T0289912

Algebraic Structures
in Integrability

Algebraic Structures in Integrability

Vladimir Sokolov

Landau Institute for Theoretical Physics, Russia

Technical Editor

Sotiris Konstantinou-Rizos

World Scientific

NEW JERSEY · LONDON · SINGAPORE · BEIJING · SHANGHAI · HONG KONG · TAIPEI · CHENNAI · TOKYO

Published by

World Scientific Publishing Co. Pte. Ltd.

5 Toh Tuck Link, Singapore 596224

USA office: 27 Warren Street, Suite 401-402, Hackensack, NJ 07601

UK office: 57 Shelton Street, Covent Garden, London WC2H 9HE

Library of Congress Control Number: 2020022361

British Library Cataloguing-in-Publication Data
A catalogue record for this book is available from the British Library.

ALGEBRAIC STRUCTURES IN INTEGRABILITY
Foreword by Victor Kac

ISBN 978-981-121-964-1 (hardcover)
ISBN 978-981-121-965-8 (ebook for institutions)
ISBN 978-981-121-966-5 (ebook for individuals)

For any available supplementary material, please visit
https://www.worldscientific.com/worldscibooks/10.1142/11809#t=suppl

To the memory of my teacher and mentor,

Alexei Shabat

Contents

Foreword

The author of this book is a leading expert in the theory of integrable systems. The book discusses the results and algebraic structures associated with the three most important concepts of the theory:

- Lax pairs, introduced by Lax in 1968;

- Compatible pairs of Hamiltonian structures, introduced by Magri in 1978;

- Generalized symmetries and conservation laws.

In all three areas, the author obtained fundamental results. For example, the Drinfeld–Sokolov reduction has reached far beyond the theory of integrable systems and is widely used in mathematical physics. Another example is the use of the symmetry approach to the classification of integrable evolution equations, developed by Mikhailov, Sokolov, Shabat and their co-authors in the 1980s.

A particularly attractive feature of the book is the abundance of examples. Therefore, although often the author omits details of the proofs or simply refers to the original source, the reader has the opportunity to understand what is happening on specific calculations.

Some algebraic structures that arose in the theory of integrable systems are not mentioned in the book. These include τ-functions and corresponding transformation groups, cluster algebras, Poisson vertex algebras, and, in particular, classical affine W-algebras. However, as Kozma Prutkov[1] said, "Nobody can embrace the immensity!"

Therefore, I unconditionally recommend the book by Vladimir Sokolov to a mathematically educated reader, who is interested in the theory of integrable systems.

Victor Kac (Massachusetts Institute of Technology)

---

[1]Fictional, but very popular Russian author.

Preface

The mathematical content of the theory of integrable systems comprises the study of hidden rich algebraic and/or analytic structures associated with such equations. In this book, algebraic structures associated with integrable ODEs and PDEs in two independent variables are considered.

The first part deals with Lax representations for differential equations. In addition, the method of factorization of Lie algebras (Adler–Konstant–Symes scheme or AKS-scheme) is formulated and generalized. The theory of classical and quantum r-matrices is not discussed in this book, because there are many excellent books on this subject (see, for example, [1, 2]). Integrable quantum systems are discussed only in Section 3.2.3.

In a small Part 2, algebraic constructions related to the bi-Hamiltonian formalism are described. The main objective here is to find a pair of compatible algebras.[1] In addition, the argument shift in quadratic Poisson brackets generating the Calogero–Moser elliptic model is discussed. For a small number of particles, commutative subalgebras of the universal enveloping algebra $\mathfrak{gl}_N$ defining the quantum elliptic Calogero–Moser systems are given.

In the third part, the basic concepts of the symmetry approach to the classification of integrable equations are formulated, and some classification results are presented. At the same time, I try as much as possible not to cross-check the well-known surveys [3]–[8] (see also [9]–[11]), dedicated to the symmetry approach. The main objectives of the study here are integrable systems in which matrices or vectors of arbitrary dimension are unknown variables. By generalizing the results obtained, deep connections between integrable polynomial multi-component systems and nonassociative algebraic structures are established (in particular, with Jordan algebras and triple Jordan systems and their deformations).

The book is addressed to experts both in algebra and in classical integrable systems. It is intelligible to undergraduate and graduate students and can serve as an introduction to separate sections of the theory of integrable systems for scientists with algebraic inclinations. The exposition is based on a series of lectures delivered by the author at MIT (Boston, 2011) and at USP (São Paulo, 2015).

---

[1]Two algebraic structures defined on a vector space are called compatible if any their linear combination is an algebraic structure of the same type.

The book was conceived as a survey of the modern theory of classical integrable nonlinear equations, in which the author is an expert. The statements are formulated in the simplest possible form. However, I try to indicate possible ways of generalization. In the proofs, I mention only essential points, while for technical details I provide references. The focus is on carefully selected examples. I hope that, for young people, they will serve as a starting point for the study of various aspects of integrable systems, while professional algebraists will be able to use some examples for wide-ranging generalizations. In addition, the book proposes many unsolved problems of various levels of complexity. I hope that this book will serve as a guide and a basis for further study, however, a deeper understanding of every chapter requires studying more rigorous and specialized literature.

In my references, I try not to point out the pioneering articles for which I apologize to their authors. Instead, survey articles are often referred, where the reader can find a description of the background and links to the original works.

I also apologize to my co-authors V. Drinfeld, I. Golubchik, A. Meshkov, A. Mikhailov, A. Odesskii, A. Shabat, S. Svinolupov, R. Hernández Heredero, A. Zhiber and T. Wolf for using certain sections of our collaborative works.

I am grateful to the first readers of parts of this book, Yu. Bagderina, S. Carpentier, D. Demskoi, C. Diemer, I. Freire, S. Startsev, and A. Zobnin who offered many suggestions, found a lot of misprints, and contributed to the improvement of the text. I particularly want to mention S. Konstantinou-Rizos, who took the trouble to read the entire book and correct many inaccuracies.

I am also grateful to V. Adler, V. Kac, I. Krasilschik, M. Kontsevich, A. Maltsev, E. Ferapontov, I. Shestakov and V. Futorny for their attention and useful discussions. The work was done as part of the state task of the Programme of the Ministry of Science and Higher Education of the Russian Federation, project No. 1.13560.2019/13. It has also benefited from the support of FAPEST and IHES, which funded my visits to Brazil and France in 2017-2019, where this book was written.

I thank my wife Olga Sokolova who provided abundant support for writing this book.

On the contents of the book

The book discusses results related to:

(i) Lax pairs;

(ii) Compatible pairs of Hamiltonian structures;

(iii) Infinitesimal symmetries and conservation laws.

In the introduction, I introduce these basic concepts of integrability theory. Here, instead of formal definitions, instructive examples are given. A more rigorous and detailed presentation is contained in the main text of the book. A systematic exposition of the concepts related to (iii) can be found in [12].

All the considered objects are invariants with respect to arbitrary local transformations (see Section 1.6) of unknown functions in the differential equations considered. Therefore, they can often be described in terms of differential geometry. Algebraic structures appear if we use a coordinate system in which the functions that we are dealing with are polynomials.

As mentioned the main text of the book is divided into three parts. The first part is devoted to Lax pairs. The hierarchies of the integrable Korteweg–de Vries (KdV) and nonlinear Schrödinger (NLS) equations are described in detail and their possible generalizations are discussed. The Lax pairs for all these systems are polynomially dependent on the spectral parameter. In trying to get rid of the polynomial hypothesis, a decomposition

$$\mathcal{G} = \mathcal{G}_+ \oplus \mathcal{G}_-$$

of infinite dimensional Lie algebras in a direct sum of subalgebras $\mathcal{G}_+$ and $\mathcal{G}_-$ in $\mathcal{G}$, plays a key role. A connection is established between these decompositions and different classes of integrable systems.

The second part presents the results related to bi-Hamiltonian systems. They are based on the study of compatible Lie and associative algebras. A class of associative linear deformations of the matrix algebra is constructed, which is associated with affine Dynkin diagrams of types A, D, E. The corresponding pairs of compatible linear Poisson brackets generate quadratic integrable ODE systems. In addition, we study a class of integrable differential operators with polynomial coefficients, associated with Olshanetsky–Perelomov quantum systems, and classical Hamiltonians corresponding to these operators.

The contributions of my co-authors, I. Golubchik, V. Drinfeld and A. Odesskii to the results presented in Sections 1 and 2 cannot be overstated.

The third part of the book is devoted to a symmetry approach to the classification of integrable systems. It has been developing since 1979 thanks to the efforts of a large group of scientists such as: A. Shabat, A. Zhiber, N. Ibragimov, A. Fokas, S. Svinolupov, A. Mikhailov, R. Yamilov, V. Adler, P. Olver, J. Sanders, J.P. Wang, V. Novikov, A. Meshkov, D. Demskoi, H. Chen, Y. Lee, C. Liu, I. Khabibullin, B. Magadeev, R. Hernández Heredero, V. Marikhin, M. Foursov, S. Startsev, M. Balakhnev, and others, including the author.

This approach is extremely effective in the case of partial differential equations with two independent variables. Under additional assumptions, it can be applied to ordinary differential equations. I do not consider equations with more than two independent variables of the Kadomtsev–Petviashvili equation type, because the vast majority of the methods used in the book do not apply to them.

From the point of view of the symmetry approach, the definition of integrability is as follows.

Definition 1. A differential equation is integrable if it has an infinite set of infinitesimal symmetries.

The reader may ask, "Why does an equation, integrable in some sense, have higher symmetries?"

The following answer is not rigorous, but rather informative. A linear differential equation has an infinite set of symmetries. Integrable nonlinear differential equations, known to date, are related to linear equations by some transformations. The same transformation generates the higher symmetries of the nonlinear equation from the linear symmetries (see the discussion after Remark 1.4.8).

It turned out that the requirement of the existence of higher symmetries makes it possible to find all the integrable equations belonging to a previously prescribed class of equations. The first classification result in the symmetry approach was obtained in [13]:

Theorem 2. *Nonlinear hyperbolic equation of the form*

$$u_{xy} = F(u)$$

has higher symmetries iff (up to scalings and shifts)[2]

$$F(u) = e^u, \qquad F(u) = e^u + e^{-u}, \qquad or \quad F(u) = e^u + e^{-2u}.$$

Statements of other classification results can be found in the surveys [3]–[8] (see also Section 4.1 of this book). Some proofs are contained in [14].

In this book, the symmetry approach is generalized to the case of special classes of multi-component systems, such as equations with matrix and vector variables and equations on certain types of nonassociative algebras.

---

[2]All constants in transformations are supposed to be complex and, therefore, the cases $F = e^u + e^{-u}$ and $F = \sin u$ are regarded as equivalent.

Chapter 1

Introduction

The purpose of this chapter is purely utilitarian. We will discuss here all the basic concepts that appear in the book, trying to restrict ourselves to the minimum of necessary information, without pretending to the rigour and completeness of presentation.

1.1 List of basic notations

1.1.1 Constants, vectors and matrices

Henceforward, the field of constants is $\mathbb{C}$; the notation $\mathbf{u}$ is used for the N-dimensional vector with the components u^i: $\mathbf{u} = (u^1, \ldots, u^N)$. The standard scalar product $\sum_{i=1}^{N} u^i v^i$ is denoted by $\langle \mathbf{u}, \mathbf{v} \rangle$. We denote the expression $\langle \mathbf{u}, \mathbf{u} \rangle$ by $|\mathbf{u}|^2$ or, simply, by $\mathbf{u}^2$.

The associative algebra of $m \times m$ square matrices is denoted by Mat_m; the matrix $(u_{ij}) \in \mathrm{Mat}_m$ is denoted by $\mathbf{U}$. The identity matrix is denoted by $\mathbf{1}_m$ or, simply, by $\mathbf{1}$. The notation $\mathbf{U}^T$ stands for the transposed matrix.

For the set of $n \times m$ matrices we use the notation $\mathrm{Mat}_{n,m}$.

1.1.2 Derivations and differential operators

In the case of ordinary differential equations, the independent variable is denoted by t, for the partial hyperbolic differential equations the variables are denoted by x and y and for the evolution equations the independent variables are denoted by t and x. By u_t we denote the partial derivative of u with respect to t. For partial derivatives with respect to x, we use the notation $u_x = u_1$, $u_{xx} = u_2$, etc.

The total derivative operator $\dfrac{d}{dx}$ is denoted by D. The associative algebra of differential operators of the form $\sum s_i D^i$ is considered. The multiplication of differential operators is uniquely defined by the relation $D \circ a = aD + D(a)$.

1

For any differential operator $L = \sum_{i=0}^{k} a_i D^i$, we define the formally adjoint operator L^+ by the formula

$$L^+ = \sum_{i=0}^{k} (-1)^i D^i \circ a_i,$$

where $\circ$ means that a_i is the multiplication operator by a_i. By L_t, we denote the operator

$$L_t = \sum_{i=0}^{k} (a_i)_t D^i.$$

1.1.3   Differential algebra

We denote by $\mathcal{F}$ the differential field of the functions [15]. In the beginning, we can assume that the elements of $\mathcal{F}$ are rational functions of a finite number of independent variables

$$u_0 = u, \ u_1, \ \ldots, \ u_i, \ \ldots. \tag{1.1}$$

However, some functions of these variables are often found by solving overdetermined systems of algebraic or partial differential equations. To add such functions, we need to expand the original field $\mathcal{F}$. We will avoid a formal description of such extensions, hoping that in each particular case it is clear what we really need from $\mathcal{F}$.

The action of derivation

$$D = \sum_{i=0}^{\infty} u_{i+1} \frac{\partial}{\partial u_i} \tag{1.2}$$

of the field $\mathcal{F}$ on u_0 generates all the variables u_i. The vector field D defined on $\mathcal{F}$ is called the *total x-derivative*.

Sometimes, we add the spatial variable x to the independent variables (1.1). In this case,

$$D = \frac{\partial}{\partial x} + \sum_{i=0}^{\infty} u_{i+1} \frac{\partial}{\partial u_i}. \tag{1.3}$$

Definition 1.1.1. For any element $a \in \mathcal{F}$ the linear differential operator

$$a_* \overset{def}{=} \sum_{k} \frac{\partial a}{\partial u_k} D^k \tag{1.4}$$

is called the *Fréchet derivative* of the element a.

By a_*^+, we denote the operator

$$a_*^+ = \sum_k (-1)^k D^k \circ \frac{\partial a}{\partial u_k}$$

formally adjoint to a_*.

Definition 1.1.2. The order of function a is defined as the order of the differential operator a_*.

By $O(k)$ we denote terms of order not exceeding k whose exact form is not important to us.

1.1.4   Algebra

We denote by $A(\circ)$ an N-dimensional algebra A over $\mathbb{C}$ with an operation $\circ$. A basis of A is denoted by $\mathbf{e}_1, \ldots, \mathbf{e}_N$, and the corresponding structural constants by C_{jk}^i:

$$\mathbf{e}_j \circ \mathbf{e}_k = \sum_i C_{jk}^i \, \mathbf{e}_i.$$

In what follows, we assume that the summation is carried out over repeated indices.

For any $a \in A$, we denote by L_a and R_a the operators of left and right multiplications by a:

$$L_a(x) = a \circ x, \qquad R_a(x) = x \circ a.$$

We will use the following notation:

$$\mathrm{As}(X, Y, Z) = (X \circ Y) \circ Z - X \circ (Y \circ Z), \qquad (1.5)$$

$$[X, Y, Z] = \mathrm{As}(X, Y, Z) - \mathrm{As}(Y, X, Z). \qquad (1.6)$$

By $\mathcal{G}$ and $\mathcal{A}$ we usually denote a Lie and an associative algebra, respectively.

The algebra of Laurent series of the form

$$M = \sum_{i=-n}^{\infty} c_i \lambda^i, \qquad c_i \in \mathbb{C}, \qquad n \in \mathbb{Z}$$

is denoted by $\mathbb{C}((\lambda))$, for the subalgebra of the Taylor series we use the notation $\mathbb{C}[[\lambda]]$, whereas $\mathbb{C}[\lambda]$ denotes polynomials in λ. By M_+ and M_- we denote

$$M_+ = \sum_{i=0}^{\infty} c_i \lambda^i, \qquad \text{and} \qquad M_- = \sum_{i=-n}^{-1} c_i \lambda^i.$$

For noncommutative Laurent series with coefficients from Lie and associative algebras, we use a similar notation.

1.2   Lax pairs

The modern theory of integrable systems was inspired by the discovery of the
inverse scattering method [16, 17]. The main ingredient of this method is the Lax
representation for the equation in question.

The Lax representation for a given differential equation is called the relation
of the form

$$L_t = [A,\, L], \tag{1.7}$$

where L and A are linear operators, which is equivalent to the equation. In order
to use the technique of the inverse scattering method, the operators L and A must
depend on an additional (complex) parameter λ.

1.2.1   Case of ODE

Lax representation for a system of ordinary differential equations

$$\mathbf{u}_t = \mathbf{F}(\mathbf{u}), \qquad \mathbf{u} = (\mathrm{u}^1, \dots, \mathrm{u}^N) \tag{1.8}$$

is the relation (1.7), where $L = L(\mathbf{u}, \lambda)$ and $A = A(\mathbf{u}, \lambda)$ are matrices.

Lemma 1.2.1. *Suppose an operator L satisfies (1.7), then*

(i) *If L_1 and L_2 satisfy (1.7), then the product $\bar{L} = L_1 L_2$ also satisfies (1.7);*

(ii) *$\bar{L} = L^n$ satisfies (1.7) for all $n \in \mathbb{N}$;*

(iii) *$\operatorname{tr} L^n$ is an integral of motion for (1.8);*

(iv) *The coefficients of the characteristic polynomial $\det(L - \mu\mathbf{1})$ are integrals
of motion.*

Proof. Item (i). We have

$$\bar{L}_t = (L_1)_t L_2 + L_1(L_2)_t = [A,\, L_1]L_2 + L_1[A,\, L_2] = A\bar{L} - \bar{L}A.$$

Item (ii) follows from (i). Item (iii): Applying the trace functional to both sides
of the identity $(L^n)_t = [A,\, L^n]$, we obtain $(\operatorname{tr} L^n)_t = 0$. Item (iv) follows from (ii)
and the formula

$$\det(L - \mu\mathbf{1}) = \exp\left(\operatorname{tr}\left(\log\left(L - \mu\mathbf{1}\right)\right)\right).$$

$\square$

Example 1.2.2. [18] Let $\mathbf{U}(t)$ be a matrix of dimension $m \times m$,

$$L = a\lambda + \mathbf{U}, \qquad A = \frac{\mathbf{U^2}}{\lambda},$$

where $a = \text{diag}(a_1, \ldots, a_m)$. Then, the relation (1.7) is equivalent to the differential equation

$$\mathbf{U}_t = [\mathbf{U}^2, a]. \qquad (1.9)$$

If

$$\mathbf{U} = \begin{pmatrix} 0 & -u_3 & u_2 \\ u_3 & 0 & -u_1 \\ -u_2 & u_1 & 0 \end{pmatrix}, \qquad a = \begin{pmatrix} a_1 & 0 & 0 \\ 0 & a_2 & 0 \\ 0 & 0 & a_3 \end{pmatrix},$$

where a_i are arbitrary parameters, then (1.9) is equivalent to Euler's top

$$(u_1)_t = (a_2 - a_3)u_2 u_3, \qquad (u_2)_t = (a_3 - a_1)u_1 u_3, \qquad (u_3)_t = (a_1 - a_2)u_1 u_2.$$

The characteristic polynomial $\det(L - \mu \mathbf{1})$ is given by the formula

$$(a_1\lambda - \mu)(a_2\lambda - \mu)(a_3\lambda - \mu) - (u_1^2 + u_2^2 + u_3^2)\mu + (a_1 u_1^2 + a_2 u_2^2 + a_3 u_3^2)\lambda.$$

The coefficients of the monomials in the variables λ and μ give two nontrivial first integrals for the top. The corresponding characteristic curve

$$\det(L - \mu \mathbf{1}) = 0$$

is elliptic. The eigenfunction $\Psi(\lambda, \mu, t)$, satisfying the equation

$$L\Psi = \mu\Psi, \qquad (1.10)$$

defines a vector bundle over this curve. The dependence of Ψ on t is determined by the linear equation

$$\Psi_t = A\,\Psi. \qquad (1.11)$$

Using (1.10) and (1.11), one can construct $\Psi(t)$ and then find the corresponding solution $\mathbf{U}(t)$.

The assumption that L and A in the relation (1.7) are functions of t and λ with values in a finite-dimensional Lie algebra $\mathcal{G}$ is a remarkable reduction of the case of generic matrices L and A which reduces the number of unknown functions in the corresponding nonlinear system of ODE.

Remark 1.2.3. *We can also assume that the operator A in (1.7) belongs to $\mathcal{G}$, whereas L belongs to a module over $\mathcal{G}$ (see, for example, [19]).*

1.2.2   Lax pairs for partial differential equations

Example 1.2.4. The Lax pair

$$L = D^2 + u + \lambda, \qquad A = 4D^3 + 6uD + 3u_x, \qquad \text{where} \quad D = \frac{d}{dx},$$

for the Korteweg–de Vries equation (KdV)

$$u_t = u_{xxx} + 6\, u\, u_x \tag{1.12}$$

was found by P. Lax in [20].

Here, unlike in Example 1.2.2, L and A are differential operators. The relations (1.10) and (1.11) allow us to construct $\Psi(x,t)$ using the inverse scattering method and, thus, to find the corresponding solution $u(x,t)$ of the KdV equation.

Example 1.2.5. A Lax representation of the nonlinear Schrödinger equation (NLS), written in the form of a system of two equations

$$u_t = -u_{xx} + 2u^2 v, \qquad v_t = v_{xx} - 2v^2 u \tag{1.13}$$

has been found by V. Zakharov and A. Shabat [21]. The Lax L-operator is given by

$$L = D + \lambda \begin{pmatrix} 1 & 0 \\ 0 & -1 \end{pmatrix} + \begin{pmatrix} 0 & u \\ v & 0 \end{pmatrix}.$$

The operator A is a polynomial in λ with matrix coefficients depending on u, v, $u_x, v_x, u_{xx}, v_{xx}, \dots$ (see Section 2.2).

In this example, the L-operator has the form $L = D - B$, where B is a matrix dependent on unknown functions and a spectral parameter λ. In this particular case, equation (1.7) can be written in the form

$$A_x - B_t = [B, A]. \tag{1.14}$$

The relation (1.14) is called *zero-curvature representation*. This is the compatibility condition of two linear systems

$$\Psi_t = A\,\Psi, \qquad \Psi_x = B\,\Psi, \qquad \Psi(x,t,\lambda),\ A(x,t,\lambda),\ B(x,t,\lambda) \in \mathrm{Mat}_N. \tag{1.15}$$

In contrast to the ODE case (see Remark 1.2.3), as a reduction, we can only assume that A and B in (1.14) are functions of x, t, λ with values in a finite-dimensional Lie algebra $\mathcal{G}$. In the NLS case we have $\mathcal{G} = \mathfrak{sl}_2$.

1.2.3 Matrix Riemann–Hilbert problem and dressing procedure

One of the possible versions of the inverse scattering method is the so-called *dressing method* [22]. The following simplified description of it shows the crucial role of the spectral parameter λ in Lax operators.

Let Γ be a simple closed contour on the complex λ-plane. The Riemann–Hilbert factorization problem is as follows. Given a matrix function G defined on the contour Γ, find matrix functions ψ_1, holomorphic outside of Γ, and ψ_2, holomorphic inside of Γ, such that

$$\psi_1(\lambda)^{-1}\,\psi_2(\lambda) = G(\lambda) \qquad (1.16)$$

on Γ. This factorization is called *regular* if $\psi_{1,2}$ and $\psi_{1,2}^{-1}$ are continuous in the closure of their domains of analyticity, including infinity.

Let a contour Γ and a matrix function $G_0(\lambda)$ on it be given. For all x and t, we define the function $G(\lambda, x, t)$ by

$$G(\lambda, x, t) = \Psi(\lambda, x, t)\, G_0(\lambda)\, \Psi(\lambda, x, t)^{-1}, \qquad (1.17)$$

where Ψ is a solution of system (1.15).

For all x and t, let us consider the factorization problem (1.16) for the function $G(\lambda, x, t)$. Differentiating (1.16) by x, we obtain

$$-\psi_1^{-1}(\psi_1)_x \psi_1^{-1} \psi_2 + \psi_1^{-1}(\psi_2)_x = B\,\psi_1^{-1}\psi_2 - \psi_1^{-1}\psi_2\,B$$

or

$$\left((\psi_1)_x + \psi_1 B\right)\psi_1^{-1} = \left((\psi_2)_x + \psi_2 B\right)\psi_2^{-1} \stackrel{def}{=} \bar{B}.$$

If follows from this formula that $\bar{B}$ has the same singularities in λ as B. In particular, if B is rational in λ, then $\bar{B}$ is also rational with the same structure of poles.

According to the definition of $\bar{B}$, we have

$$(\psi_{1,2})_x = \bar{B}\,\psi_{1,2} - \psi_{1,2}\,B.$$

This implies that the matrices $\Psi_{1,2} = \psi_{1,2}\Psi$ satisfy the linear equation $(\Psi_{1,2})_x = \bar{B}\,\Psi_{1,2}$.

Similarly, differentiating (1.16) by t, we find a matrix $\bar{A}$ such that $(\Psi_{1,2})_t = \bar{A}\,\Psi_{1,2}$. Thus, starting with a solution (A, B) of equation (1.15), we find a new solution $(\bar{A}, \bar{B})$. This solution depends on the arbitrary function G_0 in (1.17). For the initial (seed) solution (A, B) one can take, for example, any matrices $A_0(\lambda, t)$ and $B_0(\lambda, x)$ such that $[A_0, B_0] = 0$.

If (1.15) is equivalent to a system of PDEs for the coefficients of A and B (see, for instance, Example 1.2.5), then the dressing procedure generates a solution, that corresponds to $(\bar{A}, \bar{B})$, from the solution corresponding to (A, B).

1.3   Hamiltonian structures

Consider the finite-dimensional case. Suppose we have a manifold with coordinates $y_1, \ldots, y_m$. Any Poisson bracket between functions $f(y_1, \ldots, y_m)$ and $g(y_1, \ldots, y_m)$ is given by the formula

$$\{f, g\} = \sum_{i,j} P_{i,j}(y_1, \ldots, y_m) \frac{\partial f}{\partial y_i} \frac{\partial g}{\partial y_j}, \tag{1.18}$$

where $P_{i,j} = \{y_i, y_j\}$. The functions P_{ij} are not arbitrary, since by definition we must have

$$\{f, g\} = -\{g, f\}, \tag{1.19}$$

$$\{\{f, g\}, h\} + \{g, h\}, f\} + \{\{h, f\}, g\} = 0. \tag{1.20}$$

The algebra of all polynomials in the variables y_i, endowed by the operation $\{\cdot, \cdot\}$, is called *Poisson algebra*.

Formula (1.18) can be rewritten as

$$\{f, g\} = \langle \operatorname{grad} f, \mathcal{H}(\operatorname{grad} g) \rangle, \tag{1.21}$$

where $\mathcal{H} = \{P_{i,j}\}$ and $\langle \cdot, \cdot \rangle$ is a standard scalar product. The matrix $\mathcal{H}$ is called *Hamiltonian operator* or *Poisson tensor*.[1]

Definition 1.3.1. The Poisson bracket is called *degenerate* if $\det \mathcal{H} = 0$.

The Hamiltonian dynamical system is defined by the formula

$$\frac{dy_i}{dt} = \{H, y_i\}, \qquad i = 1, \ldots, m, \tag{1.22}$$

where H is a function of Hamilton. If $\{K, H\} = 0$, then K is an integral of the motion for the dynamical system. In this case, the dynamical system

$$\frac{dy_i}{d\tau} = \{K, y_i\} \tag{1.23}$$

is an infinitesimal symmetry for (1.22) (see Section 1.4 and [12]).

If $\{J, f\} = 0$ for all functions f, then J is called a *Casimir* function for the Poisson bracket. Nonconstant Casimir functions exist iff the bracket is degenerate.

---

[1]In fact, $\mathcal{H}$ is a tensor of rank (2.0), but we do not follow the procedure of raising and lowering indices here, assuming naively that $\mathcal{H}$ is a matrix, and grad is a vector column.

Example 1.3.2. Consider a symplectic manifold with coordinates q_i and p_i, $i = 1, \ldots N$. The standard (nondegenerate) Poisson constant bracket is given by formulas

$$\{p_i, p_j\} = \{q_i, q_j\} = 0, \qquad \{p_i, q_j\} = \delta_{i,j}, \tag{1.24}$$

where δ is the Kronecker symbol. The corresponding dynamical system has the usual Hamiltonian form

$$\frac{dp_i}{dt} = -\frac{\partial H}{\partial q_i}, \qquad \frac{dq_i}{dt} = \frac{\partial H}{\partial p_i}.$$

In the case of *linear Poisson brackets* we have

$$P_{ij} = \sum_k b_{ij}^k y_k, \qquad i, j, k = 1, \ldots, N.$$

It is well known that this formula defines a Poisson bracket iff b_{ij}^k are *structural constants of some Lie algebra*. Very often, the name of this Lie algebra is assigned to the corresponding Poisson bracket.

For spinning top-like systems [23, 24], the Hamiltonian structure is defined by linear Poisson brackets.

Example 1.3.3. For the models of rigid body dynamics [23] the Poisson bracket is given by

$$\{M_i, M_j\} = \sum_k \varepsilon_{ijk} M_k, \qquad \{M_i, \gamma_j\} = \sum_k \varepsilon_{ijk} \gamma_k, \qquad \{\gamma_i, \gamma_j\} = 0.$$

Here, M_i and γ_i are components of 3-dimensional vectors $\mathbf{M}$ and $\mathbf{\Gamma}$, ε_{ijk} is the totally skew-symmetric tensor. The corresponding Lie algebra $e(3)$ is the Lie algebra of the group of motions in $\mathbb{R}^3$. This bracket has two Casimir functions

$$J_1 = \langle \mathbf{M}, \mathbf{\Gamma} \rangle, \qquad J_2 = |\mathbf{\Gamma}|^2.$$

The class of quadratic Poisson brackets

$$\{y_i, y_j\} = \sum_{p,q=1,\ldots,N} r_{ij}^{pq} y_p y_q, \qquad i, j = 1, \ldots, N, \tag{1.25}$$

and related structures such as the Yang–Baxter equation are extremely important in modern mathematical physics.

All known examples of Poisson brackets (1.25) belong to two classes: rational and elliptic according to the structure of their 2-dimensional homogeneous symplectic leaves. Elliptic Poisson structures are considered as "the most nondegenerate" because rational Poisson structures can often be obtained from elliptic ones by degeneration of the corresponding elliptic curve. A wide class of examples of elliptic Poisson structures was constructed in [26] (see also [27] and references therein). These elliptic Poisson algebras (and the corresponding quantum algebras) are related to the deformation quantization, moduli spaces of holomorphic bundles, classical and quantum integrable systems and other areas of mathematics and mathematical physics.

The first example of an elliptic Poisson bracket with 4 generators was constructed in the paper [25] devoted to the R-matrix approach to quantum integrable systems. In the notation of [26, 27] this Poisson structure is denoted by $q_{4,1}(\tau)$.

More general examples constructed in [26, 27] are denoted by $q_{n,k}(\tau)$. Here $n, k \in \mathbb{Z}$, $1 \le k < n$ and n, k are coprime. An explicit expression for the coefficients of these brackets can be written in terms of theta-constants. For example, the bracket $q_{3,8}(\tau)$ is given by the formula (3.2).

Remark 1.3.4. *In the quantum case* [1], *the Hamiltonians are differential operators, and the Poisson bracket is replaced by the commutator of operators. A Hamiltonian H is called integrable if there are sufficiently many differential operators commuting with H.*

1.3.1   Hamiltonian form of evolution equations

For the evolution equations of the form

$$u_t = F(u, u_1, u_2, \ldots, u_n), \qquad u_i = \frac{\partial^i u}{\partial x^i}, \tag{1.26}$$

the Poisson brackets are also defined by the formula (1.21). However, we must use a variational derivative instead of the gradient. Furthermore, the Hamiltonian operator is not a matrix, but a differential (or even pseudo-differential) operator (see Section 4.3).

Most integrable evolution equations (1.26) can be written in Hamiltonian form

$$u_t = \mathcal{H}\left(\frac{\delta \rho}{\delta u}\right),$$

where $\mathcal{H}$ is a Hamiltonian operator. The corresponding Poisson bracket between two functions of variables (1.1) is defined by

$$\{f, g\} = \frac{\delta f}{\delta u} \mathcal{H}\left(\frac{\delta g}{\delta u}\right), \tag{1.27}$$

where

$$\frac{\delta}{\delta u} = \sum_k (-1)^k D^k \frac{\partial}{\partial u_k}$$

is the *Euler operator* or the *variational derivative.*[2]

Definition 1.3.5. Two functions ρ_1, ρ_2 of variables (1.1) are called equivalent (which is denoted $\rho_1 \sim \rho_2$) if $\rho_1 - \rho_2 \in \operatorname{Im} D$.

Proposition 1.3.1. *If $a \in \operatorname{Im} D$, then*

$$\frac{\delta a}{\delta u} = 0$$

and therefore the variational derivative is well defined on the equivalence classes.

Remark 1.3.6. *For functions $u(x), u'(x), \ldots$ which decrease rapidly as $x \mapsto \pm\infty$, two equivalent polynomial densities ρ_1 and ρ_2 with zero constant terms define the same functional of the form*

$$I(\rho) = \int_{-\infty}^{+\infty} \rho(u, u_x, \ldots)\, dx. \tag{1.28}$$

In fact, we want to define a Poisson bracket between functionals of the form (1.28). In terms of densities, this is formalized as follows. By definition, the Poisson bracket (1.27) is defined on classes of equivalent densities and must satisfy the conditions (1.19), (1.20). Finite dimensional Poisson brackets (1.18) also satisfy the Leibniz identity

$$\{f, g\,h\} = \{f, g\}\,h + g\,\{f, h\}.$$

For the brackets (1.27), the Leibniz rule does not make sense because the product of the equivalence classes is not defined.

The general theory of Poisson brackets on algebras of differential functions has been developed in [28]. In this article, algebra of differential (noncommutative) functions are defined as an associative algebra $\mathcal{D}$ with unity and commuting derivations ∂_i, $i \in \mathbb{Z}_+$ such that both following conditions are fulfilled:

- For any element $f \in \mathcal{D}$, $\partial_i(f) = 0$ for almost any i;

- $[\partial_i, D] = \partial_{i-1}$.

---

[2]It is easy to see that $\frac{\delta a}{\delta u} = a_*^+(1)$.

1.4 Infinitesimal symmetries

Consider a dynamical system of ODE

$$\frac{dy^i}{dt} = F_i(y^1, \ldots, y^n), \qquad i = 1, \ldots, n. \tag{1.29}$$

Definition 1.4.1. A dynamical system

$$\frac{dy^i}{d\tau} = G_i(y^1, \ldots, y^n), \qquad i = 1, \ldots, n \tag{1.30}$$

is called an *infinitesimal symmetry* for (1.29) if (1.29) and (1.30) are compatible.

Informally, compatibility means that for any initial data $\mathbf{y}_0$ there exists a common solution $\mathbf{y}(t, \tau)$ of the system of equations (1.29) and (1.30) such that $\mathbf{y}(0, 0) = \mathbf{y}_0$.

At the infinitesimal level, compatibility means that

$$XY - YX = 0,$$

where

$$X = \sum F_i \frac{\partial}{\partial y^i}, \quad \text{and} \quad Y = \sum G_i \frac{\partial}{\partial y^i}. \tag{1.31}$$

Remark 1.4.2. *In the case of polynomial dynamical systems, it is often said that (1.30) is a higher symmetry for (1.29) if the degree of the right-hand side of (1.30) is greater than the degree of the right-hand side of (1.29).*

We now consider evolution equations of the form (1.26). An infinitesimal (or generalized) symmetry (or commuting flow) of equation (1.26) is an evolution equation

$$u_\tau = G(u, u_x, u_{xx}, \ldots, u_m), \qquad m > 1 \tag{1.32}$$

that is compatible with (1.26). Compatibility means that

$$\frac{\partial}{\partial t} \frac{\partial u}{\partial \tau} = \frac{\partial}{\partial \tau} \frac{\partial u}{\partial t}, \tag{1.33}$$

where the partial derivatives with respect to t and τ are calculated by virtue of equations (1.26) and (1.32). In other words, for any initial data $u_0(x)$ there exists a solution $u(x, t, \tau)$ of the system of equations (1.26) and (1.32) such that $u(x, 0, 0) = u_0(x)$. A more rigorous definition in terms of commuting evolutionary vector fields is contained in Sections 2.1 and 4.2.1 (see also [12]).

Remark 1.4.3. *The infinitesimal symmetries (1.32) with $m \leq 1$ correspond to one-parameter groups of point or contact transformations (see 1.6.1 and/or [12]). They are called classical infinitesimal symmetries. These symmetries are related (see Subsection 1.6.2) to an important class of differential substitutions for evolution equations [29].*

Example 1.4.4. Any equation of the form (1.26) has the classical symmetry $u_\tau = u_x$ which corresponds to the group of shifts $x \mapsto x + \lambda$.

Example 1.4.5. For all m and n, the equation $u_\tau = u_m$ is a symmetry of the linear equation $u_t = u_n$. Symmetries with different m are compatible with each other. Thus, we have an infinite *hierarchy* of evolution linear equations such that each of the equations is a symmetry for all the others.

Example 1.4.6. The Burgers equation

$$u_t = u_{xx} + 2uu_x \tag{1.34}$$

has a third order symmetry

$$u_\tau = u_{xxx} + 3uu_{xx} + 3u_x^2 + 3u^2u_x. \tag{1.35}$$

Example 1.4.7. The simplest higher symmetry

$$u_\tau = u_5 + 10uu_3 + 20u_1u_2 + 30u^2u_1 \tag{1.36}$$

for the KdV equation (1.12) is of fifth order.

Remark 1.4.8. *As already noted, the existence of higher symmetries is a strong evidence that equation (1.26) is integrable.*

The following "explanation" of this observation may be proposed. According to Example 1.4.5, linear equations have an infinite hierarchy of higher symmetries. Usually an integrable nonlinear equation is related to a linear equation through some transformation. The same transformation generates a symmetry hierarchy of the nonlinear equation starting with the symmetries of the corresponding linear one.

For example, the Burgers equation (1.34) is integrable in particular because of the existence of the Cole–Hopf substitution

$$u = \frac{v_x}{v}, \tag{1.37}$$

which links (1.34) to the heat equation $v_t = v_{xx}$. Moreover, the same substitution connects the third order symmetry (1.35) with the equation $v_\tau = v_{xxx}$ and so on.

The transformation connecting the KdV equation with the linear equation $v_t = v_{xxx}$ is a nonlinear generalization of the Fourier transform [16, 17]. It is much more nonlocal than the Cole–Hopf substitution. However, the inverse transformation can be applied to all symmetries of the form $v_\tau = v_{2n+1}$ of the linear equation, which gives rise to a hierarchy of odd order symmetries for the KdV equation.

1.4.1   Naive symmetry test

In this section, we demonstrate how a simple classification problem for integrable homogeneous polynomial equations is posed and solved.

An evolution equation (1.26) is called λ-*homogeneous of weight* μ if it admits a one-parameter scaling group

$$(x, t, u) \longrightarrow (\tau^{-1}x,\ \tau^{-\mu}t,\ \tau^{\lambda}u).$$

For an N-component system with unknown variables $u^1, \ldots, u^N$, the corresponding scaling group has a similar form

$$(x, t, u^1, \ldots, u^N) \longrightarrow (\tau^{-1}x,\ \tau^{-\mu}t,\ \tau^{\lambda_1}u^1, \ldots, \tau^{\lambda_N}u^N). \tag{1.38}$$

Theorem 1.4.9. [30] *A scalar λ-homogeneous polynomial equation with $\lambda > 0$ can have a homogeneous polynomial higher symmetry only in one of the following cases:*

- *Case 1:* $\lambda = 2$;

- *Case 2:* $\lambda = 1$;

- *Case 3:* $\lambda = \frac{1}{2}$.

For example, the KdV equation (1.12) is homogeneous of weight 3 with $\lambda = 2$, its symmetry (1.36) is homogeneous with the same λ. The mKdV equation $u_t = u_{xxx} + u^2 u_x$ has weight 3 with $\lambda = 1$ while for the Ibragimov–Shabat equation

$$u_t = u_{xxx} + 3u^2 u_{xx} + 9uu_x^2 + 3u^4 u_x \tag{1.39}$$

the weight is 3 and $\lambda = \dfrac{1}{2}$.

The general form of the fifth order equation of weight 5 and homogeneity $\lambda = 2$ is given by the formula

$$u_t = u_5 + a_1 u u_3 + a_2 u_1 u_2 + a_3 u^2 u_1, \tag{1.40}$$

where $a_i \in \mathbb{C}$. Let us find all equations (1.40) that have a homogeneous symmetry of order 7 with $\lambda = 2$. General ansatz for such symmetry is

$$u_\tau = u_7 + c_1 u u_5 + c_2 u_1 u_4 + c_3 u_2 u_3 + c_4 u^2 u_3 + c_5 u u_1 u_2 + c_6 u_1^3 + c_7 u^3 u_1. \tag{1.41}$$

The compatibility condition (1.33) can be rewritten as $F_\tau - G_t = 0$. After eliminating the derivatives with respect to τ and t from this defining equation by means of equations (1.40) and (1.41), the left-hand side of (1.33) becomes a homogeneous polynomial P in the variables $u_1, \ldots, u_{10}$. There are no linear terms in P. Equating the coefficients of quadratic terms to zero, we find

$$c_1 = \frac{7}{5}a_1, \qquad c_2 = \frac{7}{5}(a_1 + a_2), \qquad c_3 = \frac{7}{5}(a_1 + 2a_2).$$

The vanishing conditions for cubic part allow us to express c_4, c_5 and c_6 in terms of a_1, a_2, a_3. In addition, it turns out that

$$a_3 = -\frac{3}{10}a_1^2 + \frac{7}{10}a_1 a_2 - \frac{1}{5}a_2^2.$$

Fourth degree terms lead to a formula expressing c_7 in terms of a_1, a_2 and to the basic algebraic relation

$$(a_2 - a_1)(a_2 - 2a_1)(2a_2 - 5a_1) = 0$$

between the coefficients a_1 and a_2. Solving this equation, we find that up to a scaling $u \to \lambda u$ there are only four integrable cases: the linear equation $u_t = u_5$, equations

$$u_t = u_5 + 5uu_3 + 5u_1 u_2 + 5u^2 u_1, \tag{1.42}$$

$$u_t = u_5 + 10uu_3 + 25u_1 u_2 + 20u^2 u_1, \tag{1.43}$$

and (1.36). In each of these cases, the terms of the fifth and sixth degrees in the defining equation are canceled automatically. Equations (1.42) and (1.43) are well known [31, 32].

Question 1.4.10. *The problem we solved looks rather artificial. Why did we consider the fifth order equation and the seventh order symmetry?*

Assuming that the right-hand side of equation (1.26) is polynomial and homogeneous with $\lambda > 0$, in [30] it was proved that it suffices to consider the following three cases:

- Second order equation with third order symmetry;

- Third order equation with fifth order symmetry;

- Fifth order equation with seventh order symmetry.

Other integrable equations belong to the hierarchies of such equations. This statement looks very plausible and without any assumptions about the right-hand side of the equation. In the nonpolynomial case, this is not proven, and the statement has the status of a well-known folklore hypothesis to which counterexamples are unknown.

Section 4.2 deals with an advanced version of symmetry test, where there is no requirement for the polynomiality of the right-hand side of equation (1.26), as well as fixing the order of symmetry. Only the existence of an infinite sequence of symmetries is required. A similar approach is being developed for equations with local higher conservation laws.

1.5   First integrals and local conservation laws

The concept of a first integral (or integral of motion) is one of the fundamental notions of the theory of ordinary differential equations. The function $f(y^1, \ldots, y^n)$ is called *first integral* of the system (1.29) if its value does not depend on t for any solution $\{y^1(t), \ldots, y^n(t)\}$ of (1.29). Since

$$\frac{d}{dt}\Big(f(y^1(t), \ldots, y^n(t))\Big) = X(f),$$

where the vector field X is defined by the formula (1.31), the first integral is a solution of the partial differential equation of the first order

$$X\Big(f(y^1, \ldots, y^n)\Big) = 0.$$

In the ODE case the concept of a first integral (or integral of motion) is a basic one. A function $f(y^1, \ldots, y^n)$ is called a *first integral* for the system (1.29) if its value does not depend on t for any solution $\{y^1(t), \ldots, y^n(t)\}$ of (1.29). Since

$$\frac{d}{dt}\Big(f(y^1(t), \ldots, y^n(t))\Big) = X(f),$$

where the vector field X is defined by (1.31), a first integral is, from the algebraic point of view, a solution of the first order PDE

$$Xf\Big(y^1, \ldots, y^n\Big) = 0.$$

In the case of evolution equations of the form (1.26), the integral of motion is not a function, but a functional that does not depend on t for any solution $u(x, t)$ of equation (1.26).

More rigorously, a local conservation law for equation (1.26) is a pair of functions $\rho(u, u_x, \ldots)$ and $\sigma(u, u_x, \ldots)$ such as

$$\frac{\partial}{\partial t}\Big(\rho(u, u_x, \ldots, u_p)\Big) = \frac{\partial}{\partial x}\Big(\sigma(u, u_x, \ldots, u_q)\Big) \tag{1.44}$$

for any solution $u(x, t)$ of equation (1.26). The functions ρ and σ are called the *density* and the *flow* of the conservation law (1.44), respectively. It is easy to see that $q = p + n - 1$, where n is the order of equation (1.26).

For soliton-type solutions, i.e. such that they decrease with all derivatives as $x \mapsto \pm\infty$, we obtain

$$\frac{\partial}{\partial t} \int_{-\infty}^{+\infty} \rho \, dx = 0$$

for any polynomial density ρ with a constant free term. This justifies the name *conserved density* for the function ρ. Similarly, if $u(x,t)$ is periodic in x with a period L, then the value of the functional $\int_0^L \rho\, dx$ on the solution u does not depend on time, so it is an integral of motion for equation (1.26).

Suppose ρ and σ satisfy (1.44). Then, for any function $s(u, u_x, \dots)$ the functions $\bar\rho = \rho + \dfrac{\partial s}{\partial x}$ and $\bar\sigma = \sigma + \dfrac{\partial s}{\partial t}$ also satisfy (1.44). We call the conserved densities ρ and $\bar\rho$ *equivalent*. It is clear that equivalent densities define the same functional.

Example 1.5.1. For arbitrary k and n, the function $\rho_k = u_k^2$ is a conserved density of the linear equation $u_t = u_{2n+1}$. For nonlinear evolution equations (1.26) of odd order, it can also be expected that each conserved density is a density for every symmetry as well.

Exercise 1.5.2. Verify that, for any $n \geq 1$, the equation $u_t = u_{2n}$ has only one conserved density $\rho = u$.

Example 1.5.3. The functions

$$\rho_1 = u, \qquad \rho_2 = u^2, \qquad \rho_3 = -u_x^2 + 2u^3$$

are conserved densities for the KdV equation (1.12). Indeed,

$$\frac{\partial}{\partial t}\left(\rho_1\right) = \frac{\partial}{\partial x}\left(u_{xx} + 3u^2\right),$$

$$\frac{\partial}{\partial t}\left(\rho_2\right) = \frac{\partial}{\partial x}\left(2uu_{xx} - u_x^2 + 4u^3\right),$$

$$\frac{\partial}{\partial t}\left(\rho_3\right) = \frac{\partial}{\partial x}\left(9u^4 + 6u^2 u_{xx} + u_{xx}^2 - 12uu_x^2 - 2u_x u_3\right).$$

Let us find all the conserved densities of the form $\rho(u)$ for equation (1.12). It is clear that the flow σ can only depend on u, u_x, u_{xx}. The relation (1.44) has the form

$$\rho'(u)(u_3 + 6uu_x) = \frac{\partial\sigma}{\partial u_{xx}}u_3 + \frac{\partial\sigma}{\partial u_x}u_{xx} + \frac{\partial\sigma}{\partial u}u_x. \qquad (1.45)$$

Since u is an arbitrary solution of the KdV equation, the relation (1.45) should be an identity with respect to the variables u, u_x, u_{xx}, u_3. Comparing the coefficients of u_3, we find that $\sigma = \rho'(u)u_{xx} + \sigma_1(u, u_x)$. Substituting this expression into (1.45) and equating the coefficients of u_{xx}, we obtain

$$\sigma_1 = -\frac{\rho''(u)}{2}u_x^2 + \sigma_2(u).$$

Given this and comparing the coefficients of u_x^3, we find that $\rho'''(u) = 0$, i.e. $\rho = c_2 u^2 + c_1 u + c_0$. So, up to the trivial term c_0, the density is a linear combination of the densities ρ_1 and ρ_2 from Example 1.5.3.

1.6   Transformations

1.6.1   Point and contact transformations

Let $x_1, \ldots, x_n$ be independent variables, and u be a function of these variables.
Symbols

$$x_1, \ldots, x_n, \quad u, \quad \text{and} \quad u_\alpha \tag{1.46}$$

where $\alpha = (\alpha_1, \ldots, \alpha_n)$ and $u_\alpha = \dfrac{\partial^{\alpha_1 + \cdots + \alpha_n} u}{\partial x_1^{\alpha_1} \cdots \partial x_n^{\alpha_n}}$, are treated as independent variables of the *jet space*. Let $\mathcal{F}$ be the field of functions depending on a finite number of variables (1.46).

The total derivatives with respect to x_i are given by the formulas

$$D_i = \frac{\partial}{\partial x_i} + \sum_\alpha u_{(\alpha_1, \ldots, \alpha_i + 1, \ldots, \alpha_n)} \frac{\partial}{\partial u_{(\alpha_1, \ldots, \alpha_i, \ldots, \alpha_n)}}$$

These vector fields satisfy the condition $[D_i, D_j] = 0$.

Consider transformations

$$\bar{x}_i = \phi_i, \qquad \bar{u} = \psi, \qquad \phi_i, \psi \in \mathcal{F}. \tag{1.47}$$

The derivatives of u are transformed as

$$\bar{u}_{(\alpha_1, \ldots, \alpha_n)} = \bar{D}_1^{\alpha_1} \cdots \bar{D}_n^{\alpha_n} (\bar{u}),$$

where

$$\bar{D}_i = \sum_{j=1}^n p_{ij} D_j, \qquad p_{ij} \in \mathcal{F}. \tag{1.48}$$

In order to find the coefficients p_{ij}, we apply (1.48) to $\bar{x}_k = \phi_k$ and obtain

$$\sum_{j=1}^n p_{ij} D_j(\phi_k) = \delta_k^i. \tag{1.49}$$

This means that the matrix $P = \{p_{ij}\}$ is inverse to the generalized Jacobi matrix J with elements $J_{ij} = D_i(\phi_j)$.

A transformation of the form (1.47) is called *point transformation* if the functions ψ and ϕ_i depend only on $x_1, \ldots, x_n, u$. Such a transformation is (locally) invertible if the functions ψ and ϕ_i are functionally independent. In what follows, we will assume this.

There are invertible transformations more general than point ones.

Contact transformations

Example 1.6.1. Consider the Legendre transformation

$$\bar{x}_i = \frac{\partial u}{\partial x_i}, \qquad \bar{u} = u - \sum_{k=1}^{n} x_k \frac{\partial u}{\partial x_k}. \tag{1.50}$$

Note that the functions ψ and ϕ_i depend on the first derivatives of u. For such transformations, the functions $\dfrac{\partial \bar{u}}{\partial \bar{x}_i}$ usually depend on second derivatives, etc. In this case, a transformation prolonged to derivatives of order no greater than k is not invertible for any k. However, this is not the case for the Legendre transformation (1.50). Indeed, according to (1.49), we have

$$\sum_j p_{ij} \frac{\partial^2 u}{\partial x_j \partial x_k} = \delta_k^i.$$

One can check that

$$\frac{\partial \bar{u}}{\partial \bar{x}_i} = \sum_j p_{ij} D_j \left(u - \sum_k x_k \frac{\partial u}{\partial x_k} \right) = -x_i. \tag{1.51}$$

Therefore, the transformation (1.50), after prolongation to the first derivatives by the formula (1.51), is invertible.

The Legendre transformation is an example of a *contact transformation* [9, Chapter 3, section 14]. In general, a transformation (1.47) is called *contact* if it preserves zeros of the 1-form

$$du - \sum_k \frac{\partial u}{\partial x_k} dx_k.$$

Theorem 1.6.2. [33] *If a transformation (1.47) prolonged to derivatives of order not greater than k, is invertible for some k, then this is either a point or contact transformation.*

Let us consider contact transformations in the case of $n = 1$ in more detail. They have the form

$$\bar{x} = \phi(x, u, u_x), \qquad \bar{u} = \psi(x, u, u_x), \qquad \bar{u}_1 = \chi(x, u, u_x), \tag{1.52}$$

where

$$\psi_{u_1}(\phi_u u_1 + \phi_x) = \phi_{u_1}(\psi_u u_1 + \psi_x). \tag{1.53}$$

With the assumption (1.53) the function

$$\chi = \frac{D(\psi)}{D(\phi)} = \frac{\psi_{u_1} u_2 + \psi_u u_1 + \psi_x}{\phi_{u_1} u_2 + \phi_u u_1 + \phi_x}$$

does not depend on u_2 and the transformation is invertible if the functions ϕ, ψ and χ are functionally independent.

Remark 1.6.3. *Formally, point transformations can be considered as a special case of contact ones: if the functions ϕ and ψ do not depend on u_x, then the contact condition (1.53) is obviously satisfied.*

For the Legendre transformation we have

$$\phi = u_1, \qquad \psi = u - x\,u_1. \tag{1.54}$$

In this case, the contact condition (1.53) is fulfilled and $\chi = -x$.

The formulas (1.52) and (1.53) become much easier if we enter the generating function S by the formula

$$\psi(x, u, u_1) = S(x, u, \phi(x, u, u_1)).$$

This is possible if $\phi_{u_1} \neq 0$. We have

$$\bar{u}_1 = S_\phi, \qquad S_x + u_1 S_u = 0$$

or

$$\bar{u} = S(x, u, \bar{x}), \qquad \bar{u}_1 = \frac{\partial S}{\partial \bar{x}}, \qquad u_1 = -\frac{\partial S}{\partial x}\left(\frac{\partial S}{\partial u}\right)^{-1}.$$

Exercise 1.6.4. Verify that for any function $S(x, u, \bar{x})$ the above formulas define a contact transformation.

Point and contact transformations (1.52) are important in the classification of integrable evolution equations of the form

$$u_t = F(x, u, u_x, \ldots, u_n),$$

since they preserve the form of the equation and transform the infinitesimal symmetries into symmetries, that is, they map the integrable equations again into integrable ones. Indeed, consider a contact transformation of the form

$$\bar{t} = t, \qquad \bar{x} = \phi(x, u, u_x), \qquad \bar{u} = \psi(x, u, u_x), \qquad \bar{u}_1 = \chi(x, u, u_x), \tag{1.55}$$

where the contact condition (1.53) is satisfied. The coefficients in the formulas for total derivatives

$$\bar{D}_t = \alpha D_t + \beta D, \qquad \bar{D} = \gamma D_t + \delta D$$

can be found from the conditions

$$\bar{D}_t(t) = 1, \qquad \bar{D}_t(\phi) = 0, \qquad \bar{D}(t) = 0, \qquad \bar{D}(\phi) = 1.$$

We get

$$\bar{D} = \frac{1}{D(\phi)} D, \qquad \bar{D}_t = D_t - \frac{D_t(\phi)}{D(\phi)} D \qquad (1.56)$$

and therefore

$$\bar{u}_t = \left(\psi_* - \frac{D(\psi)}{D(\phi)} \phi_* \right)(F). \qquad (1.57)$$

In this formula, it is assumed that the variables $x, u, u_x, \ldots$ in the right-hand side of the equation are replaced by new variables $\bar{x}, \bar{u}, \ldots$ with the help of the transformation inverse to (1.55). Due to the condition (1.53), the differential operator

$$Q \overset{def}{=} \psi_* - \frac{D(\psi)}{D(\phi)} \phi_* \qquad (1.58)$$

has zero order and the resulting equation has the form

$$\bar{u} = \bar{F}(\bar{x}, \bar{u}, \bar{u}_1, \ldots, \bar{u}_n)$$

with the same n.

Exercise 1.6.5. Check that the Legendre transformation (1.54) converts the heat equation $u_t = u_2$ into

$$\bar{u}_t = -\frac{1}{\bar{u}_2}.$$

Remark 1.6.6. *If we restrict ourselves to evolution equations with the right-hand side F which does not explicitly depend on x, then there exists a subgroup of contact transformations of the form $\psi = r(u, u_x)$, $\phi = x + s(u, u_x)$, which preserve this condition.*

1.6.2 Differential substitutions of Miura type

The famous Miura transformation [34]

$$\bar{u} = u_1 - u^2$$

links the mKdV equation

$$u_t = u_3 - 6u^2 u_1$$

and the KdV equation

$$\bar{u}_t = \bar{u}_3 + 6\bar{u}\bar{u}_1.$$

Indeed,

$$\bar{u}_t = D_t(u_1 - u^2) = D(u_3 - 6u^2 u_1) - 2u(u_3 - 6u^2 u_1) = u_4 - 2uu_3 - 6u^2 u_2 -$$
$$12uu_1^2 + 12u^3 u_1 = u_4 - 2uu_3 - 6u_1 u_2 + 6(u_1 - u^2)(u_2 - 2uu_1) = \bar{u}_3 + 6\bar{u}\bar{u}_1.$$

The Miura substitution is not locally invertible: for a given solution $\bar{u}$ of the KdV equation, it is necessary to solve the Riccati differential equation in order to find the solution u of the mKdV equation. Moreover, the solution u is not determined uniquely, but up to a constant.

Definition 1.6.7. A transformation of the form

$$\bar{x} = \phi(x, u, u_2, \ldots, u_k), \qquad \bar{u} = \psi(x, u, u_1, \ldots, u_k) \tag{1.59}$$

is called *differential substitution* from equation

$$u_t = F(x, u, u_1, \ldots, u_n) \tag{1.60}$$

into equation

$$\bar{u}_t = G(\bar{x}, \bar{u}, \bar{u}_1, \ldots, \bar{u}_n) \tag{1.61}$$

if for any solution $u(x, t)$ of equation (1.60) the function $\bar{u}(t, \bar{x})$ satisfies (1.61). The *order of differential substitution* is defined as the order of the corresponding differential operator (1.58).

Remark 1.6.8. *According to the definition, point and contact transformations are differential substitutions of order* 0, *while the order of the Miura substitution is equal to* 1.

The derivatives of $\bar{u}$ are expressed through the functions ϕ and ψ just as in Section 1.6.1. However, for an arbitrary function F in equation (1.60) the right-hand side of equation (1.57) cannot be expressed only in terms of the variables $\bar{x}, \bar{u}, \ldots$. If this is required, then rigid restrictions on F, ϕ and ψ appear.

Exercise 1.6.9. Check that in the case of Miura substitution, acting from the mKdV equation, the right-hand side of (1.57) is expressed only through $\bar{u}, \ldots, \bar{u}_3$ and is equal to $\bar{u}_3 + 6\bar{u}\bar{u}_1$.

Group differential substitutions

In this section, we define differential substitutions related to the classical point or contact symmetry groups [29]. For simplicity, we consider one-parameter groups. As we will see below, any such group generates some first order differential substitution. In the case of k-parameter groups, the corresponding substitutions are of order k.

Remark 1.6.10. *In Section 4.2.9 a unified approach* [35] *to differential substitutions that connect the KdV equation* (1.12) *with other integrable evolution equations is proposed. Moreover, using additional nonlocal variables, the transformations inverse to these differential substitutions are constructed.*

Section 5.4 is devoted to relations between differential substitutions for evolution equations and Darboux integrable hyperbolic equations.

Example 1.6.11. The Burgers equation

$$\bar{u}_t = \bar{u}_{xx} + 2\bar{u}\,\bar{u}_x \qquad (1.62)$$

is related to the heat equation

$$u_t = u_{xx} \qquad (1.63)$$

by the Cole–Hopf substitution

$$\bar{t} = t, \qquad \bar{x} = x, \qquad \bar{u} = \frac{u_x}{u}. \qquad (1.64)$$

This substitution admits the following group-theoretic interpretation. The one-parameter group G of dilatations $u \mapsto \tau u$ acts on solutions of the heat equation. It is easy to verify that every differential invariant of group G is a function of variables

$$\bar{t} = t, \qquad \bar{x} = x, \qquad \bar{u} = \frac{u_x}{u}, \qquad D\!\left(\frac{u_x}{u}\right), \quad \dots, \quad D^i\!\left(\frac{u_x}{u}\right), \quad \dots. \qquad (1.65)$$

Since group G preserves the equation, the total t-derivative D_t maps invariants to invariants. Consequently, the expression $D_t\!\left(\frac{u_x}{u}\right)$ must be a function of the variables (1.65). In other words, the function $\bar{u} = \frac{u_x}{u}$ must satisfy some evolution equation. A simple calculation shows that this equation has the form (1.62).

Remark 1.6.12. *The reasoning given above shows that every evolution equation with symmetry group $u \mapsto \tau u$ admits the Cole–Hopf substitution* (1.64).

Exercise 1.6.13. The heat equation (1.63) has a one-parameter symmetry group $x \mapsto x + \tau$. The simplest differential invariants of this group are

$$\bar{t} = t, \qquad \bar{x} = u, \qquad \bar{u} = u_1. \qquad (1.66)$$

The derivation $\bar{D} = \frac{1}{u_1} D$ generates a functional basis $\bar{u}_i = \bar{D}^i(u_1)$ of all differential invariants. Therefore, the differential substitution (1.66) leads to an evolution equation for $\bar{u}$. Verify that

$$\bar{u}_t = \bar{u}^2\,\bar{u}_{\bar{x}\bar{x}}.$$

In the examples considered above, equation (1.61) is a restriction of equation (1.60) to differential invariants of a certain symmetry group G. Equation (1.61) is called the *factorization* of the equation (1.60) by the group G.

Remark 1.6.14. *Recall that the one-parameter groups of contact transformations are in one-to-one correspondence with infinitesimal symmetries of the form*

$$u_\tau = G(x, u, u_1). \tag{1.67}$$

Namely, the characteristic vector field for equation (1.67) [36, Chapter 2, section 8] is given by

$$X_G = -G_{u_1}\frac{\partial}{\partial x} + (G - u_1 G_{u_1})\frac{\partial}{\partial u} + (G_x + u_1 G_u)\frac{\partial}{\partial u_1}. \tag{1.68}$$

This contact vector field generates the corresponding one-parameter group of contact transformations. This group consists of point transformations iff the function G is linear in u_1.

Remark 1.6.15. *For any one-parameter group of contact transformations, there are [29] two functionally independent invariants ϕ and ψ of order no greater than one. They generate the kernel of the vector field (1.68). A functional basis of differential invariants is*

$$\left(\frac{1}{D(\phi)}D\right)^i(\psi).$$

The corresponding differential substitution is given by the formula (1.59) with $k \leq 1$.

Consider one more important example of factorization by a one-parameter group. Every evolutionary equation of the form

$$u_t = F(u_1, u_2, \ldots, u_n) \tag{1.69}$$

admits the symmetry group $u \mapsto u + \tau$. The simplest invariants

$$\bar{t} = t, \qquad \bar{x} = x, \qquad \bar{u} = u_1$$

of this group determine the differential substitution, which leads to the equation

$$\bar{u}_t = \bar{D}\Big(F(\bar{u}, \ldots, \bar{u}_{n-1})\Big). \tag{1.70}$$

Definition 1.6.16. Equation (1.70) is said to be derived by *differentiating of u* from equation (1.69).

Remark 1.6.17. *Since every one-parameter group of point transformations is pointwise equivalent to the group $u \mapsto u + \tau$, the factorization by a one-parameter group is a composition of point transformation and differentiating of u.*

Definition 1.6.18. The transformation of $u = \int \bar{u}\,dx$ from equation (1.70) to equation (1.69), inverse to the differentiating, is called *potentiation* or *the introduction of potential.*

Remark 1.6.19. *The possibility of potentiation is related to the fact that equation (1.70) has a conserved density u. If equation (1.26) has a conserved density of order not greater than 1, then the density can be transformed into $\rho = \bar{u}$ using an appropriate point or contact transformation. The equation obtained as a result of this transformation has the form (1.70) and we can apply the potentiation to get the corresponding equation of the form (1.69).*

So, the presence of the one-parameter symmetry group (1.67) for equation (1.60) guarantees the existence of a first order differential substitution and the corresponding factor-equation (1.61), while the presence of a conserved density of the form $\rho(\bar{x}, \bar{u}, \bar{u}_1)$ for an equation of the form (1.61) leads to the existence of the corresponding equation (1.60).

Definition 1.6.20. Differential substitutions generated by infinitesimal symmetries or conservation laws of order not greater than 1 are called *quasi-local transformations.*

Question 1.6.21. *Which quasi-local transformations preserve higher symmetries?*

The answer is obvious: if (1.67) defines a symmetry group not only for equation (1.60) but also for all its symmetry hierarchy, then the corresponding factorization can be applied to the whole hierarchy. In other words, if G is a symmetry group for the entire symmetry hierarchy of an integrable equation, then the corresponding substitution preserves integrability in the sense of the symmetry approach.

Similarly, if all the equations of the hierarchy of an integrable equation (1.61) have the same conserved density $\rho(\bar{x}, \bar{u}, \bar{u}_1)$, then the corresponding (see Remark 1.6.19) equation (1.60) is integrable.

In conclusion, we give an example of factorization by a multi-parameter symmetry group.

Exercise 1.6.22. The Schwartz–KdV equation

$$\bar{u}_t = \bar{u}_3 - \frac{3\,\bar{u}_2^2}{2\,\bar{u}_1} \tag{1.71}$$

admits a three-parameter symmetry group consisting of fractional-linear transformations of the form

$$\bar{u} \to \frac{\alpha \bar{u} + \beta}{\gamma \bar{u} + \delta}, \qquad \alpha, \beta, \gamma, \delta \in \mathbb{C}, \qquad \alpha \delta - \beta \gamma = 1.$$

The simplest invariants of the group are

$$\bar{x} = x, \qquad \bar{u} = \frac{\bar{u}_3}{\bar{u}_1} - \frac{3\,\bar{u}_2^2}{2\,\bar{u}_1^2}. \tag{1.72}$$

Check that this differential substitution maps the Schwartz–KdV equation to the KdV equation $u_t = u_3 + 3uu_1$.

Remark 1.6.23. *The substitution (1.72) can be represented as a composition of differentiating (see Definition 1.6.16), the Cole–Hopf substitution, and the Miura transformation* $\bar{v} = v_1 - \frac{1}{2}v^2$.

Part 1. Lax representations for integrable systems

Chapter 2

Lax pairs and decomposition of Lie algebras

In this chapter, we discuss various types of Lax representations for integrable systems of differential equations and some constructions that allow one to derive higher symmetries and conservation laws from a Lax pair. We do not touch Hamiltonian structures related to Lax operators (see, for example, the books [1, 2] and the original articles [37, 38]).

2.0.1   Definitions of symmetries and conservation laws

For the formalization of such concepts as local symmetries and conservation laws, the language of differential algebra [15] is the most adequate.

Consider evolution equations of the form (1.26). Suppose that the right-hand side of equation (1.26) as well as all other functions of variables (1.1) belong to a certain differential field $\mathcal{F}$ (see Section 1.1.3).

As usual in differential algebra, the principle derivation

$$D \stackrel{def}{=} \frac{\partial}{\partial x} + \sum_{i=0}^{\infty} u_{i+1} \frac{\partial}{\partial u_i} \qquad (2.1)$$

generates all independent variables $u_i = D^i(u_0)$, starting with $u_0 = u$. This derivation is a formalization of the total derivative with respect to x which acts on functions of the form $g\left(x, u(x), \dfrac{\partial u}{\partial x}, \dots\right)$.

Remark 2.0.1. *We will often use that $D(f) = 0$, $f \in \mathcal{F}$, implies $f = $ const.*

Remark 2.0.2. *The variable t in the local theory of evolution equations plays the role of a parameter.*

We associate each evolution equation (1.26) with the infinite-dimensional vector field

$$D_F = F \frac{\partial}{\partial u_0} + D(F) \frac{\partial}{\partial u_1} + D^2(F) \frac{\partial}{\partial u_2} + \cdots \qquad (2.2)$$

on $\mathcal{F}$. It is easy to see that this vector field commutes with D. We call vector fields of the form (2.2) *evolutionary*. The function F is called the *generator* of the evolutionary vector field (2.2).

Remark 2.0.3. *The derivation* (2.2) *is called total derivative with respect to t by means of equation* (1.26) *and is often denoted by* D_t.

The set of all evolutionary vector fields forms a Lie algebra over $\mathbb{C}$ with respect to the commutation operation. Herewith, $[D_G, D_H] = D_K$, where

$$K = D_G(H) - D_H(G) = H_*(G) - G_*(H), \qquad (2.3)$$

and the Fréchet derivative a_* for any $a \in \mathcal{F}$ is defined by formula (1.4). Formula (2.3) defines a Lie bracket on our differential field $\mathcal{F}$.

In Section 1.4, we have defined the infinitesimal symmetry of equation (1.26) as an evolution equation (1.32),[1] compatible with (1.26). Now, we can clarify that by compatibility we mean $[D_F, D_G] = 0$. It can be rewritten as

$$G_*(F) - F_*(G) = D_t(G) - F_*(G) = 0. \qquad (2.4)$$

The concept of infinitesimal symmetry is connected with the procedure of linearization of a differential equation. For an arbitrary partial differential equation

$$Q(u, u_x, u_t, \dots) = 0$$

consider a linear equation

$$0 = \frac{\partial}{\partial \varepsilon} \Big(Q(u + \varepsilon \varphi, u_x + \varepsilon \varphi_x, u_t + \varepsilon \varphi_t, \dots) \Big) \Big|_{\varepsilon=0} \overset{def}{=} \mathcal{L}(\varphi).$$

Linear differential operator

$$\mathcal{L} = \frac{\partial Q}{\partial u} + \frac{\partial Q}{\partial u_x} D + \frac{\partial Q}{\partial u_t} D_t + \cdots$$

is called the *linearization operator* for the equation $Q = 0$. For the evolution equation (1.26), we have

$$\mathcal{L} = D_t - F_*.$$

The formula (2.4) means that the symmetry generator G is an element from the kernel of the linearization operator. Such a definition of infinitesimal symmetry can be literally transferred to the case of nonevolution equations.

---

[1] For brevity, we often call the symmetry generator G simply the symmetry G.

Definition 2.0.4. A local conservation law (see Section 1.5) for equation (1.26) is a pair of functions $\rho,\ \sigma \in \mathcal{F}$ such that

$$D_t(\rho) = D(\sigma).$$

The functions ρ and σ are called the *conserved density* and the *flux*, respectively.

Suppose that a function ρ is a conserved density. Then, for any function $s \in \mathcal{F}$, the function $\bar{\rho} = \rho + D(s)$ is also a density of the conservation law. We call the densities ρ and $\bar{\rho}$ *equivalent* and write $\rho \sim \bar{\rho}$.

2.1 Scalar Lax pairs for evolution equations

In this section, L-operators are differential operators or "ratios of differential operators". The corresponding A-operators are constructed using formal non-commutative pseudo-differential series.

2.1.1 Pseudo-differential series

Consider a skew field of (noncommutative) formal series of the form

$$A = a_m D^m + a_{m-1} D^{m-1} + \cdots + a_0 + a_{-1} D^{-1} + a_{-2} D^{-2} + \cdots \qquad a_k \in \mathcal{F}. \quad (2.5)$$

The number $m \in \mathbb{Z}$ is called the *order* of the series A and is denoted by ord A. If $a_i = 0$, for $i < 0$, then A is called a *differential operator*.

The product of two monomials is defined by the formula

$$D^k \circ b D^m = b D^{m+k} + C_k^1 D(b) D^{k+m-1} + C_k^2 D^2(b) D^{k+m-2} + \cdots ,$$

where $k, m \in \mathbb{Z}$, $b \in \mathcal{F}$ and C_n^j are the binomial coefficients:

$$C_n^j = \frac{n(n-1)(n-2)\cdots(n-j+1)}{j!}, \qquad n \in \mathbb{Z}.$$

For series this formula is extended by distributivity.

Remark 2.1.1. *For any series A and B we have* ord$(A \circ B - B \circ A) \leq$ ord $A +$ ord $B - 1$.

The series A^+, formally conjugate to A, is defined as

$$A^+ = (-1)^m D^m \circ a_m + (-1)^{m-1} D^{m-1} \circ a_{m-1} + \cdots + a_0 - D^{-1} \circ a_{-1} + D^{-2} \circ a_{-2} + \cdots .$$

Example 2.1.2. Let

$$A = uD^2 + u_1 D, \qquad B = -u_1 D^3, \qquad C = uD^{-1}.$$

Then,

$$A^+ = D^2 \circ u - D \circ u_1 = A,$$

$$B^+ = D^3 \circ u_1 = u_1 D^3 + 3u_2 D^2 + 3u_3 D + u_4,$$

$$C^+ = -D^{-1} u = -uD^{-1} + u_1 D^{-2} - u_2 D^{-3} + \cdots.$$

For any series (2.5), the inverse series

$$B = b_{-m} D^{-m} + b_{-m-1} D^{-m-1} + \cdots, \qquad b_k \in \mathcal{F}$$

such that $A \circ B = B \circ A = 1$ is uniquely defined. Indeed, multiplying A and B and equating the result to 1, we find that $a_m b_{-m} = 1$, i.e. $b_{-m} = \dfrac{1}{a_m}$. Comparing the coefficients of D^{-1}, we obtain

$$m a_m D(b_{-m}) + a_m b_{-m-1} + a_{m-1} b_{-m} = 0,$$

and, therefore,

$$b_{-m-1} = -\frac{a_{m-1}}{a_m^2} - m D\left(\frac{1}{a_m}\right), \qquad \text{and so on.}$$

In addition, we can determine the root of degree m from the series A as a series

$$C = c_1 D + c_0 + c_{-1} D^{-1} + c_{-2} D^{-2} + \cdots$$

such that $C^m = A$. This root is defined up to a constant factor ε, where $\varepsilon^m = 1$.

Example 2.1.3. Let $A = D^2 + u$. Assuming that

$$C = c_1 D + c_0 + c_{-1} D^{-1} + c_{-2} D^{-2} + \cdots,$$

we find

$$C^2 = C \circ C = c_1^2 D^2 + \left(c_1 D(c_1) + c_1 c_0 + c_0 c_1\right) D + c_1 D(c_0) + c_0^2 + c_1 c_{-1} + c_{-1} c_1 + \cdots.$$

Let us compare C^2 with A. From the coefficients of D^2 we find that $c_1^2 = 1$ or $c_1 = \pm 1$. Let $c_1 = 1$. Comparing the coefficients of D, we get $2c_0 = 0$ or $c_0 = 0$. Considering the terms with D^0, we find that $2c_{-1} = u$, the terms with D^{-1} lead to $c_{-2} = -\dfrac{u_1}{4}$, i.e.

$$C = A^{1/2} = D + \frac{u}{2} D^{-1} - \frac{u_1}{4} D^{-2} + \cdots.$$

Definition 2.1.4. The *residue* of a formal series (2.5) is the coefficient of D^{-1}:

$$\operatorname{res}(A) \overset{def}{=} a_{-1}.$$

The *logarithmic residue* of A is defined as

$$\operatorname{res} \log A \overset{def}{=} \frac{a_{m-1}}{a_m}.$$

We will use the following important statement.

Theorem 2.1.5. [39] *For any formal series A, B, the residue of the commutator belongs to* Im D:

$$\operatorname{res}[A, B] = D(\sigma(A, B)),$$

where

$$\sigma(A, B) = \sum_{p \leq \operatorname{ord}(B),\, q \leq \operatorname{ord}(A)}^{p+q+1>0} C_q^{p+q+1} \times \sum_{s=0}^{p+q}(-1)^s D^s(a_q) D^{p+q-s}(b_q).$$

Corollary 2.1.6. *For any series L and T*

$$\operatorname{res}(L - TLT^{-1}) \in \operatorname{Im} D.$$

Proof. The statement follows from the identity

$$L - TLT^{-1} = [LT^{-1}, T]$$

and Theorem 2.1.5. □

2.1.2 Korteweg–de Vries hierarchy

The Lax pair for the KdV equation (1.12) is determined by the operators [20]

$$L = D^2 + u, \qquad A = 4\left(D^3 + \frac{3}{2}uD + \frac{3}{4}u_x\right). \tag{2.6}$$

Using these L and A-operators, we will demonstrate how the scalar differential Lax pair generates higher symmetries, conservation laws, and exact solutions of the soliton type.

It is easy to verify that the commutator $[A, L]$ does not contain powers of D (i.e., it is an operator of order zero) and is equal to the left-hand side of equation (1.12). Since $L_t = u_t$, the relation (1.7) is equivalent to (1.12).

Question 2.1.7. *How to describe differential operators*

$$P_n = D^n + \sum_{i=0}^{n-1} a_i D^i$$

such that $[P_n, L]$ is a differential operator of order zero?

Remark 2.1.8. *Obviously, for any such operator P_n, relation $L_t = [P_n, L]$ is equivalent to an evolution equation of the form* (1.26).

Definition 2.1.9. For any series

$$Q = \sum_{i=-\infty}^{k} a_i D^i$$

we denote

$$Q_+ = \sum_{i=0}^{k} a_i D^i, \qquad Q_- = \sum_{i=-\infty}^{-1} a_i D^i.$$

Remark 2.1.10. *From a scientific point of view, we consider the decomposition of the associative algebra of all pseudo-differential series into a direct sum of two vector subspaces, one of which is a subalgebra of differential operators and the other is a subalgebra of all series of negative orders. The symbols $+$ and $-$ mean projections onto these subalgebras.*

Lemma 2.1.11. *If Q is a pseudo-differential series such that $[L, Q] = 0$, then*

$$[Q_+, L] = f, \qquad f \in \mathcal{F}. \tag{2.7}$$

Proof. Since $[L, Q_+ + Q_-] = 0$, we have

$$[Q_+, L] = -[Q_-, L].$$

The left-hand side of this identity is a differential operator, while, according to Remark 2.1.1, the order of the right-hand side is nonpositive. □

Lemma 2.1.12. *The following relation holds:*

$$[L, L^{\frac{1}{2}}] = 0.$$

Proof. Let

$$[L, L^{\frac{1}{2}}] = sD^p + \cdots.$$

Then, (see Example 2.1.3) we have

$$0 = [L, L] = [L, L^{\frac{1}{2}}]L^{\frac{1}{2}} + L^{\frac{1}{2}}[L, L^{\frac{1}{2}}] = 2sD^{p+1} \cdots,$$

and, therefore, $s = 0$. □

Corollary 2.1.13. *It follows from Lemmas 2.1.11 and 2.1.12 that for any $n \in \mathbb{Z}_+$ the differential operator $P = L_+^{\frac{n}{2}}$ satisfies the relation*

$$[P, L] = f_P \tag{2.8}$$

for some $f_P \in \mathcal{F}$. If n is even, then $f_P = 0$.

It is clear that the set of all differential operators satisfying (2.8) is a vector space over $\mathbb{C}$.

Lemma 2.1.14. *The differential operators $L_+^{\frac{n}{2}}$, $n \in \mathbb{Z}_+$, form a basis of this vector space.*

Proof. Suppose that $P = \sigma D^p + \cdots$ satisfies (2.8). Equating the coefficients of D^{p+1} in (2.8), we obtain that $D(\sigma) = 0$ and, therefore, $\sigma = \text{const}$. Since the operator $\sigma L_+^{\frac{p}{2}}$ has the same leading coefficient as P, the operator $P - \sigma L_+^{\frac{p}{2}}$ has strictly lower order than P. The induction over p completes the proof. $\qquad\square$

Let

$$[L_+^{\frac{n}{2}}, L] = f_n. \tag{2.9}$$

For even n we have $f_n = 0$ and the evolution equation $u_t = f_n$ that is equivalent to

$$L_t = [L_+^{\frac{n}{2}}, L] \tag{2.10}$$

is trivial. Denote

$$A_n = L_+^{\frac{2n+1}{2}}.$$

It is easy to verify that $A = 4A_1$, where A is the operator from the formula (2.6).

Remark 2.1.15. *The evolution equation corresponding to the case $n = 0$ is $u_t = u_x$. To any function f_{2n+1} one can add a "trivial" term of the form $\text{const}\, u_x$ corresponding to adding a term proportional to $A_0 = D$ to the operator A_n.*

Theorem 2.1.16. *For any $n, m \in \mathbb{N}$ the evolution equations $u_\tau = f_{2m+1}$ and $u_t = f_{2n+1}$ are compatible.[2]*

Proof. Let us rewrite these equations in the Lax form

$$L_t = [A_n, L], \qquad L_\tau = [A_m, L].$$

We have

$$(L_t)_\tau - (L_\tau)_t = [(A_n)_\tau, L] - [(A_m)_t, L] + [A_n, [A_m, L]] - [A_m, [A_n, L]].$$

---

[2]In other words, the first equation is an infinitesimal symmetry of the second equation and vice versa.

By the Jacobi identity, it suffices to prove that

$$(A_n)_\tau - (A_m)_t + [A_n, A_m] = 0.$$

Since $(L^p)_t = [A_n, L^p]$ and $(L^p)_\tau = [A_m, L^p]$ for any p, we have

$$(A_n)_\tau = \left([A_m, L^{\frac{2n+1}{2}}]\right)_+, \qquad (A_m)_t = \left([A_n, L^{\frac{2m+1}{2}}]\right)_+.$$

Therefore, it remains to prove that

$$\left([A_m, L^{\frac{2n+1}{2}}] - [A_n, L^{\frac{2m+1}{2}}] + [A_n, A_m]\right)_+ = 0.$$

Substituting

$$A_n = L^{\frac{2n+1}{2}} - \left(L^{\frac{2n+1}{2}}\right)_-$$

and

$$A_m = L^{\frac{2m+1}{2}} - \left(L^{\frac{2m+1}{2}}\right)_-$$

to the last identity, we obtain

$$\left[\left(L^{\frac{2n+1}{2}}\right)_-, \left(L^{\frac{2m+1}{2}}\right)_-\right]_+ = 0,$$

which is obviously true.   □

Corollary 2.1.17. *Any two evolution equations, defined by Lax equation* (1.7) *with different A-operators of the form*

$$A = \sum_{i \geq 0} c_i L_+^{\frac{2i+1}{2}}, \qquad c_i \in \mathbb{C},$$

are compatible.

The infinite set of pairwise compatible evolution equations $u_t = f_{2n+1}$, $n = 0, 1, \ldots$ is called the *KdV hierarchy* [40]. All of them are generated by the same L-operator, but different A-operators. This is typical for other integrable hierarchies.

Recursion operator for the KdV equation

Here, we find a recursion relation between the right-hand sides f_{2n+1} and f_{2n+3} of the equations of the KdV hierarchy. A similar method was used for the first time [41] to find the recursion operator for the Krichever–Novikov equation (see Section 4.3.3).

Since $L^{\frac{2n+3}{2}} = L\,L^{\frac{2n+1}{2}}$, we have

$$A_{n+1} = (L\,L^{\frac{2n+1}{2}})_+ = (L\,(L^{\frac{2n+1}{2}})_+)_+ + (L\,(L^{\frac{2n+1}{2}})_-)_+ = L\,(L^{\frac{2n+1}{2}})_+ + (L\,(L^{\frac{2n+1}{2}})_-)_+,$$

or

$$A_{n+1} = LA_n + R_n,$$

where $R_n = a_n D + b_n$ is some first order differential operator. Using (2.9), we obtain

$$f_{2n+3} = [A_{n+1},\,L] = L \circ f_{2n+1} + [R_n,\,L].$$

Equating coefficients of D^2, D and D^0 in this relation, we obtain

$$a_n = \frac{1}{2}D^{-1}(f_{2n+1}), \qquad b_n = \frac{3}{4}f_{2n+1}$$

and

$$f_{2n+3} = \left(\frac{1}{4}D^2 + u + \frac{1}{2}u_x D^{-1}\right) f_{2n+1},$$

what gives the standard recursion operator

$$\mathcal{R} = \frac{1}{4}D^2 + u + \frac{1}{2}u_x D^{-1} \tag{2.11}$$

for the KdV equation

$$u_t = \frac{1}{4}\left(u_{xxx} + 6uu_x\right). \tag{2.12}$$

The multiplier $\frac{1}{4}$ appeared because we took as A-operator $L_+^{\frac{3}{2}}$ instead of $4L_+^{\frac{3}{2}}$. Of course, this factor can be eliminated by rescaling $t \mapsto 4\,t$.

As shown in [42], this method of finding the recursion operator can be generalized to the case of L-operators of various types.

Exercise 2.1.18. Verify that the recursion operator (2.11) satisfies the operator relation

$$\mathcal{R}_t = [F_*,\,\mathcal{R}],$$

where

$$F_* = \frac{1}{4}\left(D^3 + 6uD + 6u_x\right)$$

is the Fréchet derivative of the right-hand side of the KdV equation (2.12).[3]

---

[3]In the language of differential geometry, this relation means that the Lie derivative of the operator R, by virtue of the KdV equation, is zero.

Exercise 2.1.19. (see [42]) Find the recursion operator for the Boussinesq equation written as a system

$$u_t = v_x, \qquad v_t = -\frac{1}{3}(u_{xxx} + 8uu_x).$$

The Lax pair for this system is given by

$$L = D^3 + 2uD + u_x + v, \qquad A = \left(L^{\frac{2}{3}}\right)_+.$$

Conservation laws for the KdV equation

Proposition 2.1.1. *For any* $n \in \mathbb{N}$, *function*

$$\rho_n = \mathrm{res}\,(L^{\frac{2n-1}{2}}), \tag{2.13}$$

where L *is defined by the formula* (2.6), *is a conserved density for the KdV equation.*

Proof. It is easy to show (cf. Lemma 1.2.1) that

$$(L^{\frac{2n-1}{2}})_t = [A,\,(L^{\frac{2n-1}{2}})].$$

Finding the residue from both sides of this identity and taking into account Theorem 2.1.5, we arrive at the desired statement. □

It is easy to verify that formula (2.13), for $n = 1, 2, 3$, gives conserved densities equivalent to those of Example 1.5.3.

Darboux transformation

The Darboux transformation for the KdV equation is defined by the following operator relation:

$$\tilde{L} = TLT^{-1}, \tag{2.14}$$

where

$$T = D^n + a_{n-1}D^{n-1} + \cdots + a_0$$

is a differential operator. In the general case, $\tilde{L}$ is a pseudo-differential series, but for special T this series turns out to be a differential operator of the form $\tilde{L} = D^2 + \tilde{u}$. In this case, we have

$$(D^n + a_{n-1}D^{n-1} + \cdots + a_0)(D^2 + u) = (D^2 + \tilde{u})(D^n + a_{n-1}D^{n-1} + \cdots + a_0). \tag{2.15}$$

Comparing the coefficients of D^n, we obtain

$$\tilde{u} = u - 2\frac{\partial}{\partial x}a_{n-1}.$$

This formula allows us to construct a new solution $\tilde{u}$ of the KdV equation based on the solution u.

From (2.15) it follows that

$$L(\operatorname{Ker} T) \subset \operatorname{Ker} T. \tag{2.16}$$

The existence of the Euclidean algorithm in the ring of differential operators [43] ensures that (2.16) is a necessary and sufficient condition that the $\tilde{L}$ is a differential operator.

Suppose that the Jordan form of the operator L, which acts on the finite-dimensional space $\operatorname{Ker} T$, is diagonal with pairwise distinct eigenvalues λ_i^2, $i = 1, \ldots n$, $\lambda_i \in \mathbb{C}$. Then, as a basis in $\operatorname{Ker} T$, we can take functions $\Psi_1, \ldots, \Psi_n$, such that

$$\frac{\partial^2}{\partial x^2}\Psi_i + u\Psi_i = \lambda_i^2\Psi_i. \tag{2.17}$$

The operator T up to a left multiplier is defined by

$$T(Y) = \mathbf{W}(\Psi_1, \ldots, \Psi_n, Y),$$

where $\mathbf{W}$ is the Wronskian. From here and from the assumption that the leading coefficient of T is equal to 1 it follows that

$$a_{n-1} = -\frac{\partial}{\partial x}\log \mathbf{W}(\Psi_1, \ldots, \Psi_n)$$

and, therefore, the Darboux transformation is given by the formula

$$\tilde{u} = u + 2\frac{\partial^2}{\partial x^2}\log \mathbf{W}(\Psi_1, \ldots, \Psi_n). \tag{2.18}$$

As usual, the t-dynamics of the functions Ψ_i is determined by the A-operator. For the KdV equation (2.12) we have

$$\frac{\partial}{\partial t}(\Psi_i) = A(\Psi_i) \equiv \frac{\partial^3}{\partial x^3}\Psi_i + \frac{3}{2}u\frac{\partial}{\partial x}(\Psi_i) + \frac{3}{4}u_x\Psi_i. \tag{2.19}$$

Theorem 2.1.20. *Let $u(x,t)$ be an arbitrary solution of the KdV equation (2.12). If functions Ψ_i satisfy linear equations (2.17) and (2.19), then the function $\tilde{u}(x,t)$ defined by the formula (2.18) satisfies the KdV equation.*

Proof. The Lax equation (1.7) can be rewritten in the commutator form $[\frac{\partial}{\partial t} - A, L] = 0$. The latter is equivalent to relation $[\frac{\partial}{\partial t} - \tilde{A}, \tilde{L}] = 0$, where the differential operator $\tilde{L}$ is defined by (2.14) and

$$\tilde{A} = TAT^{-1} + T_t T^{-1}. \tag{2.20}$$

To complete the proof it suffices to verify that the ratio (2.20) of differential operators $TA + T_t$ and T is a differential operator. This is equivalent to the inclusion

$$\operatorname{Ker} T \subset \operatorname{Ker}(TA + T_t),$$

or to the relations

$$TA(\Psi_i) + T_t(\Psi_i) = 0, \tag{2.21}$$

for $i = 1, \ldots, n$. We have $0 = (T\Psi_i)_t = T_t \Psi_i + T(\Psi_i)_t$. Substituting $-T(\Psi_i)_t$ instead of $T_t(\Psi_i)$ into (2.21) and using (2.19), we obtain the desired assertion. □

Remark 2.1.21. *The numbers λ_i from (2.17) are arbitrary parameters in the solution (2.18).*

Exercise 2.1.22. Prove that for any Jordan form of the operator L restricted onto the finite-dimensional space $\operatorname{Ker} T$ the condition

$$\left(\frac{\partial}{\partial t} - A\right) \operatorname{Ker} T \subset \operatorname{Ker} T \tag{2.22}$$

ensures that the series $\tilde{A}$ defined by (2.20) is a differential operator.

Solitons and rational solutions of the KdV equation

We apply the procedure described in the previous section to the trivial solution $u(x,t) = 0$ of the KdV equation. In this case, condition (2.16) means that $\operatorname{Ker} T$ is any finite-dimensional space of functions **V** invariant with respect to the differential operator $\frac{\partial^2}{\partial x^2}$. The dynamics of elements of **V** with respect to t is determined by the condition

$$\left(\frac{\partial}{\partial t} - \frac{\partial^3}{\partial x^3}\right) \mathbf{V} \subset \mathbf{V}.$$

In the generic case, when the Jordan form of the operator $\frac{\partial^2}{\partial x^2}$ is diagonal with distinct eigenvalues λ_i^2, a basis in V consists of the functions

$$\Psi_i(x,t) = \exp(\eta_i) + c_i \exp(-\eta_i), \qquad \text{where} \qquad \eta_i = \lambda_i x + \lambda_i^3 t, \qquad i = 1, \ldots, n.$$

The following function

$$\tilde{u} = 2 \frac{\partial^2}{\partial x^2} \log \mathbf{W}(\Psi_1, \ldots, \Psi_n)$$

is called the *n-soliton* solution of the KdV equation. The numbers λ_i, c_i are parameters of the soliton.

Example 2.1.23. In the case of $n = 1$, we have

$$\tilde{u}(x,t) = \frac{8 c_1 \lambda_1^2}{(e^{\lambda_1 x + \lambda_1^3 t} + c_1 e^{-\lambda_1 x - \lambda_1^3 t})^2}.$$

Exercise 2.1.24. Verify that the two-soliton solution of the KdV equation is given by

$$\tilde{u}(x,t) = (\lambda_2^2 - \lambda_1^2) \frac{\dfrac{8 c_1 \lambda_1^2}{(e^{\eta_1} + c_1 e^{-\eta_1})^2} - \dfrac{8 c_2 \lambda_2^2}{(e^{\eta_2} + c_2 e^{-\eta_2})^2}}{\left(\lambda_1 \dfrac{c_1 - e^{2\eta_1}}{c_1 + e^{2\eta_1}} - \lambda_2 \dfrac{c_2 - e^{2\eta_2}}{c_2 + e^{2\eta_2}} \right)^2}.$$

If the vector space $\mathbf{V}$ consists of polynomials in x, we get a rational solution of the KdV equation. In the simplest case $\dim \mathbf{V} = 1$, we have $\Psi_1 = x$ and the formula (2.18) generates a stationary rational solution of the KdV equation

$$\tilde{u}(x,t) = -\frac{2}{x^2}.$$

2.1.3 Gelfand–Dikii hierarchy and its generalizations

Let

$$L = D^n + \sum_{i=0}^{n-2} u_i D^i, \qquad A = \sum_{i=0}^{m} c_i L_+^{\frac{i}{n}}, \qquad c_i \in \mathbb{C}. \tag{2.23}$$

In the same way as Lemmas 2.1.11 and 2.1.14, one can prove that the Lax equation (1.7) is equivalent to a system of $n - 1$ evolution equations for the unknown functions $u_{n-2}, \ldots, u_0$. In this case, systems generated by the same L-operator and A-operators of the form (2.23) with different m and constants c_i are infinitesimal symmetries for each other. This infinite set of evolutionary systems is called the *Gelfand–Dikii hierarchy* [44]. For $n = 2$, we obtain the KdV hierarchy described above.

Factorization of L-operators

The relations between a factorization of scalar differential operators, substitutions of the Miura transformation type, and mKdV type systems have been discussed, for example, in [45, 46, 37]. Here, we only deal with the case of two factors.

Consider the following system of operator equations:

$$M_t = AN - MB, \qquad N_t = BN - NA, \tag{2.24}$$

where

$$M = D^r + wD^{r-1} + \sum_{i=0}^{r-2} u_i D^i, \qquad N = D^s wD^{s-1} + \sum_{i=0}^{s-2} v_i D^i,$$

$$A = \sum_{i=0}^{m} c_i\left((MN)^{\frac{i}{r+s}}\right)_+, \qquad B = (M^{-1}AM)_+.$$

System (2.24) is connected with the Lax equation (1.7). Namely, if M and N satisfy (2.24), then the product $L = MN$ satisfies (1.7).

Proposition 2.1.2. *Relations (2.24) are equivalent to a system of $r + s - 1$ evolution equations for w, u_i, v_i.*

Reductions in scalar L-operators and simple Lie algebras

We introduce the following notation:

$$Q_1(n) \stackrel{def}{=} D^{2n+1} + \sum_{i=0}^{n-1} u_i D^{2i+1} + D^{2i+1} u_i, \qquad Q_2(n) \stackrel{def}{=} D^{2n} + \sum_{i=0}^{n-1} u_i D^{2i} + D^{2i} u_i,$$

$$Q_3(n) \stackrel{def}{=} D^{2n-1} + \sum_{i=1}^{n-1} u_i D^{2i-1} + D^{2i-1} u_i + u_0 D^{-1} u_0.$$

The coefficients $u_{n-1}, \ldots, u_0$ are called *functional parameters* of the operator $Q_i(n)$. By definition,

$$Q_1(0) \stackrel{def}{=} D, \qquad Q_2(0) \stackrel{def}{=} 1, \qquad Q_3(0) \stackrel{def}{=} D^{-1}.$$

Note that the operators $Q_1(n)$ and $Q_3(n)$ are skew-symmetric (see Section 1.1.2): $Q_1(n)^+ = -Q_1(n)$ and $Q_3(n)^+ = -Q_3(n)$. The operators $Q_2(n)$ are symmetric: $Q_2(n)^+ = Q_2(n)$.

Deep connections between such operators and classical simple Lie algebras were established in [37, section 7]. It turns out that the algebra of type B_n corresponds to operator $Q_1(n)$, while the algebras of types C_n and D_n correspond to operators $Q_2(n)$ and $Q_3(n)$, respectively.

Theorem 2.1.25. (see [37, 47]) *Suppose that operators* $M = Q_i(r)$ *and* $N = Q_j(s)$, *where* $i, j \in \{1, 2, 3\}$, *have functional parameters* $u_{r-1}, \ldots, u_0$ *and* v_{s-1}, $\ldots, v_0$, *respectively. Then, the relations* (2.24), *where*

$$A = \sum_{i=0}^{m} c_i \left(L^{\frac{2i+1}{n}} \right)_+, \qquad B = \left(M^{-1} A M \right)_+,$$

are equivalent to a system of $r + s$ *evolution equations for* $u_{r-1}, \ldots, u_0, v_{s-1}, \ldots, v_0$.

Here, we give several examples of such systems with $r + s \leq 2$ and operator A of the lowest possible order. Instead of the operators M and N, we present their product $L = MN$. In the corresponding nonlinear differential equations, we produce scalings of independent variables and unknown functions to reduce the coefficients to the simplest form.

Example 2.1.26. In the cases

(a) $L = D^2 + u, \qquad A = \left(L^{\frac{3}{2}} \right)_+;$

(b) $L = (D^2 + u)D^{-1}, \qquad A = \left(L^3 \right)_+;$

(c) $L = (D^3 + 2uD + u_x)D^{-1}, \qquad A = \left(L^{\frac{3}{2}} \right)_+;$

(d) $L = (D^3 + 2uD + u_x)D, \qquad A = \left(L^{\frac{3}{4}} \right)_+$

we get the KdV equation (1.12).

Open problem 2.1.27. *Prove that any Lax pair with polynomial homogeneous coefficients for the KdV equation, where* L *is a pseudo-differential series and* A *is a third order differential operator, is equivalent[4] to one of the above four pairs.*

Example 2.1.28. The cases

(a) $L = D + uD^{-1}u, \qquad A = \left(L^3 \right)_+;$

(b) $L = (D + uD^{-1}u)D, \qquad A = \left(L^{\frac{3}{2}} \right)_+$

lead to the mKdV equation

$$u_t = u_{xxx} + 6u^2 u_x. \tag{2.25}$$

---

[4]The L-operator can be raised to a fractional power.

Example 2.1.29. The operators

$$L = (D^2 + u)D, \qquad A = \left(L^{\frac{5}{3}}\right)_+$$

produce equation (1.42).

Example 2.1.30. In the case

$$L = D^3 + 2uD + u_x, \qquad A = \left(L^{\frac{5}{3}}\right)_+$$

we arrive at equation (1.43).

Example 2.1.31. The system

$$u_t = vv_x, \qquad v_t = v_{xxx} + 2uv_x + vu_x \qquad (2.26)$$

corresponds to the operators

$$L = D^3 + 2uD + u_x + vD^{-1}v, \qquad A = L_+.$$

Example 2.1.32. The operators

$$L = (D^4 + uD^2 + D^2u + v)D^{-1}, \qquad A = L_+$$

lead to

$$u_t = w_x, \qquad w_t = w_{xxx} + wu_x + uw_x, \qquad (2.27)$$

where $w = v + \alpha v_{xx}$, and α is a properly chosen constant.

Example 2.1.33. For

$$L = (D^5 + uD^3 + D^3u + vD + Dv)D^{-1}, \qquad A = \left(L^{\frac{3}{4}}\right)_+$$

we obtain

$$u_t = -u_{xxx} + w_x - uu_x, \qquad w_t = 2w_{xxx} + uw_x, \qquad (2.28)$$

where $w = v + \beta u_{xx}$ and β is an appropriate constant. An equivalent system is generated by the Lax pair

$$L = D^4 + uD^2 + D^2u + v, \qquad A = \left(L^{\frac{3}{4}}\right)_+. \qquad (2.29)$$

Example 2.1.34. The operators

$$L = (D^3 + 2uD + u_xu + vD^{-1}v)D, \qquad A = \left(L^{\frac{3}{4}}\right)_+$$

produce the system

$$u_t = u_{xxx} + uu_x - vv_x, \qquad v_t = -2v_{xxx} - uv_x. \qquad (2.30)$$

Several more examples can be found in [37, 47].

Some further reductions in Lax operators from the above examples are possible. The most sophisticated of them defines a Lax pair for the Krichever–Novikov equation [48]

$$u_t = u_{xxx} - \frac{3}{2} \frac{u_{xx}^2}{u_x} + \frac{P(u)}{u_x},$$

where $P''''(u) = 0$. This pair is given by

$$L = (D^2 + \frac{1}{6}W)^2 + \frac{1}{18}\Big(-P'' + \frac{6\,P'}{u-k} - \frac{12\,P}{(u-k)^2}\Big), \qquad A = 4\Big(L^{\frac{3}{4}}\Big)_+,$$

where k is any root of equation $P(k) = 0$ and the function W has the form

$$W = -\frac{3u_{xxx}}{u_x} + \frac{3}{2}\frac{u_{xx}^2}{u_x^2} - \frac{P}{u_x^2} + \frac{12\,u_{xx}}{u-k} - \frac{12\,u_x^2}{(u-k)^2}.$$

This reduction of the Lax pair (2.29) was found in [41] in the case of cubic polynomial P (which corresponds to $k = \infty$). Unfortunately, there is a misprint in the expression for W in this paper (see [49] for a correct formula).

In fact, the reduction is determined by the existence of an operator of sixth order commuting with the operator L (see [48]).

Exercise 2.1.35. Find this operator explicitly.

2.2 Matrix Lax pairs

2.2.1 The NLS hierarchy

The nonlinear Schrödinger equation (NLS equation) has the form

$$Z_t = iZ_{xx} + |Z|^2 Z.$$

After (complex) scalings of t and Z the equation can be written as a system of two equations

$$u_t = \frac{1}{2}\left(u_{xx} - 2\,u^2\,v\right), \qquad v_t = \frac{1}{2}\left(-v_{xx} + 2\,v^2\,u\right), \qquad (2.31)$$

where $u = Z$, $v = \bar{Z}$. A Lax representation (1.7) for (2.31) is defined [21] by the operators

$$L = D + \begin{pmatrix} 1 & 0 \\ 0 & -1 \end{pmatrix} \lambda + \begin{pmatrix} 0 & u \\ v & 0 \end{pmatrix},$$

$$A = \begin{pmatrix} 1 & 0 \\ 0 & -1 \end{pmatrix} \lambda^2 + \begin{pmatrix} 0 & u \\ v & 0 \end{pmatrix} \lambda + \frac{1}{2} \begin{pmatrix} -uv & -u_x \\ v_x & uv \end{pmatrix}.$$

Note that the coefficients of the powers of λ in L and A belong to the Lie algebra $\mathfrak{sl}_2$.

It is easy to verify the above matrix A has the following properties:

(a) The commutator $[A, L]$ does not depend on λ;

(b) It has the following matrix structure:

$$[A, L] = \begin{pmatrix} 0 & * \\ * & 0 \end{pmatrix}.$$

We aim to find all matrices A satisfying these properties.

It is clear that if the matrix polynomial

$$A_n = \sum_{i=0}^{n} a_i \lambda^i, \qquad a_i \in \mathfrak{sl}_2, \tag{2.32}$$

obeys these properties, the Lax equation $L_t = [A_n, L]$ is equivalent to a system of two evolution equations for u and v.

Question 2.2.1. *How to describe all matrix polynomials (2.32) that satisfy these two requirements (a) and (b)?*

In the next section we answer question 2.2.1.

Formal diagonalization [37]

Theorem 2.2.2. *There is a unique series*

$$T = 1 + \begin{pmatrix} 0 & \alpha_1 \\ \beta_1 & 0 \end{pmatrix} \frac{1}{\lambda} + \begin{pmatrix} 0 & \alpha_2 \\ \beta_2 & 0 \end{pmatrix} \frac{1}{\lambda^2} + \cdots$$

such that

$$T^{-1} L T = L_0,$$

where

$$L_0 = D + \begin{pmatrix} 1 & 0 \\ 0 & -1 \end{pmatrix} \lambda + \begin{pmatrix} \rho_0 & 0 \\ 0 & -\rho_0 \end{pmatrix} + \begin{pmatrix} \rho_1 & 0 \\ 0 & -\rho_1 \end{pmatrix} \frac{1}{\lambda} + \begin{pmatrix} \rho_2 & 0 \\ 0 & -\rho_2 \end{pmatrix} \frac{1}{\lambda^2} + \cdots$$

Proof. Equating the coefficients of λ^0 in the relation $LT = TL_0$, we get

$$\left[\begin{pmatrix} 1 & 0 \\ 0 & -1 \end{pmatrix}, \begin{pmatrix} 0 & \alpha_1 \\ \beta_1 & 0 \end{pmatrix}\right] - \begin{pmatrix} \rho_0 & 0 \\ 0 & -\rho_0 \end{pmatrix} = \begin{pmatrix} 0 & u \\ v & 0 \end{pmatrix}.$$

Therefore, $\rho_0 = 0$, $\alpha_1 = \dfrac{1}{2}u$ and $\beta_1 = -\dfrac{1}{2}v$. At each next step, we obtain a similar relation of the form

$$\left[\begin{pmatrix} 1 & 0 \\ 0 & -1 \end{pmatrix}, \begin{pmatrix} 0 & \alpha_k \\ \beta_k & 0 \end{pmatrix}\right] - \begin{pmatrix} \rho_{k-1} & 0 \\ 0 & -\rho_{k-1} \end{pmatrix} = P_k,$$

where $P_k \in \mathfrak{sl}_2$ is an already known matrix. Functions $\alpha_k, \beta_k, \rho_{k-1}$ are determined from such a relation uniquely. $\square$

Proposition 2.2.1. *Let*

$$B_n = T \begin{pmatrix} 1 & 0 \\ 0 & -1 \end{pmatrix} T^{-1} \lambda^n, \tag{2.33}$$

$$A_n = (B_n)_+, \tag{2.34}$$

where, by definition,

$$\left(\sum_{i=-\infty}^{m} a_i \lambda^i\right)_+ \stackrel{def}{=} \sum_{i=0}^{m} a_i \lambda^i, \qquad \left(\sum_{i=-\infty}^{m} a_i \lambda^i\right)_- \stackrel{def}{=} \sum_{i=-\infty}^{-1} a_i \lambda^i.$$

Then, A_n satisfies requirements (a) and (b).

Proof. (cf. the proof of Lemma 2.1.11) Since $[L, B_n] = 0$, we have

$$[A_n, L] = -[(B_n)_-, L].$$

The left-hand side is a polynomial in λ, while the right-hand side has the form

$$\begin{pmatrix} 0 & * \\ * & 0 \end{pmatrix} + \sum_{i=-\infty}^{-1} b_i \lambda^i.$$

Consequently,

$$[A_n, L] = \begin{pmatrix} 0 & f_n \\ g_n & 0 \end{pmatrix}, \tag{2.35}$$

where f_n and g_n are some functions of the variables $u, v, u_x, v_x, \ldots$. $\square$

Proposition 2.2.2. *For any n and m, the system of equations*

$$u_\tau = f_m, \qquad v_\tau = g_m,$$

where f_i and g_i are defined by (2.35), *is an infinitesimal symmetry for the system* $u_t = f_n$, $v_t = g_n$.

The proof is similar to the proof of Theorem 2.1.16 for the KdV hierarchy.

Exercise 2.2.3. Prove this Proposition.

The A-operator for the NLS equation is given by the formula (2.34) with $n = 2$. Formulas (2.33), (2.34) for different n define the *NLS hierarchy*. The next system

$$u_t = -\frac{1}{4}u_{xxx} + \frac{3}{2}vuu_x, \qquad v_t = -\frac{1}{4}v_{xxx} + \frac{3}{2}uvv_x$$

of the NLS hierarchy corresponds to the operator

$$A_3 = A_2\lambda + \frac{1}{4}\left(\begin{array}{cc} vu_x - uv_x & u_{xx} - 2u^2v \\ v_{xx} - 2v^2u & uv_x - vu_x \end{array}\right).$$

The reduction $v = u$ in the above system leads to the mKdV equation

$$u_t = -\frac{1}{4}u_{xxx} + \frac{3}{2}u^2u_x.$$

Recursion operator for the NLS equation

In this section, we follow the paper [42]. Since

$$B_{n+1} = \lambda\, B_n,$$

we have

$$A_{n+1} = (\lambda\, B_n)_+ = \lambda\,(B_n)_+ + (\lambda\,(B_n)_-)_+.$$

The last formula shows that

$$A_{n+1} = \lambda\, A_n + R_n,$$

where the matrix

$$R_n = \left(\begin{array}{cc} a_n & b_n \\ c_n & -a_n \end{array}\right)$$

does not depend on λ. Substituting this expression for A_{n+1} into the Lax equation $L_{t_{n+1}} = [A_{n+1}, L]$, we obtain

$$L_{t_{n+1}} = \lambda\, L_{t_n} + [R_n, L]. \tag{2.36}$$

From the off-diagonal entries in the coefficients of λ and the diagonal entries in the coefficients of λ^0 in (2.36), we find

$$b_n = \frac{1}{2} f_n, \qquad c_n = -\frac{1}{2} g_n, \qquad a_n = \frac{1}{2} D^{-1} \left(v f_n + u g_n \right).$$

The off-diagonal entries in the coefficients of λ^0 yield the recursion operator

$$\mathcal{R} = \begin{pmatrix} -\frac{1}{2} D + u\, D^{-1} v & u\, D^{-1} u \\ -v\, D^{-1} v & \frac{1}{2} D - v\, D^{-1} u \end{pmatrix},$$

which generates the NLS hierachy:

$$\mathcal{R} \begin{pmatrix} f_n \\ g_n \end{pmatrix} = \begin{pmatrix} f_{n+1} \\ g_{n+1} \end{pmatrix}, \qquad \begin{pmatrix} f_1 \\ g_1 \end{pmatrix} = \begin{pmatrix} u_x \\ v_x \end{pmatrix}.$$

After the reduction $v = u$ the operator $\mathcal{R}^2$ leads to a recursion operator for the mKdV equation.

2.2.2   Generalizations

Consider operators of the form

$$L = D + \lambda a + q(x,t), \tag{2.37}$$

where q and a belong to a Lie algebra $\mathcal{G}$, and λ is the spectral parameter. It is assumed that the constant element a satisfies the condition

$$\mathcal{G} = \mathrm{Ker}\,(ad_a) \oplus \mathrm{Im}\,(ad_a).$$

Similar to Theorem 2.2.2, one can prove the following

Theorem 2.2.4. *There are unique series*

$$\begin{aligned} u &= u_{-1} \lambda^{-1} + u_{-2} \lambda^{-2} + \cdots, & u_i \in \mathrm{Im}\,(ad_a), \\ h &= h_0 + h_{-1} \lambda^{-1} + h_{-2} \lambda^{-2} + \cdots, & h_i \in \mathrm{Ker}\,(ad_a), \end{aligned}$$

such that

$$e^{ad_u}\,(L) \stackrel{def}{=} L + [u, L] + \frac{1}{2} [u, [u, L]] + \cdots = D_x + a\lambda + h.$$

Let b be a constant element of $\mathcal{G}$ such that

$$[b, \, \text{Ker} \, (ad_a)] = \{0\}.$$

Since

$$[b\,\lambda^n, D_x + a\lambda + h] = 0,$$

we have $[B_{b,n}, \, L] = 0$, where

$$B_{b,n} = e^{-ad_u} \, (b\,\lambda^n).$$

Then, the corresponding A-operator

$$A_{b,n} = b\,\lambda^n + a_{n-1}\,\lambda^{n-1} + \cdots + a_0$$

is determined by the formula

$$A_{b,n} = (B_{b,n})_+. \tag{2.38}$$

The set of evolution systems with Lax pairs (2.37) and (2.38) are symmetries for each other. They form an integrable hierarchy generated by the operator (2.37). For the Lie algebra $\mathcal{G} = \mathfrak{sl}_2$ and $a = \text{diag}\,(1, -1)$, we obtain the NLS hierarchy.

Example 2.2.5. Let $\mathcal{G} = \mathfrak{gl}_m$, and the L-operator have the form (2.37), where

$$a = \begin{pmatrix} \mathbf{1}_{m-1} & 0 \\ 0 & -1 \end{pmatrix}, \qquad q = \begin{pmatrix} 0 & \mathbf{u} \\ \mathbf{v}^T & 0 \end{pmatrix}.$$

Here, $\mathbf{u}$ and $\mathbf{v}$ are column vectors of dimension $m - 1$. In this case,

$$\text{Ker}\,(ad_a) = \left\{ \begin{pmatrix} \mathbf{S} & 0 \\ 0 & s \end{pmatrix} \right\}, \qquad \text{Im}\,(ad_a) = \left\{ \begin{pmatrix} 0 & \mathbf{u}_1 \\ \mathbf{u}_2^T & 0 \end{pmatrix} \right\},$$

where $\mathbf{S}$ is an $(m - 1) \times (m - 1)$ matrix, s is a scalar, and $\mathbf{u}_i$ are column vectors. Following the diagonalization procedure described above, we find that the coefficients of the operator

$$A_{a,2} = a\lambda^2 + s_1\lambda + s_2$$

are defined by the formulas

$$s_1 = \begin{pmatrix} 0 & \mathbf{u} \\ \mathbf{v}^T & 0 \end{pmatrix}, \qquad s_2 = \frac{1}{2} \begin{pmatrix} -\mathbf{u}\mathbf{v}^T & -\mathbf{u}_x \\ \mathbf{v}_x^T & \langle \mathbf{u}, \mathbf{v} \rangle \end{pmatrix}.$$

If $m = 2$, this Lax pair coincides with the Lax pair for the NLS equation from Section 2.2.1.

The corresponding nonlinear integrable system is the vector NLS equation [50]

$$\mathbf{u}_t = \mathbf{u}_{xx} + 2\langle \mathbf{u}, \mathbf{v} \rangle \, \mathbf{u}, \qquad \mathbf{v}_t = -\mathbf{v}_{xx} - 2\langle \mathbf{v}, \mathbf{u} \rangle \, \mathbf{v} \tag{2.39}$$

up to scalings of $t, \mathbf{u}, \mathbf{v}$.

Example 2.2.6. Let $\mathcal{G} = \mathfrak{gl}_m$, $a = \text{diag}\,(a_1, \ldots, a_m)$, $b = \text{diag}\,(b_1, \ldots, b_m)$, where $a_i \neq a_j$ for $i \neq j$. The system of equations corresponding to operator $A_{b,1}$ in (2.38) is called the *m-waves* equation. It has the form

$$\mathbf{Q}_t = \mathbf{P}_x + [\mathbf{Q}, \, \mathbf{P}], \tag{2.40}$$

where $\mathbf{Q}$ and $\mathbf{P}$ are $m \times m$ matrices and the entries of P and Q are related as follows:

$$p_{ij} = \frac{b_i - b_j}{a_j - a_i}\, q_{ij}.$$

Solutions of the system (2.40), independent of x, describe the dynamics of the m-dimensional rigid body [18].

Relations between scalar and matrix Lax pairs

The Gelfand–Dikii hierarchy (see Section 2.1.3) is defined by a scalar linear differential L-operator of order n. Of course, it is not difficult to replace this operator with a first order matrix differential operator of the form

$$\mathcal{L} = D + \Lambda + q, \tag{2.41}$$

where

$$\Lambda = \begin{pmatrix} 0 & 0 & \cdots & \lambda \\ 1 & 0 & \cdots & 0 \\ . & . & \cdots & . \\ 0 & \cdots & 1 & 0 \end{pmatrix}, \qquad q = \begin{pmatrix} 0 & 0 & \cdots & u_1 \\ 0 & 0 & \cdots & u_2 \\ . & . & \cdots & u_{n-1} \\ 0 & 0 & \cdots & 0 \end{pmatrix}. \tag{2.42}$$

Consider operators (2.41), where q is an arbitrary upper triangular matrix. The gauge transformation $\tilde{\mathcal{L}} = N \mathcal{L} N^{-1}$, where N is an arbitrary function with values in the group of upper triangular matrices with units on the diagonal, preserves the class of such operators. It turns out that the matrix q, defined by formula (2.42), is one of the possible canonical forms of the upper triangular potential q with respect to the gauge action.

The approach from [37], based on this observation, allows us to construct an analogue of the Gelfand–Dikii hierarchy for any Kac–Moody algebra G. We briefly recall it.

Let e_i, f_i, h_i, where $i = 0, \ldots, r$, be the canonical generators of a Kac–Moody algebra G satisfying the relations

$$[h_i, \, h_j] = 0, \qquad [e_i, \, f_j] = \delta_{ij}\, h_i, \qquad [h_i, \, e_j] = A_{ij}\, e_j, \qquad [h_i, \, f_j] = -A_{ij}\, f_j,$$

where $A = \{A_{ij}\}$ is the Cartan matrix of the algebra G.

Let us take $\sum_{i=0}^{r} e_i$ for the element $\Lambda \in G$ in operator (2.41). The potential q depends on the choice of the vertex c_m of the Dynkin diagram of the algebra G. We consider a grading $G = \oplus G_i$ such that $e_m \in G_1$, $f_m \in G_{-1}$ and the remaining canonical generators belong to G_0. It is well known that the component $\mathcal{G} \stackrel{def}{=} G_0$ is a semisimple finite-dimensional Lie algebra. As a potential q we take the generic element of the Borel subalgebra $\mathcal{B} \subset \mathcal{G}$ generated by f_i, h_i, where $i \neq m$.

If $\mathcal{L}$ is an operator of the form (2.41) and S belongs to the nilpotent subalgebra $\mathcal{N} \subset \mathcal{B}$, then the operator

$$\bar{\mathcal{L}} = e^{ad\,S}(\mathcal{L}) \tag{2.43}$$

has the same form (2.41) (but with a different potential q). This follows from the fact that $[\mathcal{N}, \mathcal{B}] \subset \mathcal{N}$, $[\mathcal{N}, e_m] = \{0\}$, $[\mathcal{N}, e_i] \subset \mathcal{B}$.

Any canonical form with respect to the gauge group of transformations (2.43) leads to a system of r evolution equations. Systems corresponding to different canonical forms are connected by invertible triangular differential-polynomial transformations of unknown functions.

As usual, the $\mathcal{L}$-operator (2.41) generates a commutative hierarchy of integrable nonlinear evolution systems. The corresponding A-operators can be constructed using the formal diagonalization procedure, which generalizes the construction of Theorem 2.2.4.

In [37], it was shown that integrable systems defined by L-operators of the form (2.41) include the systems from Theorem 2.1.25.

Further generalizations can be found in [51, 52, 53].

2.3 Decompositions of loop algebras and Lax pairs

For all classes of Lax representations discussed above, L and A-operators depend on the spectral parameter λ polynomially. However, there are important examples in which λ is a parameter on an elliptic curve or on its degenerations [54, 55].

An algebraic curve of genus $g > 1$ arises in the following example.

Example 2.3.1. [56] Consider a vector equation

$$\mathbf{u}_t = \left(\mathbf{u}_{xx} + \frac{3}{2}\langle \mathbf{u}_x, \mathbf{u}_x \rangle \mathbf{u}\right)_x + \frac{3}{2}\langle \mathbf{u}, \mathbf{R}\,\mathbf{u} \rangle \mathbf{u}_x, \qquad |\mathbf{u}| = 1, \tag{2.44}$$

where $\mathbf{u} = (u^1, \ldots, u^N)$, $\mathbf{R} = \mathrm{diag}\,(r_1, \ldots, r_N)$, and $\langle \cdot, \cdot \rangle$ denotes the standard scalar product. In the case $N = 3$, this equation is a higher symmetry of the famous integrable Landau–Lifschitz equation

$$\mathbf{u}_t = \mathbf{u} \times \mathbf{u}_{xx} + (\mathbf{R}\,\mathbf{u}) \times \mathbf{u}, \qquad |\mathbf{u}| = 1. \tag{2.45}$$

Here × stands for the cross product. It is interesting to note that for $N \neq 3$ all the symmetries of equation (2.44) have odd orders. In particular, equation (2.44) does not have a symmetry of order 2.

Equation (2.44) has a Lax representation (see Section 2.3.6) with the operator

$$L = D + \left(\begin{array}{cc} 0, & \Lambda\mathbf{u} \\ \mathbf{u}^T\Lambda, & 0 \end{array} \right). \qquad (2.46)$$

Here,

$$\Lambda = \mathrm{diag}(\lambda_1, \lambda_2, \cdots, \lambda_N)$$

is a matrix satisfying the relation

$$\Lambda^2 = \frac{1}{\lambda^2} - \mathbf{R}.$$

It is clear that

$$\lambda_1^2 + r_1 = \lambda_2^2 + r_2 = \cdots = \lambda_N^2 + r_N.$$

For generic constants r_i, this algebraic curve has genus $g = 1 + (N - 3)\, 2^{N-2}$. In the case of $N = 3$, this form of an elliptic spectral curve was used in [54, 1].

Some class of Lax operators related to algebraic curves of genus $g > 1$ was considered in [57, 58].

Factoring subalgebras

If we do not want to fix *a priori* the nature of the dependence of the L-operator on λ, we can assume that L is a Laurent series in λ with coefficients from some finite-dimensional Lie algebra $\mathcal{G}$.

The Lie algebra $\mathcal{G}((\lambda))$ of formal series of the form

$$\sum_{i=-n}^{\infty} g_i\lambda^i \quad | \quad g_i \in \mathcal{G}, \quad n \in \mathbb{Z} \qquad (2.47)$$

is called the (extended) *loop algebra* over $\mathcal{G}$.

If the algebra $\mathcal{G}$ is simple, then the formula

$$\langle X(\lambda), Y(\lambda) \rangle = \mathrm{res}\Big(X(\lambda), Y(\lambda) \Big), \qquad X(\lambda), Y(\lambda) \in \mathcal{G}((\lambda)) \qquad (2.48)$$

defines an invariant nondegenerate bilinear form on $\mathcal{G}((\lambda))$. Here, $(\cdot, \cdot)$ is the Killing form on $\mathcal{G}$. The invariance of the form (2.48) means that

$$\langle [a, b], c \rangle = -\langle b, [a, c] \rangle \qquad (2.49)$$

for any $a, b, c \in \mathcal{G}((\lambda))$.

If we assume that L and A in the Lax equation (1.7) are arbitrary elements of $\mathcal{G}((\lambda))$, then (1.7) is equivalent to an infinite collection of evolution equations. In order to obtain a finite system of PDEs, we need additional assumptions of the structure of L and A.

The basic ingredient for constructing Lax pairs in $\mathcal{G}((\lambda))$ is a decomposition (see [59, 60])

$$\mathcal{G}((\lambda)) = \mathcal{G}[[\lambda]] \oplus \mathcal{U} \tag{2.50}$$

of the loop algebra into a direct sum of vector subspaces, the first of which is the Lie subalgebra $\mathcal{G}[[\lambda]]$ of all Taylor series, and the second one is a Lie subalgebra. The Lie algebra $\mathcal{U}$ is called *factoring*, or *complementary*.

It turns out that if A-operator belongs to $\mathcal{U}$, and L belongs to $\mathcal{U}$ or to the orthogonal complement to $\mathcal{U}$ with respect to the form (2.48), then the Lax equation (1.7) is equivalent to a finite system of nonlinear differential equations. Indeed, the difference $L_t - [A, L]$ is identically equal to zero iff all coefficients of λ^{-i}, $i > 0$ in this difference vanish (see Item (a) of Lemma 2.3.2 and Exercise 2.3.21).

The following statement is obvious:

Lemma 2.3.2. *Let $\mathcal{U}$ be a factoring subalgebra. Then:*

(a) *It does not contain nonzero Taylor series;*

(b) *For any principle part $P = \sum_{i=-n}^{-1} g_i \lambda^i$, where $g_i \in \mathcal{G}$, there exists a unique element $\bar{P} \in \mathcal{U}$ of the form $\bar{P} = P + O(0)$.*

Example 2.3.3. The simplest factoring subalgebra consists of polynomials in $\frac{1}{\lambda}$ with a zero free term:

$$\mathcal{U}^{st} = \Big\{ \sum_{i=1}^{n} g_i \lambda^{-i} \ \Big| \ g_i \in \mathcal{G}, \quad n \in \mathbb{N} \Big\}. \tag{2.51}$$

This factoring subalgebra is called *standard*.

Definition 2.3.4. Factoring subalgebras are called *equivalent* if they are connected by a transformation of the parameter λ of the form

$$\lambda \to \lambda + k_2 \lambda^2 + k_3 \lambda^3 + \cdots, \qquad k_i \in \mathbb{C}, \tag{2.52}$$

or by an automorphism of the form

$$\exp\left(\mathrm{ad}_{g_1 \lambda + g_2 \lambda^2 + \dots}\right), \qquad g_i \in \mathcal{G}. \tag{2.53}$$

Remark 2.3.5. *Clearly, transformations (2.52) and (2.53) preserve the subalgebra $\mathcal{G}[[\lambda]]$.*

Suppose that an r-dimensional Lie algebra $\mathcal{G}$ is semisimple. Let $\mathbf{e}_1, \ldots, \mathbf{e}_r$ be a basis in $\mathcal{G}$. According to Item (b) of Lemma 2.3.2 for any i there exists a unique element $\mathbf{E}_i \in \mathcal{U}$ such that

$$\mathbf{E}_i = \frac{\mathbf{e}_i}{\lambda} + O(0), \tag{2.54}$$

where $O(0)$ means a Taylor series.

Proposition 2.3.1. *The elements* $\mathbf{E}_i$, $i = 1, \ldots, r$ *generate* $\mathcal{U}$.

Proof. We have to show that for any i, k the element $\mathbf{E}_{ik}$ of the form

$$\mathbf{E}_{ik} = \frac{\mathbf{e}_i}{\lambda^k} + O(-k+1)$$

can be obtained as a commutator of length k of elements $\mathbf{E}_j$, where $j = 1, \ldots, r$. It can be proved by induction over k. The induction step follows from the well-known property of semisimple Lie algebras: $[\mathcal{G}, \mathcal{G}] = \mathcal{G}$. □

If we take arbitrary series with the property (2.54), then the Lie subalgebra generated by them will contain nonzero Taylor series. Item (a) of Lemma 2.3.2 imposes severe restrictions on the generators (2.54).

Proposition 2.3.2. *The subalgebra* $\mathcal{U}$ *generated by the elements* $\mathbf{E}_i$, $i = 1, \ldots, r$ *of the form* (2.54) *is a factoring one iff for any* m, *the dimension* d_m *of the vector space spanned by commutators of length not greater than* m, *is equal to* $r\,m$.

Proof. Since the elements $\mathbf{E}_{ik}$, $k \leq m$ are linearly independent, we have $d_m \geq r\,m$. If $d_m > r\,m$, then $\mathcal{U}$ contains nonzero Taylor series. □

Conjecture 2.3.6. *Let* $\mathcal{G}$ *be a simple Lie algebra of dimension* r, *nonisomorphic to* $\mathfrak{sl}_2$. *Then, elements of* $\mathbf{E}_i \in \mathcal{G}((\lambda))$ *of the form* (2.54) *generate a factoring subalgebra if* $d_2 = 2r$. *In the case of* $\mathcal{G} = \mathfrak{sl}_2$, *it suffices to satisfy the conditions* $d_2 = 6$ *and* $d_3 = 9$.

Obviously, the subalgebra $\mathcal{G}[[\lambda]]$ is isotropic with respect to the form (2.48). The description of isotropic factoring subalgebras is closely related to the classification of Yang–Baxter r-matrices [61]. Without the assumption of isotropy, the problem of classifying factoring subalgebras was not seriously considered. In Section 2.3.1, we solve it [62] for $\mathcal{G} = \mathfrak{so}_3$.

Multiplicands

Definition 2.3.7. A scalar Laurent series

$$\mathbf{m} = \sum_{i=-n}^{\infty} c_i \lambda^i, \qquad c_i \in \mathbb{C},$$

is called a *multiplicand* of a factoring subalgebra $\mathcal{U}$ if $\mathbf{m}\mathcal{U} \subset \mathcal{U}$. The number n is called *the order* of the multiplicand $\mathbf{m}$.

Definition 2.3.8. If $\mathcal{U}$ has a multiplicand $\mathbf{m}$ of order $n = 1$, then $\mathcal{U}$ is called *homogeneous*.

Remark 2.3.9. *For any homogeneous subalgebra the first order multiplicand can be reduced to the form $m = \dfrac{1}{\lambda}$ by a transformation* (2.52).

Obviously, the set of all multiplicands forms a (commutative) associative algebra with unity over $\mathbb{C}$.

Remark 2.3.10. *In the case of a semisimple algebra $\mathcal{G}$, a series $\mathbf{m}$ is a multiplicand iff $\mathbf{m}\,\mathbf{E}_i \in \mathcal{U}$ for all generators* (2.54).

Let $\mathcal{G}$ be a simple Lie algebra. The following construction allows us to associate an algebraic curve with each factoring subalgebra $\mathcal{U}$.

Theorem 2.3.11. [63] *Let $\mathcal{U}$ be a factoring subalgebra in $\mathcal{G}((\lambda))$, then:*

(i) *$\mathcal{U}$ has no multiplicands of negative order;*

(ii) *The complement from the set of orders of all multiplicands to the set of natural numbers is finite.*

It follows from the statement (ii) that any two multiplicands are related by an algebraic relation and, therefore, the set of all multiplicands is isomorphic to the coordinate ring of some algebraic curve. For the homogeneous factoring subalgebras each multiplicand is a polynomial in a first order multiplicand and this curve is a straight line. Examples of multiplicands and corresponding algebraic curves are given in the next section.

The established canonical connection between factoring subalgebras and algebraic curves allows the use of methods of algebraic geometry for the study of factoring subalgebras.

2.3.1 Factoring subalgebras for $\mathcal{G} = \mathfrak{so}_3$

In this section, we follow the paper [62].

Consider the standard basis

$$\mathbf{e}_1 = \begin{pmatrix} 0 & 1 & 0 \\ -1 & 0 & 0 \\ 0 & 0 & 0 \end{pmatrix}, \qquad \mathbf{e}_2 = \begin{pmatrix} 0 & 0 & 1 \\ 0 & 0 & 0 \\ -1 & 0 & 0 \end{pmatrix}, \qquad \mathbf{e}_3 = \begin{pmatrix} 0 & 0 & 0 \\ 0 & 0 & 1 \\ 0 & -1 & 0 \end{pmatrix}$$

in $\mathfrak{so}_3$. Let $\mathcal{U}$ be a factoring subalgebra. Define the elements $\mathbf{E}_i \in \mathcal{U}$, $i = 1, 2, 3$, using the condition (2.54).

Automorphisms (2.53) are orthogonal transformations depending on λ as Taylor series. The functions

$$|\mathbf{E}_1|^2, \quad |\mathbf{E}_2|^2, \quad |\mathbf{E}_3|^2, \quad (\mathbf{E}_1, \mathbf{E}_2), \quad (\mathbf{E}_1, \mathbf{E}_3), \quad (\mathbf{E}_2, \mathbf{E}_3), \qquad (2.55)$$

where

$$\left(\sum_i x_i \mathbf{e_i}, \sum_j y_j \mathbf{e_j} \right) = \sum_i x_i y_i,$$

are invariant with respect to these transformations.

Proposition 2.3.3. *For any factoring subalgebra the following relations hold:*

$$\begin{pmatrix} [\mathbf{E}_1, [\mathbf{E}_2, \mathbf{E}_3]] \\ [\mathbf{E}_3, [\mathbf{E}_1, \mathbf{E}_2]] \\ [\mathbf{E}_2, [\mathbf{E}_3, \mathbf{E}_1]] \end{pmatrix} = \mathbf{A} \begin{pmatrix} [\mathbf{E}_3, \mathbf{E}_1] \\ [\mathbf{E}_1, \mathbf{E}_2] \\ [\mathbf{E}_2, \mathbf{E}_3] \end{pmatrix} + \mathbf{B} \begin{pmatrix} \mathbf{E}_2 \\ \mathbf{E}_3 \\ \mathbf{E}_1 \end{pmatrix},$$

$$(2.56)$$

$$\begin{pmatrix} [\mathbf{E}_3, [\mathbf{E}_2, \mathbf{E}_3]] + [\mathbf{E}_1, [\mathbf{E}_1, \mathbf{E}_2]] \\ [\mathbf{E}_1, [\mathbf{E}_3, \mathbf{E}_1]] + [\mathbf{E}_2, [\mathbf{E}_2, \mathbf{E}_3]] \\ [\mathbf{E}_2, [\mathbf{E}_1, \mathbf{E}_2]] + [\mathbf{E}_3, [\mathbf{E}_3, \mathbf{E}_1]] \end{pmatrix} = \mathbf{C} \begin{pmatrix} [\mathbf{E}_3, \mathbf{E}_1] \\ [\mathbf{E}_1, \mathbf{E}_2] \\ [\mathbf{E}_2, \mathbf{E}_3] \end{pmatrix} + \mathbf{D} \begin{pmatrix} \mathbf{E}_2 \\ \mathbf{E}_3 \\ \mathbf{E}_1 \end{pmatrix},$$

where

$$\mathbf{A} = \begin{pmatrix} -u & w & 0 \\ u & 0 & -v \\ 0 & -w & v \end{pmatrix}, \qquad \mathbf{B} = \begin{pmatrix} -\alpha & \beta & 0 \\ \alpha & 0 & -\gamma \\ 0 & -\beta & \gamma \end{pmatrix},$$

$$(2.57)$$

$$\mathbf{C} = \begin{pmatrix} x & v & -w \\ -v & y & u \\ w & -u & z \end{pmatrix}, \qquad \mathbf{D} = \begin{pmatrix} \varepsilon & \gamma & -\beta \\ -\gamma & \tau & \alpha \\ \beta & -\alpha & \delta \end{pmatrix}$$

are spme constant matrices. In addition, $\operatorname{tr} \mathbf{C} = \operatorname{tr} \mathbf{D} = 0$ *and*

$$c_1 \mathbf{A} + c_2 \mathbf{B} = 0, \qquad c_1 \mathbf{C} + c_2 \mathbf{D} = 0 \qquad (2.58)$$

for some constants c_1, c_2.

Proof. The coefficients of λ^{-3} in the expressions appearing in the left-hand side of (2.56) are equal to zero. Therefore, these expressions must be linear combinations of the elements

$$\mathbf{E}_1, \quad \mathbf{E}_2, \quad \mathbf{E}_3, \quad [\mathbf{E}_1, \mathbf{E}_2], \quad [\mathbf{E}_3, \mathbf{E}_1], \quad [\mathbf{E}_2, \mathbf{E}_3].$$

The structure of the coefficients in these linear combinations is determined by Lemma 2.3.2 and Proposition 2.3.2. $\qquad\square$

Example 2.3.12. For the standard factoring subalgebra (2.51), the conditions (2.56) are satisfied with $\mathbf{A} = \mathbf{B} = \mathbf{C} = \mathbf{D} = 0$. In this case

$$\mathbf{E}_i = \frac{\mathbf{e}_i}{\lambda}, \qquad i = 1, 2, 3.$$

It is clear that the algebra of multiplicands is generated by the multiplicand $x = \dfrac{1}{\lambda}$ and, therefore, the corresponding algebraic curve is a straight line.

Example 2.3.13. Suppose that

$$(\mathbf{E}_3, \mathbf{E}_1) = -\alpha, \qquad (\mathbf{E}_1, \mathbf{E}_2) = -\beta, \qquad (\mathbf{E}_2, \mathbf{E}_3) = -\gamma, \tag{2.59}$$

$$|\mathbf{E}_3|^2 - |\mathbf{E}_1|^2 = \varepsilon, \qquad |\mathbf{E}_1|^2 - |\mathbf{E}_2|^2 = \tau, \qquad |\mathbf{E}_2|^2 - |\mathbf{E}_3|^2 = \delta, \tag{2.60}$$

where $\alpha, \beta, \gamma, \delta, \varepsilon, \tau$ are fixed constants such that $\varepsilon + \tau + \delta = 0$. From (2.60) it follows that we can parametrize the relations (2.59) by entering the spectral parameter λ:

$$|\mathbf{E}_1| = \frac{\sqrt{1 - p\lambda^2}}{\lambda}, \qquad |\mathbf{E}_2| = \frac{\sqrt{1 - q\lambda^2}}{\lambda}, \qquad |\mathbf{E}_3| = \frac{\sqrt{1 - r\lambda^2}}{\lambda},$$

where $\varepsilon = p - r$, $\tau = q - p$, $\delta = r - q$. The elements $\mathbf{E}_i$ of the form

$$\mathbf{E}_1 = c_1 \mathbf{e}_1, \qquad \mathbf{E}_2 = c_2 \mathbf{e}_1 + c_3 \mathbf{e}_2, \qquad \mathbf{E}_3 = c_4 \mathbf{e}_1 + c_5 \mathbf{e}_2 + c_6 \mathbf{e}_3, \qquad c_i \in \mathbb{C}((\lambda)),$$

can be easily recovered from (2.59).

In particular, if $\alpha = \beta = \gamma = 0$, then

$$\mathbf{E}_1 = \frac{\sqrt{1 - p\lambda^2}}{\lambda} \mathbf{e}_1, \qquad \mathbf{E}_2 = \frac{\sqrt{1 - q\lambda^2}}{\lambda} \mathbf{e}_2, \qquad \mathbf{E}_3 = \frac{\sqrt{1 - r\lambda^2}}{\lambda} \mathbf{e}_3. \tag{2.61}$$

One can verify that such elements $\mathbf{E}_i$ satisfy (2.56), (2.57) with $\mathbf{A} = \mathbf{C} = 0$ and generate a factoring subalgebra that is isotropic with respect to the form (2.48). The expressions $X_i(\lambda) = |\mathbf{E}_i|$ are functions on the elliptic curve

$$X_1^2 + p = X_2^2 + q = X_3^2 + r. \tag{2.62}$$

The functions

$$x = \frac{1}{\lambda^2}, \qquad y = \frac{\sqrt{(1 - p\lambda^2)(1 - q\lambda^2)(1 - r\lambda^2)}}{\lambda^3}$$

are multiplicands of $\mathcal{U}$ of orders 2 and 3, respectively. For the generators (2.61) we have

$$x\,\mathbf{E}_2 = [[\mathbf{E}_1, \mathbf{E}_2], \mathbf{E}_1] + p\,\mathbf{E}_2, \qquad y\,\mathbf{E}_2 = [[\mathbf{E}_2, \mathbf{E}_3], [\mathbf{E}_1, \mathbf{E}_2]]$$

and so on. The corresponding algebraic curve is elliptic:

$$y^2 = (x - p)(x - q)(x - r).$$

Example 2.3.14. Let

$$|\mathbf{E}_1|^2 = (\mu - r)(\mu - q) - u^2, \qquad |\mathbf{E}_2|^2 = (\mu - r)(\mu - p) - v^2,$$

$$|\mathbf{E}_3|^2 = (\mu - q)(\mu - p) - w^2, \qquad (\mathbf{E}_1, \mathbf{E}_2) = w(\mu - r) + uv, \qquad (2.63)$$

$$(\mathbf{E}_1, \mathbf{E}_3) = v(\mu - q) + uw, \qquad (\mathbf{E}_2, \mathbf{E}_3) = u(\mu - p) + vw,$$

where $\mu = \lambda^{-1}$, p, q, r, u, v, w are arbitrary parameters. Then, elements $\mathbf{E}_i$, $i = 1, 2, 3$ satisfy conditions (2.56), (2.57) with $x = p - r, y = q - p, z = r - q$, $\mathbf{B} = \mathbf{D} = 0$ and generate a factoring subalgebra.

If $u = v = w = 0$, then the generators $\mathbf{E}_i$ are given by the formulas

$$\mathbf{E}_1 = \frac{\sqrt{(1 - r\lambda)(1 - q\lambda)}}{\lambda} \, \mathbf{e}_1, \qquad \mathbf{E}_2 = \frac{\sqrt{(1 - r\lambda)(1 - p\lambda)}}{\lambda} \, \mathbf{e}_2,$$

$$\mathbf{E}_3 = \frac{\sqrt{(1 - q\lambda)(1 - p\lambda)}}{\lambda} \, \mathbf{e}_3 \, . \qquad (2.64)$$

At first glance, we have functions

$$X_1 = \sqrt{\frac{(1 - r\lambda)}{\lambda}}, \qquad X_2 = \sqrt{\frac{(1 - q\lambda)}{\lambda}}, \qquad X_3 = \sqrt{\frac{(1 - p\lambda)}{\lambda}}$$

on the elliptic curve (2.62), but in fact the generators $\mathbf{E}_i$ depend only on the products

$$Z_1 = X_2 X_3, \qquad Z_2 = X_1 X_3, \qquad Z_3 = X_1 X_2.$$

The algebraic curve connecting the variables Z_i can be written as

$$\frac{Z_1 Z_2}{Z_3} + p = \frac{Z_1 Z_3}{Z_2} + q = \frac{Z_2 Z_3}{Z_1} + r.$$

This curve is rational. Indeed, substituting

$$Z_3 = \frac{(q - r) Z_2 Z_1}{Z_2^2 - Z_1^2}$$

into the curve equation, we obtain

$$(Z_2^2 - Z_1^2)^2 + a^2 Z_2^2 - b^2 Z_1^2 = 0, \qquad (2.65)$$

where $a^2 = (r - q)(q - p)$, $b^2 = (r - p)(q - p)$. The curve (2.65) allows the rational parametrization

$$Z_1 = \frac{a(t^3 + St)}{t^4 + Kt^2 + S^2}, \qquad Z_2 = \frac{b(t^3 - St)}{t^4 + Kt^2 + S^2},$$

where

$$S = \frac{(a^2 - b^2)^2}{4a^2b^2}, \qquad K = \frac{a^4 - b^4}{2a^2b^2}.$$

The function $x = \dfrac{1}{\lambda}$ is a multiplicand. For example,

$$x\mathbf{E}_1 = [\mathbf{E}_3, \, \mathbf{E}_2] + p\,\mathbf{E}_1.$$

The algebra of all multiplicands is generated by the multiplicand x.

Example 2.3.15. Let

$$\mathbf{E}_i = \frac{\mathbf{e}_i}{\lambda} + \nu\,[\mathbf{V}, \, \mathbf{e}_i] + \frac{1}{2}[\mathbf{V}, \, [\mathbf{V}, \, \mathbf{e}_i]],$$

where

$$\mathbf{V} = v_1\mathbf{e}_1 + v_2\mathbf{e}_2 + v_3\mathbf{e}_3, \tag{2.66}$$

ν and v_i are parameters. The constants in (2.56)–(2.58) are given by

$$u = v_1 v_3, \quad v = v_2 v_3, \quad w = v_1 v_2, \quad x = v_1^2 - v_3^2, \quad y = v_2^2 - v_1^2, \quad z = v_3^2 - v_2^2,$$

$$c_1 = -1, \qquad c_2 = \nu^2 + \frac{\Delta}{4}, \qquad \Delta = v_1^2 + v_2^2 + v_3^2.$$

The elements $\mathbf{E}_i$ generate a factoring subalgebra.

Exercise 2.3.16. Check that $x = \dfrac{1}{\lambda}$ is a multiplicand.

Theorem 2.3.17. *Any factoring subalgebra for $\mathcal{G} = \mathfrak{so}_3$ is equivalent to one of the subalgebras from Examples 2.3.12–2.3.15.*

Proof. The proof is given in [62]. It is based on the necessary conditions from Proposition 2.3.3. $\qquad\square$

Corollary 2.3.18. *If the generators $\mathbf{E}_i$ satisfy the conditions (2.56)–(2.58), then the corresponding subalgebra $\mathcal{U} \subset \mathfrak{so}_3((\lambda))$ is factoring.*[5]

Conjecture 2.3.19. *If the generators satisfy the conditions (2.56), then the corresponding subalgebra $\mathcal{U}$ is factoring.*

---

[5]This proves the part of Conjecture 2.3.6 concerning $\mathcal{G} = \mathfrak{sl}_2$.

On factoring subalgebras for $\mathcal{G} = \mathfrak{so}_4$

The classification of factoring subalgebras for the semisimple algebra $\mathcal{G} = \mathfrak{so}_4$ is very important for applications. In particular, any such subalgebra leads to an integrable system of the form

$$\mathbf{u}_t = \mathbf{u} \times (\mathbf{J}\,\mathbf{u} + \mathbf{G}\,\mathbf{v}), \qquad \mathbf{v}_t = \mathbf{v} \times (\bar{\mathbf{J}}\,\mathbf{v} + \bar{\mathbf{G}}\,\mathbf{u}),$$

where $\mathbf{u}$ and $\mathbf{v}$ are three-dimensional vectors such that $|\mathbf{u}| = |\mathbf{v}| = 1$, $\mathbf{J}, \bar{\mathbf{J}}, \mathbf{G}, \bar{\mathbf{G}}$ are some constant matrices and $\times$ means a vector product.

Some class of factoring subalgebras is described in [64].

Open problem 2.3.20. *Describe all factoring subalgebras for $\mathcal{G} = \mathfrak{so}_4$.*

2.3.2 Integrable top-like systems

The standard reduction in the Lax equation (1.7) is based on the assumption that both operators L and A take values in the same Lie algebra $\mathcal{U}$. But, as noted in Remark 1.2.3, we can assume that the A-operator in (1.7) belongs to the Lie algebra $\mathcal{U}$, while L lies in some module over it.

In this section, we assume that $\mathcal{G}$ is semisimple. Then, on the loop algebra there exists the nondegenerate invariant form (2.48). It turns out [19], that the Hamiltonian top-like systems, which are the most interesting for applications, arise if L belongs to the orthogonal complement $\mathcal{U}^\perp$ to $\mathcal{U}$ with respect to this form. From its invariance it follows that $\mathcal{U}^\perp$ is a module over $\mathcal{U}$.

Exercise 2.3.21. Prove that $\mathcal{U}^\perp$ does not contain nonzero Taylor series.

We say that the operator L is of order k if $L \in \mathcal{O}_k$, where

$$\mathcal{O}_k \overset{def}{=} \lambda^{-k}\mathcal{G}[[\lambda]] \cap \mathcal{U}^\perp.$$

To construct A-operators, we generalize the scheme of Sections 2.1.2 and 2.2.1. Namely, we find elements of $\mathcal{G}((\lambda))$, that commute with L, and project them onto $\mathcal{U}$. Denote by π_+ and π_- the projection operators on $\mathcal{U}$ and $\mathcal{G}[[\lambda]]$, corresponding to the decomposition (2.50).

For the sake of simplicity, we assume that the algebra $\mathcal{G}$ is embedded into a matrix algebra. The main objects in constructing A-operators are elements of the form $B_{ij} = \lambda^i L^j$ that belong to $\mathcal{G}((\lambda))$.

Proposition 2.3.4. *Let $L \in \mathcal{O}_k$. Then,*

(i) $[\pi_+(B_{ij}), L] \in \mathcal{O}_k$;

(ii) *For any i, j, p, q the Lax equations*

$$L_t = [\pi_+(B_{ij}),\, L]$$

and

$$L_\tau = [\pi_+(B_{pq}),\, L]$$

are infinitesimal symmetries for each other.

Exercise 2.3.22. Prove this proposition (see the proof of Theorem 2.1.16).

A general theory of similar Lax pairs and corresponding Hamiltonian structures for semisimple Lie algebras $\mathcal{G}$ was proposed in [19]. Below, we consider the simplest case of $\mathcal{G} = \mathfrak{so}_3$.

2.3.3   Classical $\mathfrak{so}_3$ spinning tops

In this section, we show that in the case $L \in \mathcal{O}_2$ the factoring subalgebras in $\mathfrak{so}_3$ described in Section 2.3.1 are in one-to-one correspondence with the classical integrable cases in the Kirchhoff problem of the motion of a rigid body in an ideal fluid [23]. The equations of motion in this problem have the form

$$\frac{d\boldsymbol{\Gamma}}{dt} = \boldsymbol{\Gamma} \times \frac{\partial H}{\partial \mathbf{M}}, \qquad \frac{d\mathbf{M}}{dt} = \mathbf{M} \times \frac{\partial H}{\partial \mathbf{M}} + \boldsymbol{\Gamma} \times \frac{\partial H}{\partial \boldsymbol{\Gamma}}, \qquad (2.67)$$

where $\mathbf{M} = (M_1, M_2, M_3)$ is the total angular momentum, $\boldsymbol{\Gamma} = (\gamma_1, \gamma_2, \gamma_3)$ is the gravitational vector, $\times$ stands for the cross product,

$$\frac{\partial H}{\partial \mathbf{M}} = \left(\frac{\partial H}{\partial M_1}, \frac{\partial H}{\partial M_2}, \frac{\partial H}{\partial M_3} \right), \qquad \frac{\partial H}{\partial \boldsymbol{\Gamma}} = \left(\frac{\partial H}{\partial \gamma_1}, \frac{\partial H}{\partial \gamma_2}, \frac{\partial H}{\partial \gamma_3} \right),$$

and $H(\mathbf{M}, \boldsymbol{\Gamma})$ is a homogeneous quadratic Hamiltonian.

Structure of the orthogonal complement to $\mathcal{U}$

Proposition 2.3.5. *The orthogonal complement to $\mathcal{U}$ has the following properties:*

(i) *There exist unique elements $\mathbf{R}_i \in \mathcal{U}^\perp$ of the form*

$$\mathbf{R}_i = \frac{\mathbf{e}_i}{\lambda} + O(0), \qquad i = 1, 2, 3. \qquad (2.68)$$

They generate $\mathcal{U}^\perp$ as a $\mathcal{U}$-module;

(ii) *The following commutator relations hold:*

$$
\begin{pmatrix} [\mathbf{E}_1, \mathbf{R}_1] \\ [\mathbf{E}_3, \mathbf{R}_3] \\ [\mathbf{E}_2, \mathbf{R}_2] \end{pmatrix} = \mathbf{A} \begin{pmatrix} \mathbf{R}_2 \\ \mathbf{R}_3 \\ \mathbf{R}_1 \end{pmatrix}, \qquad \begin{pmatrix} [\mathbf{E}_3, \mathbf{R}_1] + [\mathbf{E}_1, \mathbf{R}_3] \\ [\mathbf{E}_1, \mathbf{R}_2] + [\mathbf{E}_2, \mathbf{R}_1] \\ [\mathbf{E}_2, \mathbf{R}_3] + [\mathbf{E}_3, \mathbf{R}_2] \end{pmatrix} = \mathbf{C} \begin{pmatrix} \mathbf{R}_2 \\ \mathbf{R}_3 \\ \mathbf{R}_1 \end{pmatrix},
$$

where $\mathbf{A}$ and $\mathbf{C}$ are the same constant matrices as in (2.56), (2.57).

Remark 2.3.23. *The elements $\mathbf{R}_i \in \mathfrak{so}_3((\lambda))$ of the form* (2.68) *are determined from the commutator relations[6] uniquely up to a term of the form $S(\lambda)\,\mathbf{e}_i$, where S is a scalar Taylor series.*

Open problem 2.3.24. *Prove that for any scalar series $S(\lambda)$, the $\mathcal{U}$-module generated by the corresponding elements $\mathbf{R}_i$, $i = 1, 2, 3$, is the orthogonal complement to $\mathcal{U}$ with respect to a form*

$$
\langle X(\lambda), Y(\lambda) \rangle_P = \operatorname{res} P(\lambda)\Big(X(\lambda), Y(\lambda) \Big), \qquad X(\lambda), Y(\lambda) \in \mathfrak{so}_3((\lambda)),
$$

where P is a properly selected scalar Taylor series.

The simplest choice $L \in \mathcal{O}_1$, $A = \pi_+(L)$ corresponds to the Euler top for the vector $\mathbf{M} = (M_1, M_2, M_3)$. In this case, we have

$$
L = M_1 \mathbf{R}_1 + M_2 \mathbf{R}_2 + M_3 \mathbf{R}_3, \qquad A = M_1 \mathbf{E}_1 + M_2 \mathbf{E}_2 + M_3 \mathbf{E}_3. \tag{2.69}
$$

Note that if $\mathcal{U}$ is isotropic, then $\mathbf{R_i} = \mathbf{E_i}$ and the corresponding nonlinear system is trivial.

Lax pairs for integrable systems of Kirchhoff type have the following structure:

$$
L = \gamma_1[\mathbf{R}_3, \mathbf{E}_2] + \gamma_2[\mathbf{R}_1, \mathbf{E}_3] + \gamma_3[\mathbf{R}_2, \mathbf{E}_1] + m_1 \mathbf{R}_1 + m_2 \mathbf{R}_2 + m_3 \mathbf{R}_3,
$$

$$
A = \pi_+\Big(\lambda L\Big) = \gamma_1 \mathbf{E}_1 + \gamma_2 \mathbf{E}_2 + \gamma_3 \mathbf{E}_3, \tag{2.70}
$$

where $m_i = M_i + c_i \gamma_i$ for some constants c_i. From Proposition 2.3.5 it follows that any operator L from $\mathcal{O}_2$ has this form. The only nontrivial symmetry for the system (2.70) corresponds to the choice of $A = \pi_+(L)$.

Clebsch integrable case

The factoring subalgebra $\mathcal{U}$ from Example 2.3.13, generated by the elements (2.61), is isotropic and therefore $\mathbf{R}_i = \mathbf{E}_i$, $i = 1, 2, 3$. The Lax equation (1.7), (2.70) with $m_i = M_i$, is equivalent to (2.67), where

$$
H = -\frac{1}{2}\Big(M_1^2 + M_2^2 + M_3^2 - (q + r)\gamma_1^2 - (p + r)\gamma_2^2 - (p + q)\gamma_3^2 \Big).
$$

---

[6]Without the condition that $\mathbf{R}_i$ is orthogonal to elements from $\mathcal{U}$.

This is exactly the integrable Clebsch case in the Kirchhoff problem of the motion of a rigid body in an ideal fluid.

Since the subalgebra $\mathcal{U}$ is isotropic, the Lax pair (2.69) leads to a trivial system of nonlinear equations.

Euler and Steklov–Lyapunov cases

For the factoring subalgebra from Example 2.3.14, generated by generators (2.64), we have

$$\mathbf{R}_1 = \mathbf{e}_1 \frac{1}{\sqrt{(1-r\lambda)(1-q\lambda)}\,\lambda}, \qquad \mathbf{R}_2 = \mathbf{e}_2 \frac{1}{\sqrt{(1-r\lambda)(1-p\lambda)}\,\lambda},$$

$$\mathbf{R}_3 = \mathbf{e}_3 \frac{1}{\sqrt{(1-q\lambda)(1-p\lambda)}\,\lambda}.$$

The Lax pair (2.69) leads to the Euler equations (see also Example 1.2.2)

$$\mathbf{M}_t = \mathbf{M} \times \mathbf{V}\,\mathbf{M},$$

where $\mathbf{V} = \mathrm{diag}\,(p,q,r)$. This Lax pair is different from that discussed in Example 1.2.2.

The Lax equation (1.7), (2.70) is equivalent to (2.67), where

$$H = -\frac{1}{2}\left(M_1^2 + M_2^2 + M_3^2 + (r+q)M_1\gamma_1 + (r+p)M_2\gamma_2 + (q+p)M_3\gamma_3\right) - \frac{1}{8}\left((r-q)^2\gamma_1^2 + (p-r)^2\gamma_2^2 + (q-p)^2\gamma_3^2\right),$$

and

$$m_1 = M_1 + \frac{r-q}{2}\gamma_1, \qquad m_2 = M_2 + \frac{p-r}{2}\gamma_2, \qquad m_3 = M_3 + \frac{q-p}{2}\gamma_3.$$

This is nothing but the Steklov–Lyapunov integrable case.

Kirchhoff integrable case

For the factoring subalgebra described in Example 2.3.15, the elements

$$\mathbf{R}_i = \frac{\mathbf{e}_i}{\lambda} + \nu\,[\mathbf{V}, \mathbf{e}_i] - \frac{1}{2}[\mathbf{V}, [\mathbf{V}, \mathbf{e}_i]] + (\mathbf{V}, \mathbf{V})\,\mathbf{e}_i, \qquad i = 1,2,3$$

satisfy the commutator relations from Proposition 2.3.5 and therefore generate a $\mathcal{U}$-module that does not contain nonzero Taylor series. Any such module can be used to construct a Lax pair.

Remark 2.3.25. *This module is slightly different from* $\mathcal{U}^{\perp}$ *(see Remark 2.3.23).*

Exercise 2.3.26. Find an integrable system corresponding to the Lax pair (2.69). Answer: this is the Euler top with a pair of coinciding eigenvalues of the V matrix.

Exercise 2.3.27. Verify that the Lax pair (2.70) leads to the integrable Kirchhoff case (see [19]).

Open problem 2.3.28. *Find the generators* $\mathbf{R}_i$ *for the module* $\mathcal{U}^{\perp}$.

2.3.4   Generalization of Euler and Steklov–Lyapunov models to $\mathfrak{so}_n$-case

The factoring subalgebra from Example 2.3.14 can be described as follows:

$$\mathcal{U} = (1 + \lambda \mathbf{V})^{1/2} \mathcal{U}^{st} (1 + \lambda \mathbf{V})^{1/2}, \tag{2.71}$$

where $\mathbf{V} = \text{diag}\,(p, q, r)$,

$$(1 + \lambda \mathbf{V})^{1/2} = 1 + \frac{1}{2} \mathbf{V} \lambda - \frac{1}{8} \mathbf{V}^2 \lambda^2 + \cdots,$$

and $\mathcal{U}^{st}$ is defined by formula (2.51). It is easy to verify that formula (2.71), where $\mathbf{V}$ is an arbitrary diagonal matrix, defines a factoring subalgebra also for $\mathcal{G} = \mathfrak{so}_n$ in the case of arbitrary n. The orthogonal complement to $\mathcal{U}$ has the form

$$\mathcal{U}^{\perp} = (1 + \lambda \mathbf{V})^{-1/2} \mathcal{U}^{st} (1 + \lambda \mathbf{V})^{-1/2}.$$

The simplest Lax pair $L \in \mathcal{O}_1$, $A = \pi_+(L)$ is given by the formulas

$$L = (1 + \lambda \mathbf{V})^{-1/2} \frac{\mathbf{M}}{\lambda} (1 + \lambda \mathbf{V})^{-1/2}, \qquad A = (1 + \lambda \mathbf{V})^{1/2} \frac{\mathbf{M}}{\lambda} (1 + \lambda \mathbf{V})^{1/2},$$

where $\mathbf{M} \in \mathfrak{so}_n$. It leads to the Euler equations

$$\mathbf{M}_t = [\mathbf{V}, \mathbf{M}^2],$$

on $\mathfrak{so}_n$.

The system of equations

$$\mathbf{M}_t = [\mathbf{V}, \mathbf{M}^2] + [\mathbf{M}, \boldsymbol{\Gamma}], \qquad \boldsymbol{\Gamma}_t = \mathbf{V} \mathbf{M} \boldsymbol{\Gamma} - \boldsymbol{\Gamma} \mathbf{M} \mathbf{V}, \qquad \mathbf{M}, \boldsymbol{\Gamma} \in \mathfrak{so}_n$$

has the Lax pair

$$L = (1 + \lambda \mathbf{V})^{-1/2} \left(\frac{\mathbf{M}}{\lambda^2} + \frac{\boldsymbol{\Gamma}}{\lambda} \right) (1 + \lambda \mathbf{V})^{-1/2}, \qquad A = (1 + \lambda \mathbf{V})^{1/2} \frac{\mathbf{M}}{\lambda} (1 + \lambda \mathbf{V})^{1/2},$$

which corresponds to the orbit $L \in \mathcal{O}_2$. It can be regarded as a generalization [19] of the Steklov–Lyapunov top to the algebra $\mathfrak{so}_n$.

2.3.5   Factoring subalgebras for Kac–Moody algebras

For the Clebsch, Steklov–Lyapunov, and Kirchhoff integrable cases there exist first integrals of second degree.[7] In order to obtain more complex integrable models like the Kovalevskaya top, one has to consider a decomposition problem for the Kac–Moody algebras (see [19]).

Let $\mathcal{G}$ be a semisimple Lie algebra and ϕ its automorphism of finite order k. The infinite-dimensional $\mathbb{Z}$-graded Lie algebra

$$\mathcal{G}((\lambda, \phi)) = \Big\{ \sum_{i=-n}^{\infty} g_i \Big\}, \qquad g_i \in \mathcal{G}_i, \quad n \in \mathbb{Z},$$

is called *the Kac–Moody algebra*[8] associated with the pair $\mathcal{G}$, ϕ. Here,

$$\mathcal{G}_i \overset{def}{=} \{ a \in \mathcal{G} \mid \phi(a) = \varepsilon^i\, a \}\, \lambda^i, \qquad i \in \mathbb{Z},$$

where ε is a primitive root of unity of degree k.

Some interesting integrable systems are related to Kac–Moody algebras generated by the Lie algebra

$$\mathcal{G} = \{ \mathbf{A} \in \mathrm{Mat}_{n+m} \mid \mathbf{A}^T = -\mathbf{SAS} \},$$

where

$$\mathbf{S} = \begin{pmatrix} \mathbf{1}_n & 0 \\ 0 & -\mathbf{1}_m \end{pmatrix}.$$

It is clear that over $\mathbb{C}$ the algebra $\mathcal{G}$ is isomorphic to $\mathfrak{so}_{n+m}$.

Consider the subalgebra $\mathcal{A}$ in the loop algebra over $\mathcal{G}$ consisting of Laurent series such that the coefficients for even (respectively, odd) powers of λ belong to $\mathcal{G}_1$ (respectively, $\mathcal{G}_{-1}$). Here, by $\mathcal{G}_{\pm 1}$ we denote the eigenspaces of the inner automorphism of the second order

$$\phi : \mathcal{G} \mapsto \mathbf{S}\mathcal{G}\mathbf{S}^{-1},$$

corresponding to the eigenvalues ± 1. More explicitly, this means that the coefficients of even powers of λ have the following block structure:

$$\begin{pmatrix} v_1 & 0 \\ 0 & v_2 \end{pmatrix},$$

where $v_1 \in \mathfrak{so}_n$, $v_2 \in \mathfrak{so}_m$, while the coefficients of odd powers have the form

$$\begin{pmatrix} 0 & w \\ w^T & 0 \end{pmatrix},$$

where $w \in \mathrm{Mat}_{n,m}$.

---

[7]For the Kirchhoff case, there is also a first degree integral.
[8]Its central expansion is often considered, but the center is not needed for our purposes.

Choose $\mathrm{res}(\lambda^{-1}\mathrm{tr}(X\,Y))$ as the nondegenerate invariant form on $\mathcal{A}$. Note that in this case the form $\mathrm{res}(\mathrm{tr}(X\,Y))$ is degenerate.

Let $\mathcal{T}$ be the subalgebra of all Taylor series in $\mathcal{A}$ and

$$\mathcal{U} \overset{def}{=} (1+\lambda r)^{1/2}\mathcal{U}^{st}(1+\lambda r)^{1/2},$$

where $\mathcal{U}^{st}$ is the subalgebra of polynomials in λ^{-1} from $\mathcal{A}$ and r is an arbitrary constant matrix of the form

$$r = \begin{pmatrix} 0 & r_1 \\ -r_1^T & 0 \end{pmatrix}, \qquad r_1 \in \mathrm{Mat}_{n,m}. \tag{2.72}$$

One can check that in the decomposition

$$\mathcal{A} = \mathcal{T} + \mathcal{U} \tag{2.73}$$

the sum is direct. The factoring subalgebra $\mathcal{U}$ is a natural generalization of the subalgebra (2.71) to the case when the block structure of the coefficients of the series from $\mathcal{U}$ is determined by an additional automorphism of second order.

The orthogonal complement of $\mathcal{U}$ with respect to the form

$$\langle X,\,Y \rangle = \mathrm{res}\ \lambda^{-1}\mathrm{tr}\,(XY)$$

is given by the formula

$$\mathcal{U}^{\perp} = (1+\lambda r)^{-1/2}\,\mathcal{U}^{st}\,(1+\lambda r)^{-1/2}.$$

The Lax equation $L_t = [\pi_+(L),\,L]$ with the operator L of the form

$$L = (1+\lambda r)^{-1/2}(\lambda^{-1}w + v + \lambda u)(1+\lambda r)^{-1/2}$$

is equivalent to the following system of nonlinear matrix equations:

$$w_t = [w,\,wr+rw-v], \qquad v_t = [u,\,w]+vwr-rwv, \qquad u_t = uwr-rwu. \tag{2.74}$$

It is easy to see that this system allows the reduction (cf. formula (2.72))

$$u = \begin{pmatrix} 0 & r_1 \\ r_1^T & 0 \end{pmatrix},$$

which leads to the model found in [65]. When $n = 3, m = 2$, by further reductions we arrive at the Lax pair of the integrable case [66] in the Kirchhoff problem with the Hamiltonian

$$H = \frac{1}{2}|\mathbf{u}|^2|\mathbf{M}|^2 + \frac{1}{2}\big(\mathbf{u},\,\mathbf{M}\big)^2 + \big(\mathbf{u} \times \mathbf{v},\,\mathbf{M} \times \mathbf{\Gamma}\big),$$

where **u** and **v** are arbitrary constant vectors such that $\langle \mathbf{u}, \mathbf{v} \rangle = 0$. This integrable case has similar properties to the Kovalevskaya case in the problem of the motion of a rigid body around a fixed point. In particular, the additional integral of motion has a fourth degree.

Open problem 2.3.29. *Study systematically the decomposition problem for Kac–Moody algebras. Find nontrivial examples of factoring subalgebras for Kac–Moody algebras of small dimension.*

2.3.6   Integrable systems of Landau–Lifschitz type

Landau–Lifschitz equations related to $\mathfrak{so}_3$

Any factoring subalgebra $\mathcal{U}$ in $\mathfrak{so}_3$ generates a Lax pair of the form

$$L = \frac{d}{dx} + U, \qquad U = \sum_{i=1}^{3} s_i \, \mathbf{E}_i, \qquad s_1^2 + s_2^2 + s_3^2 = 1, \qquad (2.75)$$

$$A = \sum s_i \, [\mathbf{E}_j, \, \mathbf{E}_k] + \sum t_i \, \mathbf{E}_i, \qquad (2.76)$$

leading to a nonlinear integrable PDE of the Landau–Lifschitz type.

Let us rewrite the Lax equation in the form

$$U_t - A_x + [U, A] = 0. \qquad (2.77)$$

The Laurent expansion of the left-hand side of (2.77) contains terms with λ^k, where $k \geq -2$. If the coefficients of λ^{-2} and of λ^{-1} vanish, then the left-hand side of (2.77) is identically equal to zero. Indeed, the element $U_t - A_x + [U, A] \in \mathcal{U}$ is a Taylor series. According to Lemma 2.3.2 the subalgebra $\mathcal{U}$ does not contain any nonzero Taylor series.

The standard method for constructing all A-operators of a hierarchy defined by a given L-operator of the form (2.75) is based on the diagonalization procedure proposed in [67] (cf. Theorem 2.2.4). In particular, we could find the coefficients t_i in the A-operator using this procedure. However, we do not discuss it here. Instead, we find the coefficients t_i and the corresponding nonlinear system of the form

$$\mathbf{s}_t = \vec{F}(\mathbf{s}, \mathbf{s}_x, \mathbf{s}_{xx}), \qquad \text{where} \quad \mathbf{s} = (s_1, s_2, s_3), \qquad \mathbf{s}^2 = 1,$$

using a direct calculation. Namely, comparing the coefficients of λ^{-2} in the relation (2.77), we express t_i in terms of $\mathbf{s}, \mathbf{s}_x$. Then, equating the coefficients of λ^{-1}, we obtain a system of evolution equations for **s**.

Consider the case of Example 2.3.13, where the generators $\mathbf{E}_i$ are defined by formula (2.61). Equating to zero the coefficient of λ^{-2} in (2.77), we get $\mathbf{s}_x = \mathbf{s} \times \mathbf{t}$, where $\mathbf{t} = (t_1, t_2, t_3)$. Since $\mathbf{s}^2 = 1$, we find that $\mathbf{t} = \mathbf{s}_x \times \mathbf{s} + \mu \mathbf{s}$. Comparing the coefficients of λ^{-1}, we arrive at the equation $\mathbf{s}_t = \mathbf{t}_x - \mathbf{s} \times \mathbf{V}\mathbf{s}$, where $\mathbf{V} = \operatorname{diag}(p, q, r)$. Substituting the expression for $\mathbf{t}$, we obtain

$$\mathbf{s}_t = \mathbf{s}_{xx} \times \mathbf{s} + \mu_x \mathbf{s} + \mu \mathbf{s}_x - \mathbf{s} \times \mathbf{V}\mathbf{s}.$$

Since the scalar product $\langle \mathbf{s}, \mathbf{s}_t \rangle$ has to be zero, we find that $\mu = \text{const}$. The resulting equation coincides with (2.45) up to the involution $t \mapsto -t$, a trivial additional term $\mu \mathbf{s}_x$ (see Remark 2.1.15) and a change of notation.

The factoring subalgebra from Example 2.3.14 leads to the equation

$$\mathbf{s}_t = \mathbf{s} \times \mathbf{s}_{xx} + \langle \mathbf{s}, \mathbf{V}\mathbf{s} \rangle \mathbf{s}_x + 2\mathbf{s} \times (\mathbf{s} \times \mathbf{V}\mathbf{s}_x).$$

The subalgebra described in Example 2.3.15 corresponds to the equation

$$\mathbf{s}_t = \mathbf{s} \times \mathbf{s}_{xx} + \langle \mathbf{s}, \mathbf{Z}\mathbf{s} \rangle \mathbf{s}_x + 2\mathbf{s} \times (\mathbf{s} \times \mathbf{Z}\mathbf{s}_x) + c\,\mathbf{s} \times \mathbf{Z}\mathbf{s},$$

where

$$\mathbf{Z} = \begin{pmatrix} r_1^2 & r_1 r_2 & r_1 r_3 \\ r_1 r_2 & r_2^2 & r_2 r_3 \\ r_1 r_3 & r_2 r_3 & r_3^2 \end{pmatrix}, \qquad c = \nu^2 + \frac{r_1^2 + r_2^2 + r_3^2}{4}.$$

In this equation, $\mathbf{Z}$ is an arbitrary symmetric matrix of rank 1, and c is an arbitrary constant.

The systems obtained above were found in [68] using the symmetry approach to integrability [5].

Perelomov model and vector equation (2.44)

Consider the special case $n = N$, $m = 1$ of the Kac–Moody algebra from Section 2.3.5. It is easy to verify that the subalgebra

$$\mathcal{U} = \left\{ \sum_{i=-n}^{0} \lambda^{2i} \begin{pmatrix} \Lambda \mathbf{A}_i \Lambda & \Lambda \mathbf{u}_i \\ \mathbf{u}_i^T \Lambda & 0 \end{pmatrix}, \quad n \in \mathbb{N} \right\}$$

is factoring. Here,

$$\Lambda = \frac{1}{\lambda} \sqrt{1 - \lambda^2 \mathbf{R}} = \frac{1}{\lambda} - \frac{\mathbf{R}}{2} \lambda - \frac{\mathbf{R}^2}{8} \lambda^3 + \cdots, \tag{2.78}$$

$\mathbf{R} = \operatorname{diag}(r_1, \ldots, r_N)$, $\mathbf{A}_i$ are $N \times N$ skew-symmetric matrices, and $\mathbf{u}_i$ are column vectors. The orthogonal complement of $\mathcal{U}$ with respect to the form $\langle X, Y \rangle = \operatorname{res}(\lambda^{-1}(X, Y))$ is given by

$$\mathcal{U}^\perp = \left\{ \sum_{i=-n}^{-1} \lambda^{2i} \begin{pmatrix} \Lambda^{-1} \mathbf{A}_i \Lambda^{-1} & \Lambda^{-1} \mathbf{u}_i \\ \mathbf{u}_i^T \Lambda^{-1} & 0 \end{pmatrix}, \quad n \in \mathbb{N} \right\}. \tag{2.79}$$

The simplest L-operator belongs to $\mathcal{O}_1$ and has the form

$$L = \frac{1}{\lambda^2} \begin{pmatrix} \Lambda^{-1}\mathbf{V}\Lambda^{-1} & \Lambda^{-1}\mathbf{u} \\ \mathbf{u}^T\Lambda^{-1} & 0 \end{pmatrix}.$$

The Lax equation (1.7) with the A-operator of the form

$$A = \pi_+(\lambda^{-2}L) = \frac{1}{\lambda^2} \begin{pmatrix} 0 & \Lambda\mathbf{u} \\ \mathbf{u}^T\Lambda & 0 \end{pmatrix} + \begin{pmatrix} \Lambda\mathbf{V}\Lambda & \Lambda\mathbf{R}\mathbf{u} \\ \mathbf{u}^T\mathbf{R}\Lambda & 0 \end{pmatrix}$$

leads to the following generalization of the Clebsch top:

$$\mathbf{V}_t = [\mathbf{V}^2, \mathbf{R}] + [\mathbf{u}\mathbf{u}^T, \mathbf{R}^2], \qquad \mathbf{u}_t + (\mathbf{V}\mathbf{R} + \mathbf{R}\mathbf{V})\mathbf{u} = 0$$

to the case of the algebra $\mathfrak{so}_N$, found by Perelomov [69].

For equation (2.44), the L-operator is given by the formula (2.46), and the A-operator is

$$A = \frac{1}{\lambda^2} \begin{pmatrix} 0, & \Lambda\mathbf{u} \\ \mathbf{u}^T\Lambda, & 0 \end{pmatrix} + \begin{pmatrix} \Lambda\mathbf{V}\Lambda & \Lambda\mathbf{y} \\ \mathbf{y}^T\Lambda & 0 \end{pmatrix},$$

where the entries $v_{i,j}$ of the matrix $\mathbf{V}$ are $v_{i,j} = u_i(u_j)_x - u_j(u_i)_x$, and

$$\mathbf{y} = \mathbf{u}_{xx} + \left(\frac{3}{2}\langle \mathbf{u}_x, \mathbf{u}_x \rangle + \frac{1}{2}\langle \mathbf{u}, \mathbf{R}\mathbf{u} \rangle\right)\mathbf{u}.$$

2.3.7 Integrable hyperbolic models of chiral type

A special class of factoring subalgebras for $\mathcal{G} = \mathfrak{so}_4$ and their connection with integrable $\mathfrak{so}_4$-tops was studied in [64] (see Problem 2.3.20). These subalgebras also generate [70, 71] integrable hyperbolic systems of the form

$$\mathbf{u}_\xi = \mathbf{A}\mathbf{v} \times \mathbf{u}, \qquad \mathbf{v}_\eta = \bar{\mathbf{A}}\mathbf{u} \times \mathbf{v}, \tag{2.80}$$

where

$$\mathbf{A} = \mathrm{diag}(a_1, a_2, a_3), \qquad \bar{\mathbf{A}} = \mathrm{diag}(\bar{a}_1, \bar{a}_2, \bar{a}_3),$$

$\mathbf{u}, \mathbf{v}$ are three-dimensional vectors,[9] and the constants $a_i, \bar{a}_j$ satisfy the following relations:

$$a_1\bar{a}_1(a_3^2 - a_2^2) + a_2\bar{a}_2(a_1^2 - a_3^2) + a_3\bar{a}_3(a_2^2 - a_1^2) = 0,$$
$$a_1\bar{a}_1(\bar{a}_3^2 - \bar{a}_2^2) + a_2\bar{a}_2(\bar{a}_1^2 - \bar{a}_3^2) + a_3\bar{a}_3(\bar{a}_2^2 - \bar{a}_1^2) = 0.$$

The Cherednik model [72] corresponds to the choice of $a_i = \bar{a}_i$, $i = 1, 2, 3$. In the Golubchik–Sokolov case [70], we have $a_i = \bar{a}_i^{-1}$, $i = 1, 2, 3$. The integrable case, for which $a_1 = a_2$, $\bar{a}_1 = \bar{a}_2$ and $a_3, \bar{a}_3$ are arbitrary, was found in [71].

In the case of an arbitrary factoring subalgebra for $\mathcal{G} = \mathfrak{so}_4$ the matrices $\mathbf{A}$ and $\bar{\mathbf{A}}$ in the system (2.80) are not necessarily diagonal.

---

[9]Clearly, system (2.80) is consistent with the condition $|\mathbf{u}| = |\mathbf{v}| = 1$, which is usually applied when it comes to applications in geometry or in physics.

Open problem 2.3.30. *Construct examples of integrable systems* (2.80) *with nondiagonal matrices* **A** *and* **Ā**.

Summary

A description of factoring subalgebras is one of the fundamental problems in the theory of integrable systems. Each factoring subalgebra generates several different integrable nonlinear systems of ordinary differential and partial differential equations.

2.4　Finite-dimensional factorization method, reductions and nonassociative algebras

The factorization method (or, AKS-scheme [73]) is a finite-dimensional analogue of the matrix Riemann–Hilbert problem (see Section 1.2.3 and [17]). Just as in Section 2.3, we deal with decomposing a Lie algebra into a direct sum of vector spaces being Lie subalgebras. However, the Lie algebra in the AKS-scheme is finite-dimensional and the spectral parameter λ is absent.

At first glance, the factorization method is related to a very special class of nonlinear ordinary differential equations. However, various reductions and generalizations significantly expand the field of its applications. When discussing reductions, nonassociative algebras arise for the first time in this book. Most of these reductions can also be described in terms of $\mathbb{Z}_2$-graded Lie algebras.

2.4.1　Factorization method

Let $\mathcal{G}$ be a finite-dimensional Lie algebra, $\mathcal{G}_+$ and $\mathcal{G}_-$ be subalgebras in $\mathcal{G}$ such that

$$\mathcal{G} = \mathcal{G}_+ \oplus \mathcal{G}_-. \tag{2.81}$$

The simplest example is the Gauss decomposition of the matrix algebra into the sum of upper and law triangular matrices.

The standard factorization method is used to integrate rather special quadratic systems of the form

$$X_t = [\pi_+(X), X], \qquad X(0) = x_0. \tag{2.82}$$

Here, $X(t) \in \mathcal{G}$, and π_+ denotes the projector onto $\mathcal{G}_+$ parallel to $\mathcal{G}_-$. Very often we denote by X_+ and X_- the projections of the element X on $\mathcal{G}_+$ and on $\mathcal{G}_-$, respectively. For simplicity, we assume that the algebra $\mathcal{G}$ is embedded in a matrix algebra.

Remark 2.4.1. *It follows from Lemma 1.2.1 that, for any k, the function* $\operatorname{tr} X^k$ *is an integral of motion of system* (2.82). *Since X does not depend on the spectral parameter λ, in general, the number of these integrals is not sufficient for the Liouville integrability. However, additional integrals of motion sometimes appear somewhat mysteriously (see, for example, a construction of rational integrals of motion for the complete Toda lattice in* [74]).

Proposition 2.4.1. *The solution of the Cauchy problem* (2.82) *is given by the formula*

$$X(t) = A(t)\, x_0\, A^{-1}(t), \tag{2.83}$$

where the function $A(t)$ is defined as the solution of the following factorization problem:

$$A^{-1} B = \exp\left(-x_0\, t\right), \qquad A \in G_+, \quad B \in G_-. \tag{2.84}$$

Here, G_+ and G_- are the Lie groups of the algebras $\mathcal{G}_+$ and $\mathcal{G}_-$, respectively.

Proof. Differentiating (2.83) by t, we obtain

$$X_t = A_t\, x_0\, A^{-1} - A\, x_0\, A^{-1}A_t\, A^{-1} = [A_t\, A^{-1},\, X].$$

From (2.84) it follows that

$$-A^{-1}A_t A^{-1}B + A^{-1}B_t = -x_0 A^{-1}B.$$

The latter relation is equivalent to

$$-A_t A^{-1} + B_t B^{-1} = -A x_0 A^{-1}.$$

Projecting it onto $\mathcal{G}_+$, we obtain $A_t A^{-1} = X_+$, which proves the equality (2.82). $\qquad\square$

If the groups G_+ and G_- are algebraic, then the condition

$$A \in G_+, \qquad A\exp(-x_0\, t) \in G_-$$

is equivalent to a system of algebraic equations from which (for small t) the matrix $A(t)$ is determined uniquely.

The factorization problem (2.84) can also be reduced to a system of linear differential equations with variable coefficients for $A(t)$. We define the linear operator $L(t) : \mathcal{G}_+ \to \mathcal{G}_+$ by the formula

$$L(t)(v) = \Big(\exp(x_0\, t)\, v\, exp(-x_0\, t)\Big)_+.$$

Since $L(0)$ is the identical operator, $L(t)$ is invertible for small t.

Proposition 2.4.2. *Let $A(t)$ be a solution to the Cauchy problem*

$$A_t = A \, L(t)^{-1} \Big((x_0)_+ \Big), \qquad A(0) = 1. \tag{2.85}$$

We set $B = A \exp(-x_0 \, t)$. Then, the pair $(A, \, B)$ is the solution of the factorization problem (2.84).

Proof. Since $A^{-1} A_t \in \mathcal{G}_+$ and $A(0) = I$, we have $A \in \mathcal{G}_+$. It suffices for us to verify that $B^{-1} B_t \in \mathcal{G}_-$. From the definition of B, we find

$$B^{-1} B_t = \exp(x_0 \, t) \, A^{-1} \Big(A_t \exp(-x_0 \, t) - A x_0 \exp(-x_0 \, t) \Big) =$$
$$\exp(x_0 \, t) \Big(L(t)^{-1} (x_0)_+ \Big) \exp(-x_0 \, t) - x_0.$$

Projecting this identity on $\mathcal{G}_+$ and using the definition of the operator $L(t)$, we obtain $(B^{-1} B_t)_+ = 0.$ □

2.4.2 Reductions

From formula (2.83) it follows that, if the initial data x_0 for the system (2.82) belongs to some $\mathcal{G}_+$-module $\mathcal{M}$, then $X(t) \in \mathcal{M}$ for any t. This specialization of equation (2.82) can be written as

$$M_t = [\pi_+(M), \, M], \qquad M \in \mathcal{M}. \tag{2.86}$$

Introducing the product

$$M_1 \circ M_2 = [\pi_+(M_1), \, M_2], \qquad M_i \in \mathcal{M}, \tag{2.87}$$

we endow $\mathcal{M}$ with a structure of (generally speaking, noncommutative and non-associative) algebra. The system (2.86) is called $\mathcal{M}$-*reduction*, and the operation (2.87) is called $\mathcal{M}$-*product*.

Some classes of modules $\mathcal{M}$ correspond to interesting nonassociative algebras defined by the formula (2.87). These algebras are often similar to those appearing in [75].

In particular, algebras from the following classes arise.[10]

Definition 2.4.2. Algebras defined by the identity $[X, Y, Z] = 0$ are called *left-symmetric* [76].

Definition 2.4.3. Algebras with the identity

$$[V, X, Y \circ Z] - [V, X, Y] \circ Z - Y \circ [V, X, Z] = 0 \tag{2.88}$$

are called *SS-algebras* [77, 78].

---

[10]Below, we use the notation from Section 1.1.4.

Remark 2.4.4. *From* (2.88) *it follows that for any SS-algebra $\mathcal{A}$ the operator*

$$K_{YZ} \overset{def}{=} [L_Y, L_Z] - L_{Y \circ Z} + L_{Z \circ Y}$$

is a derivation of $\mathcal{A}$ for any Y, Z.

Definition 2.4.5. An algebra with identities

$$[X, Y, Z] + [Y, Z, X] + [Z, X, Y] = 0, \qquad (2.89)$$

and

$$V \circ [X, Y, Z] = [V \circ X, Y, Z] + [X, V \circ Y, Z] + [X, Y, V \circ Z] \qquad (2.90)$$

is called a *G-algebra* [79].

Remark 2.4.6. *The identity* (2.89) *means that the operation $X \circ Y - Y \circ X$ is a Lie bracket.*

Reductions in the case of $\mathbb{Z}_2$-graded Lie algebras

Let

$$\mathcal{G} = \mathcal{G}_0 \oplus \mathcal{G}_1 \qquad (2.91)$$

be a $\mathbb{Z}_2$-graded Lie algebra, i.e.

$$[\mathcal{G}_0, \mathcal{G}_0] \subset \mathcal{G}_0, \qquad [\mathcal{G}_0, \mathcal{G}_1] \subset \mathcal{G}_1, \qquad [\mathcal{G}_1, \mathcal{G}_1] \subset \mathcal{G}_0.$$

Suppose that we have a decomposition (2.81), where $\mathcal{G}_+ = \mathcal{G}_0$. Consider $\mathcal{G}_1$-reductions.

It is clear that

$$\mathcal{G}_- = \{m - R(m) \mid m \in \mathcal{G}_1\}, \qquad (2.92)$$

where $R = \pi_+$ is the projection operator onto $\mathcal{G}_+ = \mathcal{G}_0$ parallel to $\mathcal{G}_-$.

Theorem 2.4.7. [79] *The vector space* (2.92) *is a Lie subalgebra in $\mathcal{G}$ iff the operator $R : \mathcal{G}_1 \to \mathcal{G}_0$ satisfies the modified Yang–Baxter equation*

$$R\Big([R(X), Y] - [R(Y), X]\Big) - [R(X), R(Y)] - [X, Y] = 0, \qquad X, Y \in \mathcal{G}_1.$$

Remark 2.4.8. *It is important to note that in our case R is an operator defined on $\mathcal{G}_1$ and acting from $\mathcal{G}_1$ in $\mathcal{G}_0$, while as usual (see* [80]*) R is assumed to be an operator on $\mathcal{G}$.*

Proposition 2.4.3. *If $[\mathcal{G}_1, \mathcal{G}_1] = \{0\}$, then $\mathcal{G}_1$ is a left-symmetric algebra with respect to product* (2.87).

Proof. Let $X, Y, Z \in \mathcal{G}_1$. We should check that

$$[X, Y, Z] = [[X, Y], Z]. \tag{2.93}$$

According to (2.87) and (1.6), the left-hand side of (2.93) is given by

$$[[X_+, Y]_+, Z] - [[Y_+, X]_+, Z] + [Y_+, [X_+, Z]] - [X_+, [Y_+, Z]].$$

We have

$$[[X_+, Y]_+, Z] - [[Y_+, X]_+, Z] = [[X, Y]_+, Z] - [[X_-, Y]_+, Z] + [[X, Y_+]_+, Z] =$$

$$[[X, Y], Z] + [[X_+, Y_+]_+, Z] = [[X, Y], Z] + [[X_+, Y_+], Z].$$

Now, (2.93) follows from the Jacobi identity for X_+, Y_+, Z. Since $[\mathcal{G}_1, \mathcal{G}_1] = \{0\}$, the statement of Proposition follows from (2.93). ☐

Without the assumption $[\mathcal{G}_1, \mathcal{G}_1] = \{0\}$, we arrive at G-algebras (see Definition 2.4.5).

Theorem 2.4.9.   (i) *The vector space $\mathcal{G}_1$ is a G-algebra with respect to the operation* (2.87)*;*

(ii) *Any G-algebra can be obtained from an appropriate $\mathbb{Z}_2$-graded Lie algebra using this construction.*

Proof. To prove identity (2.89), it suffices to project the Jacobi identity for X_-, Y_-, Z_- on $\mathcal{G}_1$. Rewriting (2.90) in terms of $\mathcal{G}$-bracket with the use of (2.93), we see that (2.90) follows from the Jacobi identity for $\mathcal{G}$.

It remains to prove the second part of the theorem. Let $\mathcal{G}_1$ be a G-algebra. We define $\mathcal{G}$ by the formula (2.91), where $\mathcal{G}_0$ is a Lie algebra generated by all left multiplication operators (see Section 1.1.4) in $\mathcal{G}_1$. The bracket on $\mathcal{G}$ is defined as follows:

$$[(A, X), (B, Y)] = \Big([A, B] - [L_X, L_Y] + L_{X \circ Y} - L_{Y \circ X}, \; A(Y) - B(X)\Big). \tag{2.94}$$

Its skew-symmetry is obvious. It is easy to verify that the identities (2.89), (2.90) are equivalent to the Jacobi identity for (2.94). From (2.94) it follows that the decomposition (2.91) defines a $\mathbb{Z}_2$-grading. To define a decomposition (2.81), we take the set $\{(-L_X, X)\}$ for $\mathcal{G}_-$, and $\mathcal{G}_0$ for $\mathcal{G}_+$. Formula (2.94) implies that $\mathcal{G}_-$ is a subalgebra in $\mathcal{G}$. For $\mathcal{G}_\pm$, defined in this way, (2.87) has the form $(0, X) \circ (0, Y) = [(L_X, 0), (0, Y)]$ which follows from (2.94). ☐

Example 2.4.10. Let $\mathcal{G}_0 = \mathcal{G}_+$, $\mathcal{G}_1 = \mathcal{M}$, $\mathcal{G}_-$ and $\mathcal{G}$ be the sets of skew-symmetric, symmetric, upper triangular, and all matrices, respectively. Then, formula (2.87) defines the structure of a G-algebra on the set of symmetric matrices.

The first part of Theorem 2.4.9 allows us to construct quadratic systems of ODE integrable by the factorization method.

Example 2.4.11. Consider the algebra $\mathcal{G} = \mathfrak{sl}_3$, and set

$$\mathcal{G}_+ = \left\{ \begin{pmatrix} a, & c, & 0 \\ d, & b, & 0 \\ 0, & 0, & -a-b \end{pmatrix} \right\}, \qquad \mathcal{G}_1 = \left\{ \begin{pmatrix} 0, & 0, & P \\ 0, & 0, & Q \\ R, & S, & 0 \end{pmatrix} \right\}.$$

Let us choose a complementary subalgebra $\mathcal{G}_-$ as follows:

$$\mathcal{G}_- = \left\{ \begin{pmatrix} -Y+X+\alpha U, & X, & Y-X \\ -Z+(2-3\alpha)U, & (1-2\alpha)U, & Z+(3\alpha-2)U \\ -Y+X+U, & X, & Y-X+(\alpha-1)U \end{pmatrix} \right\},$$

where α is an arbitrary parameter.

The operation (2.87) turns the vector space $\mathcal{M} = \mathcal{G}_1$ into a G-algebra. The corresponding system (2.86) for the entries of the matrix

$$\mathbf{M} = \begin{pmatrix} 0, & 0, & P \\ 0, & 0, & Q \\ R, & S, & 0 \end{pmatrix}$$

has the form

$$\begin{cases} P_t = P^2 - RP - QS, \\ Q_t = (\beta - 2)RQ + \beta PQ, \\ R_t = R^2 - RP - QS, \\ S_t = (3 - \beta)RS + (1 - \beta)PS, \end{cases} \tag{2.95}$$

where $\beta = 3\alpha$. From (2.86) it follows that $I_1 = \operatorname{tr} \mathbf{M}^2 = RP + QS$ is the first integral of the system. Other integrals of the form $\operatorname{tr} \mathbf{M}^k$ are trivial. Nevertheless, it is not difficult to integrate the system (2.95) in quadratures. Two additional integrals have the form

$$I_2 = \frac{P-R}{QS}, \qquad I_3 = Q^{1-\beta} S^{-\beta} (R^2 - RP - QS).$$

For a generic value of the parameter β, the integral I_3 is a multi-valued function. This shows that the system (2.95) is *nonintegrable* from the viewpoint of the Painlevé approach (see, for example, [81]).

Exercise 2.4.12. Write down in explicit form the linear system (2.85) solving the factorization problem for system (2.95). Find out for which values of the parameter β system (2.95) satisfies the Kovalevskaya–Lyapunov test (see Section 6.1.5).

The singularities in the general solution of the system (2.82) are determined by the singularities of the coefficients of the corresponding linear system (2.85). Since the coefficients of this system depend on the initial data, the singularities are "movable" in the sense of the Painlevé test. First integrals of (2.82) can also be obtained from the first integrals of (2.85).

Open problem 2.4.13. *Obtain the rational integrals for the complete A_n-Toda lattice [74] using the corresponding linear system (2.85). Generalize the results obtained to the case of other series of simple Lie algebras.*

2.4.3   Generalized factorization method

Suppose that

$$\mathcal{G} = V_1 \oplus V_2 \tag{2.96}$$

for some vector subspaces V_i. Let

$$\mathcal{H} \stackrel{def}{=} [V_1, V_1]_- + [V_2, V_2]_+, \tag{2.97}$$

where $+$ and $-$ denote the projections on V_1 and on V_2, respectively.

Additionally, assume that V_1 and V_2 satisfy the following conditions:

$$[\mathcal{H}, V_1] \subset V_1, \qquad [\mathcal{H}, V_2] \subset V_2. \tag{2.98}$$

If V_i are subalgebras, then $\mathcal{H} = \{0\}$ and the conditions (2.98) are trivial.

It turns out [82] that the construction of a solution of equation (2.82), where π_+ denotes the projector on V_1 parallel to V_2, can be reduced to solving a system of linear differential equations with variable coefficients (cf. Proposition 2.4.2).

Remark 2.4.14. *If the conditions (2.98) are fulfilled, then the vector subspaces $\mathcal{H}$, $\mathcal{G}_+ \stackrel{def}{=} V_1 + [V_1, V_1]_-$ and $\mathcal{G}_- \stackrel{def}{=} V_2 + [V_2, V_2]_+$ are Lie subalgebras in $\mathcal{G}$. Moreover, $\mathcal{H} = \mathcal{G}_+ \cap \mathcal{G}_-$.*[11]

Theorem 2.4.15.   (i) *Let $\mathcal{G} = \mathcal{G}_0 \oplus \mathcal{G}_1$ be a $\mathbb{Z}_2$-graded Lie algebra such that $[\mathcal{G}_1, \mathcal{G}_1] = 0$. Having a decomposition (2.96) of the algebra $\mathcal{G}$ into a direct sum of vector spaces, where $V_1 = \mathcal{G}_0$ and V_2 satisfies the condition (2.98), we turn V_2 into an algebra using the operation (2.87). Then, V_2 is a SS-algebra with respect to this operation $\circ$ (see Definition 2.4.3);*

(ii) *Any SS-algebra $\mathcal{A}$ can be obtained from a suitable $\mathbb{Z}_2$-graded Lie algebra using this construction.*

---

[11] This implies that $\mathcal{G}$ is a sum of two subalgebras with a nontrivial intersection.

Proof. The first assertion can be proved similarly to the first part of Theorem 2.4.9. We will only indicate how to construct $\mathcal{G}, \mathcal{G}_+, V_2$ for a given SS-algebra. As $\mathcal{G}_+$ we take the Lie algebra $\text{End}\,\mathcal{A}$ of all endomorphisms of $\mathcal{A}$. The vector space

$$\mathcal{G} = (\text{End}\,\mathcal{A}) \oplus \mathcal{A}$$

becomes a $\mathbb{Z}_2$-graded Lie algebra if we define the bracket by the formula

$$[(A, X),\, (B, Y)] = \Big([A,\, B],\, A(Y) - B(X)\Big).$$

It is easy to show that the identity (2.88) implies that: (a) the vector space $\mathcal{H}$ generated by all elements of the form

$$\Big([L_Y,\, L_Z] - L_{Y \circ Z} + L_{Z \circ Y},\, 0\Big)$$

is a Lie subalgebra in $\mathcal{G}$, and (b) the vector spaces $V_2 = \{(-L_X,\, X)\}$, $V_1 = \mathcal{G}_+$ and the subalgebra $\mathcal{H}$ satisfy the conditions (2.97) and (2.98). $\square$

Example 2.4.16. Let us choose

$$\mathcal{G} = \left\{ \begin{pmatrix} *, & *, & *, & * \\ *, & *, & *, & * \\ *, & *, & *, & * \\ 0, & 0, & 0, & 0 \end{pmatrix} \right\}$$

as a $\mathbb{Z}_2$-graded Lie algebra. Clearly, $\mathcal{G} = \mathcal{G}_0 \oplus \mathcal{G}_1$, where

$$\mathcal{G}_0 = \left\{ \begin{pmatrix} *, & *, & *, & 0 \\ *, & *, & *, & 0 \\ *, & *, & *, & 0 \\ 0, & 0, & 0, & 0 \end{pmatrix} \right\}, \qquad \mathcal{G}_1 = \left\{ \begin{pmatrix} 0, & 0, & 0, & P \\ 0, & 0, & 0, & Q \\ 0, & 0, & 0, & R \\ 0, & 0, & 0, & 0 \end{pmatrix} \right\}.$$

Let $\mathcal{G}_+ = \mathcal{G}_0$ and

$$\mathcal{G}_- = \left\{ \begin{pmatrix} c, & \mu c, & a, & a \\ -\mu c, & c, & b, & b \\ a, & b, & c, & c \\ 0, & 0, & 0, & 0 \end{pmatrix} \right\},$$

where μ is an arbitrary parameter. A simple calculation shows that

$$\mathcal{H} = \left\{ \begin{pmatrix} 0, & -d, & 0, & 0 \\ d, & 0, & 0, & 0 \\ 0, & 0, & 0, & 0 \\ 0, & 0, & 0, & 0 \end{pmatrix} \right\}$$

and that the conditions (2.98) are fulfilled. The corresponding system (2.82) (up to a scaling) has the form

$$\begin{cases} P_t = 2PR + \mu QR, \\ Q_t = 2QR - \mu PR, \\ R_t = P^2 + Q^2 + R^2. \end{cases}$$

Exercise 2.4.17. Find first integrals and the general solution of this system.

Part 2. Algebraic structures in theory of bi-Hamiltonian systems

Chapter 3

Bi-Hamiltonian formalism

Definition 3.0.1. [83] Two Poisson brackets $\{\cdot, \cdot\}_1$ and $\{\cdot, \cdot\}_2$ defined on the same manifold are called *compatible* if their linear combination

$$\{\cdot, \cdot\}_\lambda = \{\cdot, \cdot\}_1 + \lambda\{\cdot, \cdot\}_2$$

is a Poisson bracket for any $\lambda \in \mathbb{C}$.

General results on the properties and geometric structure of a pencil of compatible brackets $\{\cdot, \cdot\}_\lambda$ can be found in [84, 85]. In particular, if the bracket $\{\cdot, \cdot\}_\lambda$ is degenerate, then the set of integrals commuting with respect to both brackets can be constructed as follows.

Theorem 3.0.2. [83] *Let*

$$C(\lambda) = C_0 + \lambda C_1 + \lambda^2 C_2 + \cdots$$

be the Casimir function for the pencil $\{\cdot, \cdot\}_\lambda$. Then, the coefficients C_i commute with respect to each of the brackets $\{\cdot, \cdot\}_1$ and $\{\cdot, \cdot\}_2$.

This method of constructing a family of commuting Hamiltonians is called the *Lenard–Magri scheme*.

From Theorem 3.0.2 it follows that

$$\{C_{k+1}, y\}_1 = -\{C_k, y\}_2$$

for any function y and integer k. If we take the function C_{k+1} as the Hamiltonian H, then the dynamical system (1.22) can be written in Hamiltonian form in two different ways:

$$\frac{dy_i}{dt} = \{C_{k+1}, y_i\}_1 = -\{C_k, y_i\}_2.$$

Moreover, all functions C_j are commuting integrals of motion for this system.

We see that the same dynamical system can sometimes admit two different compatible Hamiltonian structures with different Hamiltonians and integrals that commute with respect to both structures. In this case, we say that this system has a *bi-Hamiltonian representation* [83].

Integrable top-like systems are usually bi-Hamiltonian with respect to two compatible linear Poisson brackets. The corresponding algebraic object is a pair of compatible Lie algebras (see Section 3.1).

3.0.1   Shift argument method

Let us discuss the standard method for constructing compatible Poisson brackets.

Let $\mathbf{a} = (a_1, \ldots, a_N)$ be a constant vector. Then, any linear Poisson bracket generates a constant Poisson bracket compatible with the original linear one by shifting the coordinates $x_i \to x_i + \lambda a_i$ (see [18, 86]).

Now, consider a quadratic Poisson bracket (1.25). The shift $x_i \mapsto x_i + \lambda a_i$ leads to the Poisson bracket $\{\cdot,\cdot\}_\lambda = \{\cdot,\cdot\} + \lambda\{\cdot,\cdot\}_1 + \lambda^2\{\cdot,\cdot\}_2$. If the coefficient of λ^2 is equal to zero, then this formula defines the linear Poisson bracket $\{\cdot,\cdot\}_1$, compatible with (1.25).

Thus, in the case of quadratic brackets, the shift vector $\mathbf{a} = (a_1, \ldots, a_N)$ is not arbitrary. Its components must satisfy the overdetermined system of algebraic equations

$$\sum_{p,q} r_{ij}^{pq} a_p a_q = 0, \qquad i,j = 1, \ldots, N. \tag{3.1}$$

Such a vector $\mathbf{a}$ is called *admissible*. Admissible vectors are nothing but 0-dimensional symplectic leaves of the Poisson bracket (1.25).

Any p-dimensional vector space of admissible vectors generates p pairwise compatible linear Poisson brackets, each of which is compatible with the original quadratic bracket (1.25).

Many interesting integrable models can be obtained [87, 88] by shifting the argument from *elliptic quadratic Poisson brackets* [27, 89]. In particular, [90], the classical elliptic Calogero–Moser system with three or four particles can be obtained in this way (see Section 3.2.4).

Example 3.0.3. Consider the quadratic Poisson bracket which is defined on the generators $x_0, \ldots, x_7$ by the following formulas (indexes are taken modulo 8):

$$\{x_i, x_{i+1}\} = p_1 x_i x_{i+1} + k_1 x_{i+2} x_{i+7} - 2k_2 x_{i+3} x_{i+6} + p_2 x_{i+4} x_{i+5},$$

$$\{x_i, x_{i+2}\} = p_3(x_{i+1}^2 - x_{i+5}^2),$$

$$\{x_i, x_{i+3}\} = p_1 x_i x_{i+3} + k_1 x_{i+5} x_{i+6} - 2k_2 x_{i+1} x_{i+2} + p_2 x_{i+4} x_{i+7},$$

$$\{x_i, x_{i+4}\} = p_4(x_{i+1} x_{i+3} - x_{i+5} x_{i+7}).$$

(3.2)

Here,

$$p_1 = -\frac{1}{2}k_1^{1/2}k_2^{-1/2}(4k_2^2 + k_1^2)^{1/2}, \qquad p_2 = k_2^{1/2}k_1^{-1/2}(4k_2^2 + k_1^2)^{1/2},$$

$$p_3 = k_2^{1/4}k_1^{1/4}(4k_2^2 + k_1^2)^{1/4}, \qquad p_4 = k_2^{-1/4}k_1^{-1/4}(4k_2^2 + k_1^2)^{3/4},$$

where k_1, k_2 are arbitrary parameters. This bracket depends on one essential parameter $\tau = \dfrac{k_1}{k_2}$. In the paper [27] the Poisson algebra (3.2) is denoted by $q_{8,3}(\tau)$.

The bracket (3.2) has the following four Casimir functions:

$$C_i = k_2(x_i^2 + x_{i+4}^2) + p_3(x_{i+3}x_{i+5} + x_{i+1}x_{i+7}) + k_1 x_{i+2}x_{i+6}, \qquad i = 0, 1, 2, 3. \quad (3.3)$$

The admissible vectors are given by

$$\mathbf{a}_\pm = (t_1, 0, t_2, 0, \pm t_1, 0, \pm t_2, 0), \qquad \mathbf{b}_\pm = (0, t_1, 0, t_2, 0, \pm t_1, 0, \pm t_2),$$

where t_1, t_2 are arbitrary parameters. We see that admissible vectors form four 2-dimensional vector spaces such that $\mathbb{R}^8$ is their direct sum.

Consider the coordinate shift by the vector $\mathbf{a}_+$. As a result, we get a linear bracket $\{\cdot, \cdot\}_a = t_1\{\cdot, \cdot\}_1 + t_2\{\cdot, \cdot\}_2$ or, in other words, a pair of compatible linear Poisson brackets $\{\cdot, \cdot\}_{1,2}$. For generic parameters t_1 and t_2 the bracket $\{\cdot, \cdot\}_a$ is isomorphic to $\mathfrak{gl}_2 \oplus \mathfrak{gl}_2$. It is easy to verify that the bracket has two linear Casimir functions $K_1 = x_0 + x_4$ and $K_2 = x_2 + x_6$. After reducing the linear brackets to the surface $K_1 = K_2 = 0$, we obtain a pair of compatible $\mathfrak{sl}_2 \oplus \mathfrak{sl}_2$ brackets. It is important to note that the initial quadratic bracket (3.2) cannot be reduced to the surface $K_1 = K_2 = 0$ because K_i are not Casimir functions for (3.2). It can be verified that the Lenard–Magri scheme applied to the reduced linear brackets $\{\cdot, \cdot\}_{1,2}$ leads to the $\mathfrak{so}_4$-Schottky–Manakov top.

Theorem 3.0.4. *For a quadratic Poisson algebra of the type* $q_{mn^2, kmn-1}(\tau)$ *(see [89] for the definition), the set of admissible vectors is the union of n^2 copies of m-dimensional vector space. The space of generators of this algebra is the direct sum of these spaces.*

Conjecture 3.0.5. *Linear Poisson brackets obtained from the* $q_{mn^2, kmn-1}(\tau)$ *bracket by shifting the argument by generic admissible vectors are isomorphic to* $\oplus_{i=1}^m \mathfrak{gl}_n$. *All these Lie algebras have a common center. After quotienting by this center, m pairwise compatible $\oplus_{i=1}^m \mathfrak{sl}_n$ Poisson brackets are obtained.*

Remark 3.0.6. *For the algebra from Example 3.0.3 we have $m = n = 2$, $k = 1$.*

Open problem 3.0.7. *Find the integrable systems defined by Theorem 3.0.2 which are generated by the shift of the argument in the elliptic Poisson brackets* $q_{mn^2, kmn-1}(\tau)$.

3.0.2   Bi-Hamiltonian form for KdV equation

We will not discuss here the bi-Hamiltonian formalism and the Lenard–Magri scheme for evolution equations (1.26). We only note that the KdV equation (1.12) is a bi-Hamiltonian system [83]. The two compatible Poisson brackets are given by the formula (1.27), where the Hamiltonian operators $\mathcal{H}_i$ are the following differential operators:

$$\mathcal{H}_1 = D, \qquad \mathcal{H}_2 = D^3 + 4uD + 2u_x. \tag{3.4}$$

Note that operator $\mathcal{H}_1$ can be obtained from $\mathcal{H}_2$ by shifting the argument $u \mapsto u+\lambda$.

Exercise 3.0.8. Verify that the KdV equation (1.12) can be written in bi-Hamiltonian form

$$u_t = \mathcal{H}_1 \frac{\delta \rho_1}{\delta u} = \mathcal{H}_2 \frac{\delta \rho_2}{\delta u},$$

where

$$\rho_1 = -\frac{u_x^2}{2} + u^3, \qquad \rho_2 = \frac{u^2}{2}.$$

3.1   Bi-Hamiltonian formalism and pairs of compatible algebras

In this section, we consider pairs of compatible linear Poisson brackets. As mentioned in the Introduction, each of these brackets corresponds to a Lie algebra. Denote by $[\cdot,\cdot]_1$ and $[\cdot,\cdot]_2$ the operations of these algebras. It is clear that the Poisson brackets are compatible iff the operation $\lambda_1[\cdot,\cdot]_1 + \lambda_2[\cdot,\cdot]_2$ is a Lie bracket for arbitrary $\lambda_i \in \mathbb{C}$. Without loss of generality, we set $\lambda_1 = 1$.

Definition 3.1.1. Two Lie brackets $[\cdot,\cdot]$ and $[\cdot,\cdot]_1$ defined on the same vector space **V**, are called *compatible* if the operation

$$[\cdot,\cdot]_\lambda = [\cdot,\cdot] + \lambda[\cdot,\cdot]_1 \tag{3.5}$$

is a Lie bracket for any λ.

In what follows, we will assume that the Lie algebra $\mathcal{G}$ with the bracket $[\cdot,\cdot]$ is semisimple. The following classification problem arises:

Open problem 3.1.2. *Describe all possible Lie brackets $[\cdot,\cdot]_1$, compatible with a given semisimple Lie bracket $[\cdot,\cdot]$.*

The Lie algebra with the bracket (3.5) can be considered as a linear deformation of the algebra $\mathcal{G}$. Since any semisimple Lie algebra is rigid (that is, $H^2[\mathcal{G}, \mathcal{G}] = 0$), the bracket (3.5) is isomorphic to the original bracket $[\cdot, \cdot]$. This means that there exists a formal series of the form

$$A_\lambda = 1 + R\lambda + S\lambda^2 + \cdots, \tag{3.6}$$

such that

$$A_\lambda^{-1}[A_\lambda(X), A_\lambda(Y)] = [X, Y] + \lambda[X, Y]_1. \tag{3.7}$$

Here, the coefficients $R, S, \ldots$ are (constant) linear operators on $\mathcal{G}$, and 1 is the identity operator.

Equating the coefficients of λ in (3.7), we obtain

$$[X, Y]_1 = [R(X), Y] + [X, R(Y)] - R([X, Y]), \tag{3.8}$$

where R is the coefficient of λ in A_λ.

Lemma 3.1.3. (i) *The operation $[\cdot, \cdot]_1$ is a Lie bracket iff there exists a linear operator $S : \mathcal{G} \to \mathcal{G}$ such that*

$$R\big([R(X), Y] - [R(Y), X]\big) - [R(X), R(Y)] - R^2([X, Y]) =$$

$$[S(X), Y] - [S(Y), X] - S([X, Y]);$$

(ii) *If this relation is satisfied, then the bracket $[\cdot, \cdot]_1$ is compatible with $[\cdot, \cdot]$.*

Proof. We note only that Item (ii) follows from the fact that if the condition from Item (i) is fulfilled for a pair of (R, S), then it holds for the pair $(R + \mu 1, S)$ for any constant μ. $\square$

In the special case $S = 0$, the relation from Lemma 3.1.3 takes the form

$$R\big([R(X), Y] - [R(Y), X]\big) - [R(X), R(Y)] - R^2([X, Y]) = 0. \tag{3.9}$$

In each specific case, the series A_λ is an operator-valued meromorphic function of the variable λ which changes the basis such that the bracket (3.5) is reduced to the bracket $[\cdot, \cdot]$. This function can be regarded as a kind of Yang–Baxter r-matrix (see Theorem 3.1.21). Like solutions of the Yang–Baxter equation, A_λ-matrices are usually rational, trigonometric, or elliptic functions of λ. Examples of elliptic A_λ-matrices can be found in [88] (see also Section 3.1.2).

Open problem 3.1.4. *Find all elliptic A_λ-matrices.*

While Problem 3.1.2 for Lie algebras is almost hopelessly complex, a similar problem for associative algebras looks quite amenable to study (see [91, 92, 93]).

88 *Algebraic Structures in Integrability*

Definition 3.1.5. Two associative algebras with multiplications $\star$ and $\circ$, defined on the same vector space $\mathbf{V}$ are called compatible if the multiplication

$$a \bullet b = a \star b + \lambda \, a \circ b \qquad (3.10)$$

is associative for any constant λ.

Remark 3.1.6. *For compatible associative algebras with multiplications $\star$ and $\circ$, associated Lie algebras with brackets $[X, Y]_1 = X \star Y - Y \star X$ and $[X, Y]_2 = X \circ Y - Y \circ X$ are compatible.*

Example 3.1.7. Let $\mathbf{V}$ be the vector space of polynomials in one variable of degree $\leq k - 1$, μ_1 and μ_2 be polynomials of degree k without common roots. Any polynomial Z, where $\deg Z \leq 2k - 1$, can be uniquely represented as $Z = \mu_1 P + \mu_2 Q$, where $P, Q \in \mathbf{V}$. The explicit form of P and Q can be found using the Lagrange interpolation formula. We define the multiplications $\circ$ and $\star$ on $\mathbf{V}$ by the formula

$$X Y = \mu_1 (X \circ Y) + \mu_2 (X \star Y), \qquad X, Y \in \mathbf{V}.$$

It can be verified that associative algebras with multiplications $\circ$ and $\star$ are compatible.

Example 3.1.8. Let $\mathbf{e}_1, \ldots, \mathbf{e}_m$ be a basis in $\mathbf{V}$ and the multiplication $\star$ is given by

$$\mathbf{e}_i \star \mathbf{e}_j = \delta_j^i \mathbf{e}_i. \qquad (3.11)$$

Let

$$r_{ii} = q_0 - \sum_{k \neq i} r_{ki}, \qquad r_{ij} = \frac{q_i \lambda_i}{\lambda_j - \lambda_i}, \qquad i \neq j,$$

where λ_i, q_j are arbitrary constants. Then, the multiplication

$$\mathbf{e}_i \circ \mathbf{e}_j = r_{ij} \mathbf{e}_j + r_{ji} \mathbf{e}_i - \delta_j^i \sum_{k=1}^{m} r_{ik} \mathbf{e}_k$$

is associative and compatible with $\star$. Since it is linear with respect to the parameters q_i, in fact, we have a family of $m + 1$ pairwise compatible multiplications.

Suppose that an associative algebra $\mathcal{A}$ with multiplication $\star$ is semisimple.[1] Since such algebras are rigid, an associative algebra with multiplication (3.10) is isomorphic to the algebra $\mathcal{A}$ for almost all values of λ. Consequently, on $\mathbf{V}$ there exists a linear operator S_λ such that

$$S_\lambda(X) \star S_\lambda(Y) = S_\lambda \Big(X \star Y + \lambda \, X \circ Y \Big). \qquad (3.12)$$

---

[1]This means that A is the direct sum of several matrix algebras Mat_{m_i}.

If

$$S_\lambda = 1 + R\lambda + O(\lambda^2), \tag{3.13}$$

then the multiplication $\circ$ is given by the formula (cf. (3.8))

$$X \circ Y = R(X) \star Y + X \star R(Y) - R(X \star Y). \tag{3.14}$$

One can check that if the multiplication $\circ$ of the form (3.14) is associative, then it is compatible with $\star$.

3.1.1 Compatible Lie algebras. Examples and applications

We present here examples of compatible Lie brackets [70] along with the corresponding operators A_λ and R (see formulas (3.7) and (3.8)).

Example 3.1.9. Let $\mathcal{G}$ be a Lie algebra associated with an associative algebra $\mathcal{A}$. We fix an element $r \in \mathcal{A}$ and choose as R an operator of left multiplication by r. Then, the bracket

$$[X, Y]_1 = XrY - YrX$$

is compatible with the bracket $[X, Y] = XY - YX$. In this case $A_\lambda = 1 + \lambda R$.

Example 3.1.10. Let $\mathcal{A}$ be an associative algebra with an involution $\ast$, $\mathcal{G}$ a Lie algebra consisting of all skew-symmetric elements of $\mathcal{A}$, and r an element symmetric with respect to $\ast$. In this case, the following operator:

$$A_\lambda : g \to \sqrt{1 + r\lambda} \; g \; \sqrt{1 + r\lambda}$$

can be selected as A_λ. Then, $R(X) = \dfrac{1}{2}(rX + Xr)$ and $[X, Y]_1 = XrY - YrX$.

Example 3.1.11. Let φ be an automorphism of a Lie algebra $\mathcal{G}$ of order n, $\mathcal{G}_i$ the eigenspace of the operator φ corresponding to its eigenvalue ε^i, where $\varepsilon^n = 1$. Then, the vector space $\mathcal{G}_0$ is a Lie subalgebra. Suppose

$$\mathcal{G}_0 = \mathcal{G}_+ \oplus \mathcal{G}_-,$$

where the vector subspaces $\mathcal{G}_+$ and $\mathcal{G}_-$ are some subalgebras in $\mathcal{G}_0$. Consider the operator A_λ, acting on $\mathcal{G}_+$, $\mathcal{G}_-$, and $\mathcal{G}_i$, $i > 0$, by multiplying by

$$1 + \alpha\lambda, \quad 1 + \beta\lambda, \quad \sqrt[n]{(1 + \alpha\lambda)^i (1 + \beta\lambda)^{n-i}},$$

respectively. Operator R has the form

$$R(g) = \alpha\, g_+ + \beta\, g_- + \sum_{i=1}^{n-1} \left(\frac{i}{n}\alpha + \frac{n-i}{n}\beta \right) g_i,$$

where $g_\pm$ means the projection of g onto $\mathcal{G}_\pm$.

Example 3.1.12. This class of compatible brackets is associated with finite-dimensional $\mathbb{Z}_2 \times \mathbb{Z}_2$-graded Lie algebras. Recall the definition. Let φ and ψ be two commuting automorphisms of $\mathcal{G}$ such that $\varphi^2 = \psi^2 = 1$. The decomposition $\mathcal{G} = \oplus \mathcal{G}_{ij}$, $i, j = \pm 1$ to a direct sum of the following four invariant vector subspaces

$$\mathcal{G}_{ij} = \{g \in \mathcal{G} \mid \varphi(g) = i\,g, \quad \psi(g) = j\,g\}$$

is called a $\mathbb{Z}_2 \times \mathbb{Z}_2$-grading.

Let us define the operator A_λ on the homogeneous components $\mathcal{G}_{ij}$ by the formulas

$$A_\lambda(g_{1,1}) = (1 + \gamma\lambda)\,g_{1,1}, \qquad A_\lambda(g_{-1,1}) = \sqrt{(1 + \beta\lambda)(1 + \gamma\lambda)}\,g_{-1,1},$$

$$A_\lambda(g_{1,-1}) = \sqrt{(1 + \alpha\lambda)(1 + \gamma\lambda)}\,g_{1,-1},$$

$$A_\lambda(g_{-1,-1}) = \sqrt{(1 + \alpha\lambda)(1 + \beta\lambda)}\,g_{-1,-1} + \lambda\sqrt{(\gamma - \alpha)(\gamma - \beta)}\,\rho(g_{-1,-1}).$$

Here, α, β and γ are arbitrary constants, and operator $\rho : \mathcal{G}_{-1,-1} \to \mathcal{G}_{1,1}$ is an arbitrary solution of the modified Yang–Baxter equation

$$\rho\Big([\rho(X), Y] - [\rho(Y), X]\Big) - [\rho(X), \rho(Y)] - [X, Y] = 0.$$

In other words (see Theorem 2.4.7), it is assumed that the Lie algebra $\mathcal{G}_{1,1} \oplus \mathcal{G}_{-1,-1}$ is decomposed into the direct sum of $\mathcal{G}_{1,1}$ and some additional subalgebra $\mathcal{B}$ and r denotes the projection on $\mathcal{G}_{1,1}$ parallel to $\mathcal{B}$.

The corresponding operator R has the form

$$R(g) = \gamma\,g_{1,1} + \frac{1}{2}(\alpha + \gamma)\,g_{1,-1} + \frac{1}{2}(\beta + \gamma)\,g_{-1,1} +$$

$$\frac{1}{2}(\alpha + \beta)\,g_{-1,-1} + \sqrt{(\gamma - \alpha)(\gamma - \beta)}\,\rho(g_{-1,-1}).$$

Remark 3.1.13. *Operator A_λ in Example 3.1.12 can be parametrized by points of the elliptic curve*

$$\lambda_1^2 - \alpha = \lambda_2^2 - \beta = \lambda_3^2 - \gamma = \frac{1}{\lambda}.$$

Remark 3.1.14. *A wide class of $\mathbb{Z}_2 \times \mathbb{Z}_2$-graded Lie algebras can be constructed as follows. Let $\mathcal{G} = \oplus \mathcal{G}_i$ be a $\mathbb{Z}$-graded Lie algebra possessing an involution ψ such that $\psi(\mathcal{G}_i) = \mathcal{G}_{-i}$. In particular, such an involution exists for any standard grading of a simple Lie algebra. Choosing as φ the involution $\varphi(X) = (-1)^i X$, where $X \in \mathcal{G}_i$, we define on $\mathcal{G}$ a structure of $\mathbb{Z}_2 \times \mathbb{Z}_2$-graded algebra.*

Applications

Very often, a Lax representation for the corresponding bi-Hamiltonian model can be written in terms of A_λ [19, 94].

Application 3.1.15. Let $\mathbf{V}$ be a vector space endowed with a pair of compatible Lie brackets $[\cdot, \cdot]$ and $[\cdot, \cdot]_1$, where $[\cdot, \cdot]$ is semisimple, and A_λ be an operator of the form (3.6) that satisfy (3.7). Consider the following system of ordinary differential equations for three vectors $u, v, w \in \mathbf{V}$:

$$w_t = [w, v] + w * w, \qquad v_t = [w, u] + w * v, \qquad u_t = w * u,$$

where

$$X * Y \overset{def}{=} [R(X), Y] - [X, R^*(Y)] + R^*([X, Y]),$$

and R^* denotes the operator adjoint to R with respect to the Killing form. Then, operators

$$\mathcal{L} = (A_\lambda^{-1})^*(\lambda u + v + \lambda^{-1} w), \qquad \mathcal{A} = \lambda^{-1} A_\lambda(w) \qquad (3.15)$$

form a Lax pair for this system.

As usual (see Lemma 1.2.1), the integrals of motion can be found from the expressions $\operatorname{tr} \mathcal{L}^k$, $k = 1, 2 \dots$. Each of them depends on λ, which plays the role of a spectral parameter in the Lax pair, and therefore defines several first integrals.

In the case of Example 3.1.9, the bracket $[\cdot, \cdot]$ is the usual matrix commutator, the bracket $[X, Y]_1$ is given by $[X, Y]_1 = XrY - YrX$, and the product is defined by the formula $X * Y = rXY - YXr$. Then,

$$R(w) = r w, \qquad A_\lambda(w) = (1 + \lambda r) w, \qquad (A_\lambda^{-1})^*(w) = w (1 + \lambda r)^{-1}.$$

In the case of Lax pair (3.15), we arrive at (2.74), where u, v and w are matrices of a general form. In the particular case $u = v = 0$, we have

$$\mathcal{L} = w(1 + \lambda r)^{-1}, \qquad \mathcal{A} = \lambda^{-1}(1 + \lambda r) w.$$

The corresponding Lax equation is equivalent to (1.9), where $\mathbf{U} = w$, $a = -r$.

Application 3.1.16. Consider a hyperbolic system of equations

$$u_x = [u, v], \qquad v_y = [v, u]_1, \qquad (3.16)$$

where u and v belong to the vector space $\mathbf{V}$ endowed with a pair of Lie brackets $[\cdot, \cdot]$ and $[\cdot, \cdot]_1$. For the well-known integrable system of the principle chiral field

$$u_x = [u, v], \qquad v_y = [v, u]$$

the brackets $[\cdot, \cdot]$ and $[\cdot, \cdot]_1$ coincide with the matrix commutator.

Theorem 3.1.17. *If the brackets $[\cdot,\cdot]$ and $[\cdot,\cdot]_1$ are compatible, then the system (3.16) has the Lax pair*

$$\mathcal{L} = \frac{d}{dy} + \frac{1}{\lambda} A_\lambda(u), \qquad \mathcal{A} = \frac{d}{dx} + A_\lambda(v).$$

The compatible Lie brackets and the corresponding operator A_λ are also closely related to the different versions of the Yang–Baxter equation [95, 91].

Application 3.1.18. Consider the classical Yang–Baxter equation

$$[r^{1,2}(\lambda,\mu),\, r^{1,3}(\lambda,\nu)] + [r^{1,2}(\lambda,\mu),\, r^{2,3}(\mu,\nu)] + [r^{1,3}(\lambda,\nu),\, r^{2,3}(\mu,\nu)] = 0,$$

where

$$r(x,y) = \sum_i a_i(x,y) \otimes b_i(x,y)$$

is a function of two complex variables with values in $\mathfrak{gl}_N \otimes \mathfrak{gl}_N$, a $r^{i,j}$, where $1 \le i < j \le 3$, are the following functions with values in the tensor cube $\mathfrak{gl}_N \otimes \mathfrak{gl}_N \otimes \mathfrak{gl}_N$:

$$r^{1,2}(\lambda,\mu) = \sum_i a_i(\lambda,\mu) \otimes b_i(\lambda,\mu) \otimes 1,$$

$$r^{1,3}(\lambda,\nu) = \sum_i a_i(\lambda,\nu) \otimes 1 \otimes b_i(\lambda,\nu),$$

$$r^{2,3}(\mu,\nu) = \sum_i 1 \otimes a_i(\mu,\nu) \otimes b_i(\mu,\nu).$$

We assume, as usual, that the unitary condition

$$r^{1,2}(\lambda,\mu) = -r^{2,1}(\mu,\lambda),$$

where

$$r^{2,1}(\mu,\lambda) = \sum_i b_i(\mu,\lambda) \otimes a_i(\mu,\lambda) \otimes 1,$$

is fulfilled.

Theorem 3.1.19. *Let $[\cdot\cdot]_1$ and $[\cdot,\cdot]_2$ be a pair of compatible Lie brackets on N-dimensional vector space $\mathbf{V}$. Suppose that there exists a nondegenerate symmetric form $\omega(X,Y)$ on $\mathbf{V}$, invariant under both brackets $[\cdot\cdot]_{1,2}$. Let $\mathbf{e}_1,\ldots,\mathbf{e}_N$ be a basis in $\mathbf{V}$, orthonormal with respect to ω. Then*

$$r(x,y) = \sum_{i=1}^N \frac{(x\, ad_1 \mathbf{e_i} + ad_2 \mathbf{e_i}) \otimes (y\, ad_1 \mathbf{e_i} + ad_2 \mathbf{e_i})}{x - y}$$

satisfies the classical Yang–Baxter equation and the unitary condition. In the formula for r, the linear operators $ad_i q$ are defined by formula

$$ad_i q(p) = [q,p]_i, \qquad i = 1,2.$$

Application 3.1.20. The classical Yang–Baxter equation can be written in operator form [80] as

$$[r(u, w)x, \, r(u, v)y] = r(u, v)[r(v, w)x, \, y] + r(u, w)[x, \, r(w, v)y].$$

Here, $r(u, v) \in \text{End}(\mathcal{G})$. The solution is called unitary if

$$\langle x, \, r(u, v)y \rangle = -\langle r(v, u)x, \, y \rangle$$

for the Killing form $\langle \cdot, \cdot \rangle$ on $\mathcal{G}$.

Theorem 3.1.21. [91] *Let $[\cdot, \cdot]$ and $[\cdot, \cdot]_1$ be compatible Lie brackets, where the algebra $\mathcal{G}$ with the bracket $[\cdot, \cdot]$ is semisimple, and an operator A_λ satisfies (3.7). Then*

$$r(u, v) = \frac{1}{u - v} A_u A_v^{-1} \tag{3.17}$$

is a solution of the operator classical Yang–Baxter equation.

Remark 3.1.22. *The constructed r-matrix (3.17) is unitary with respect to the Killing form $\langle \cdot, \cdot \rangle$ if the operator A_λ is orthogonal. In this case, the formula (3.7) implies also the invariance of the form $\langle \cdot, \cdot \rangle$ with respect to the bracket $[\cdot, \cdot]_1$.*

Application 3.1.23. As it is known (see Section 2.3), the decomposition (2.50) of the loop algebra over a Lie algebra $\mathcal{G}$ into a direct sum of the Lie subalgebra of Taylor series and a factoring subalgebra $\mathcal{U}$ leads to Lax representations for various integrable models.

In the case when the factoring subalgebra $\mathcal{U}$ is homogeneous (see Definition 2.3.8), we have

$$\frac{1}{\lambda} \mathcal{U} \subset \mathcal{U}.$$

It turns out [94] that for any semisimple Lie algebra $\mathcal{G}$ with the bracket $[\cdot, \cdot]$ there is a one-to-one correspondence between homogeneous factoring subalgebras in the loop algebra over $\mathcal{G}$ and Lie brackets $[\cdot, \cdot]_1$, compatible with $[\cdot, \cdot]$.

Theorem 3.1.24. (i) *Any homogeneous factoring subalgebra $\mathcal{U}$ has the form*

$$\mathcal{U} = \Big\{ \sum_{i=1}^{k} \lambda^{-i} A_\lambda(g_i), \, | \, g_i \in \mathcal{G}, \, k \in \mathbb{N} \Big\}, \tag{3.18}$$

where A_λ is a series of the form (3.6);

(ii) *The vector space (3.18) is a factoring Lie subalgebra iff A_λ satisfies the relation (3.7) for some Lie bracket $[\cdot, \cdot]_1$, compatible with the bracket on $\mathcal{G}$.*

3.1.2   Compatible Lie brackets associated with θ-functions

In this section, we consider an elliptic analogue of Example 3.1.7 replacing polynomials with scalar θ-functions on an elliptic curve $\mathcal{E}$ with values in $\mathfrak{sl}_n$. A generalization to the case of vector θ-functions appears in [88].

Let $\Gamma \subset \mathbb{C}$ be a lattice generated by the unity and τ, where $\operatorname{Im} \tau > 0$, and $m \in \mathbb{N}$. Denote by $\Theta_m(\tau)$ the vector space of holomorphic functions $\phi : \mathbb{C} \mapsto \mathbb{C}$ such that

$$\phi(z+1) = \phi(z), \qquad \phi(z+\tau) = (-1)^m \exp\left(-2\pi i m \, z\right) \phi(z).$$

The elements of this vector space are called θ-*functions of order* m. It is clear that the product of θ-functions of orders m_1 and m_2 is a θ-function of order $m_1 + m_2$.

The properties of θ-functions are described, for example, in the appendix of survey [27]. In particular, the dimension of the space $\Theta_m(\tau)$ is equal to m. Let us fix a generator of the space $\Theta_1(\tau)$ and denote it by $\theta(z)$. It is known that any element $f \in \Theta_m(\tau)$ has m zeros modulo Γ and the sum of them equals zero modulo Γ. In particular, $\theta(z)$ has zero at $z = 0$ modulo Γ. If $x_1, \ldots, x_m$ are all zeros of an element $f \in \Theta_m(\tau)$, then $f(z) = c\,\theta(z-x_1)\cdots\theta(z-x_m)$, where c is some constant.

We fix relative prime positive integers k and n such that $1 \leq k < n$. Let $\mathbf{A}$ and $\mathbf{B}$ be $n \times n$ matrices with the following properties:

$$\mathbf{A}^n = \mathbf{B}^n = 1, \qquad \mathbf{B}\,\mathbf{A} = \exp\left(\frac{2\pi i k}{n}\right) \mathbf{A}\,\mathbf{B}. \tag{3.19}$$

Note that such a pair of matrices $(\mathbf{A}, \mathbf{B})$ is an important ingredient of several "elliptic" constructions related to $\mathfrak{sl}_n$ (see [61, 2, 27]). In this book they will appear again in Example 3.1.48.

It is clear that matrices $\mathbf{A}^\alpha \mathbf{B}^\beta$, where $\alpha, \beta = 0, \ldots, n-1$, $(\alpha, \beta) \neq (0,0)$, form a basis in $\mathfrak{sl}_n$. The commutation relations between these matrices are of the form

$$[\mathbf{A}^{\alpha_1} \mathbf{B}^{\beta_1}, \, \mathbf{A}^{\alpha_2} \mathbf{B}^{\beta_2}] = \left[\exp\left(\frac{2\pi i k \beta_1 \alpha_2}{n}\right) - \exp\left(\frac{2\pi i k \beta_2 \alpha_1}{n}\right)\right] \mathbf{A}^{\alpha_1 + \alpha_2} \mathbf{B}^{\beta_1 + \beta_2}. \tag{3.20}$$

Denote by $\mathbf{V}_m$ the vector space of holomorphic functions $f : \mathbb{C} \mapsto \mathfrak{sl}_n$ satisfying the following quasiperiodicity conditions:

$$f(z+1) = \mathbf{A} f(z) \mathbf{A}^{-1}, \qquad f(z+\tau) = (-1)^m \exp\left(-2\pi i m \, z\right) \mathbf{B} f(z) \mathbf{B}^{-1}. \tag{3.21}$$

Note that if $f_1 \in \mathbf{V}_{m_1}$ and $f_2 \in \mathbf{V}_{m_2}$, then $f_1 f_2 \in \mathbf{V}_{m_1+m_2}$. From (3.21) it follows that if

$$f(z) = \sum f_{\alpha,\beta}(z) \, \mathbf{A}^\alpha \, \mathbf{B}^\beta,$$

then

$$f_{\alpha,\beta}(z+1) = \exp\left(-\frac{2\pi i k\,\beta}{n}\right) f_{\alpha,\beta}(z),$$

$$f_{\alpha,\beta}(z+\tau) = (-1)^m \exp\left(-2\pi i m z + \frac{2\pi i k\,\alpha}{n}\right) f_{\alpha,\beta}(z). \tag{3.22}$$

These identities imply that

$$f_{\alpha,\beta}(z) = \exp\left(-\frac{2\pi i k\,\beta}{n}\,z\right) g_{\alpha,\beta}\left(z - \frac{k\alpha}{mn} - \frac{k\beta}{mn}\tau\right), \tag{3.23}$$

where $g_{\alpha,\beta}(z)$ belongs to $\Theta_m(\tau)$.

Lemma 3.1.25. *Let $\mu_1, \mu_2 \in \Theta_m(\tau)$ be a pair of θ-functions that do not have common zeros. Then, any element $Z \in \mathbf{V}_{2m}$ can be uniquely represented as*

$$Z = \mu_1 P + \mu_2 Q, \qquad P, Q \in \mathbf{V}_m.$$

Proof. Consider the linear map $L : \mathbf{V}_m \oplus \mathbf{V}_m \mapsto \mathbf{V}_{2m}$ given by the formula

$$L(P,Q) = \mu_1 P + \mu_2 Q.$$

We will prove that L is an isomorphism. Since $\dim(\mathbf{V}_m \oplus \mathbf{V}_m) = \dim\mathbf{V}_{2m} = 2m(n^2 - 1)$, it suffices to show that $\mathrm{Ker}\, L = 0$. Substituting

$$P = \sum P_{\alpha,\beta}(z)\,\mathbf{A}^\alpha\,\mathbf{B}^\beta, \qquad Q = \sum Q_{\alpha,\beta}(z)\,\mathbf{A}^\alpha\,\mathbf{B}^\beta$$

into $L(P,Q) = 0$, we find that $\mu_1(z)P_{\alpha,\beta}(z) + \mu_2(z)Q_{\alpha,\beta}(z) = 0$ for all $(\alpha,\beta) \neq 0$. Since $\mu_1(z)$ and $\mu_2(z)$ have no common zeros, we see that every zero of $\mu_1(z)$ is a zero of $Q_{\alpha,\beta}(z)$. We know that $\mu_1(z) \in \Theta_m(\tau)$ has exactly m zeros modulo Γ and that the sum of all these zeros is 0. From (3.23) it follows that if $Q_{\alpha,\beta}(z) \not\equiv 0$, then $Q_{\alpha,\beta}(z)$ also has exactly m zeros, but their sum is $\dfrac{k\alpha}{n} + \dfrac{k\beta}{n}\tau \neq 0$. Since $Q_{\alpha,\beta}(z) \equiv 0$, we get $Q \equiv P \equiv 0$. $\square$

Using Lemma 3.1.25, we define for any $f_1, f_2 \in \mathbf{V}_m$ brackets $[f_1, f_2]_1$ and $[f_1, f_2]_2$ by the formula (cf. Example 3.1.7)

$$[f_1, f_2] = \mu_1[f_1, f_2]_1 + \mu_2[f_1, f_2]_2. \tag{3.24}$$

Theorem 3.1.26. *The bilinear operations $[f_1, f_2]_1$ and $[f_1, f_2]_2$ are compatible Lie brackets on $\mathbf{V}_m$.*

Proof. Obviously, the usual commutator bracket $[f_1, f_2] = f_1 f_2 - f_2 f_1$ is a Lie bracket on the vector space $\oplus_{p>0} \mathbf{V}_p$. Lemma 3.1.25 shows that $[\cdot, \cdot]_{1,2}$ are well defined brackets on $\mathbf{V}_m$. Substituting (3.24) into the antisymmetricity condition $[f_1, f_2] + [f_2, f_1] = 0$ for the usual bracket, we obtain that

$$\mu_1([f_1, f_2]_1 + [f_2, f_1]_1) + \mu_2([f_1, f_2]_2 + [f_2, f_1]_2) = 0.$$

It follows from Lemma 3.1.25 that $[f_1, f_2]_i + [f_2, f_1]_i = 0$ for $i = 1, 2$. Substituting (3.24) into the Jacobi identity for the usual bracket, we get the identity of the form $\mu_1^2 P + \mu_1 \mu_2 Q + \mu_2^2 R = 0$ for some $P, Q, R \in \mathbf{V}_m$. Using the same arguments as in the proof of Lemma 3.1.25, it can be shown that $P \equiv Q \equiv R \equiv 0$. It is easy to verify that the identities $P = 0$ and $R = 0$ coincide with the Jacobi identity for the brackets $[\cdot, \cdot]_1$ and $[\cdot, \cdot]_2$, respectively. The identity $Q = 0$ is equivalent to the compatibility of the brackets $[\cdot, \cdot]_{1,2}$. $\square$

Let $x_i(\lambda)$, $i = 1, \ldots, m$ be zeros of the θ-function $\mu_2(z) - \lambda \mu_1(z)$. Then,

$$\mu_2(z) - \lambda \mu_1(z) = c(\lambda) \theta(z - x_1(\lambda)) \cdots \theta(z - x_m(\lambda)) \tag{3.25}$$

If $x_1(\lambda), \ldots, x_m(\lambda)$ are pairwise distinct modulo Γ, we say that the point λ is regular.

Proposition 3.1.1. *The linear combination*

$$[f_1, f_2]_\lambda = [f_1, f_2]_1 + \lambda [f_1, f_2]_2 \tag{3.26}$$

of the brackets $[\cdot, \cdot]_i$ is isomorphic to the Lie bracket $\oplus_{i=1}^m \mathfrak{sl}_n$ for all regular λ.

Remark 3.1.27. *It is not difficult to construct the "trigonometric" and "rational" degenerations of the elliptic brackets described above. Namely, in the trigonometric case, $\theta(z)$ should be replaced by $1 - \exp(2\pi i z)$, $\Theta_m(\tau)$ by the space of functions of the form*

$$a_0 + a_1 \exp(2\pi i z) + \cdots + a_m \exp(2\pi i m z),$$

such that $a_m = (-1)^m a_0$, and $\mathbf{V}_m$ by the space of $\mathfrak{sl}_n$-valued functions of the form

$$\sum c_{\alpha,\beta} \exp\left(2\pi i \frac{\beta}{n} z\right) \mathbf{A}^\alpha \mathbf{B}^\beta, \qquad 0 \le \alpha \le n, \quad 0 \le \beta \le mn,$$

where $c_{\alpha,0} = (-1)^m c_{\alpha,mn}$. In this formula, we assume that $\mathbf{A}^n = \mathbf{B}^n = 1$, $\mathbf{B}\mathbf{A} = \exp\left(\dfrac{2\pi i}{n}\right) \mathbf{A}\mathbf{B}$.

In the rational case, $\theta(z)$ is replaced by z, $\Theta_m(\tau)$ by the space of polynomials of the form $\sum\limits_{i=0}^m c_\alpha z^\alpha$, $c_{m-1} = 0$, and $\mathbf{V}_m$ by the space of matrix polynomials of the form $\sum\limits_{i=0}^m g_\alpha z^\alpha$, $g_\alpha \in \mathfrak{sl}_n$, $g_{m-1} = 0$.

Open problem 3.1.28. *Verify that the bi-Hamiltonian system corresponding to the compatible elliptic brackets constructed above is (after a change of variables) the elliptic Gaudin model* [96].

3.1.3 Associative algebras compatible with Mat_m

Consider the case when the associative algebra with a multiplication of $\star$ coincides with Mat_m. Then, $R : \mathrm{Mat}_m \to \mathrm{Mat}_m$ is a linear operator on the space of $m \times m$ matrices. In other words, we study *linear associative deformations of matrix multiplication*.[2] We will omit the sign $\star$ everywhere.

Example 3.1.29. (see Example 3.1.9) Let r be a fixed element of Mat_m, and $R : X \to X$ is the operator of left multiplication by r. Then, the corresponding multiplication (3.14), which has the form $X \circ Y = X\,r\,Y$, is associative and compatible with the matrix product in Mat_m.

Example 3.1.30. Let $a, b \in \mathrm{Mat}_2$ be arbitrary fixed matrices. Then, the multiplication

$$X \circ Y = (aX - Xa)\,(bY - Yb) \tag{3.27}$$

is associative and compatible with the usual product in Mat_2. The corresponding operator R is given by the formula

$$R(X) = a\,(Xb - bX).$$

Remark 3.1.31. *Since any matrix $A \in \mathrm{Mat}_2$ can be written in the form $A = c_1 a + c_2 1$, where $a^2 = 1$, $c_i \in \mathbb{C}$, without loss of generality we may assume that $a^2 = b^2 = 1$ in the formula* (3.27).

Proposition 3.1.2. *In the case of the algebra Mat_2, any linear deformation of the matrix product is equivalent[3] to one of the two given in Examples 3.1.29 and 3.1.30.*

Remark 3.1.32. *The operator*

$$R(X) = aXb + baX, \qquad \text{where} \quad a^2 = b^2 = 1$$

defines the multiplication

$$X \circ Y = a\,X\,b\,Y + X\,a\,Y\,b + X\,b\,a\,Y - a\,X\,Y\,b, \qquad a^2 = b^2 = 1 \tag{3.28}$$

equivalent to (3.27). *Indeed, if $a^2 = b^2 = 1$, where $a, b \in \mathrm{Mat}_2$, then $ab + ba$ is proportional to* **1**.

It turns out that formula (3.28) *defines an associative multiplication, compatible with the matrix product in Mat_m for an arbitrary m.*

---

[2]Linear deformations of arbitrary semisimple associative algebras were investigated in [93].

[3]Multiplications $\circ$ and $\star$ are equivalent if $X \circ Y = k_1 X \star Y + k_2 XY$, where $k_1 \neq 0$.

Exercise 3.1.33. Verify this fact.

It is easy to see that any operator $R : \mathrm{Mat}_m \to \mathrm{Mat}_m$ can be written as

$$R(x) = \mathbf{a}_1 \, x \, \mathbf{b}^1 + \cdots + \mathbf{a}_{p+1} \, x \, \mathbf{b}^{p+1},$$

where $\mathbf{a}_i, \mathbf{b}^i \in \mathrm{Mat}_m$. We will always assume that p is the minimum possible number for a given R. In the case of the minimum p, matrices $\mathbf{a}_1, \ldots, \mathbf{a}_{p+1}$ are linearly independent as well as the matrices $\mathbf{b}^1, \ldots, \mathbf{b}^{p+1}$.

Since the transformation

$$R \mapsto R + \mathrm{ad}_s \tag{3.29}$$

does not change the multiplication (3.14) for any $s \in \mathrm{Mat}_m$, and the shift $R \mapsto R + \mu \mathbf{1}$ leads to an equivalent multiplication, we can represent $R(x)$ in the following form:

$$R(x) = \mathbf{a}_1 \, x \, \mathbf{b}^1 + \cdots + \mathbf{a}_p \, x \, \mathbf{b}^p + \mathbf{c} \, x, \tag{3.30}$$

where matrices $\mathbf{a}_1, \ldots, \mathbf{a}_p, \mathbf{1}$ and $\mathbf{b}^1, \ldots, \mathbf{b}^p, \mathbf{1}$ are linearly independent.

Integrable matrix differential equations associated with the operator R

Let $R : \mathrm{Mat}_m \to \mathrm{Mat}_m$ be a linear operator such that the multiplication (3.14) is associative. Consider the following matrix differential equation [19, 92]:

$$\frac{dx}{dt} = [x, \, R(x) + R^*(x)], \qquad x(t) \in \mathrm{Mat}_m, \tag{3.31}$$

where R^* denotes the operator adjoint to R with respect to the bilinear form $\langle x, \, y \rangle = \mathrm{tr}\,(x\,y)$. For an operator written in the form (3.30), we have

$$R^*(x) = \mathbf{b}^1 \, x \, \mathbf{a}_1 + \cdots + \mathbf{b}^p \, x \, \mathbf{a}_p + x \, \mathbf{c}.$$

Theorem 3.1.34. *If the multiplication (3.14) is associative, then equation (3.31) has the Lax pair*

$$L = \left(S_\lambda^{-1} \right)^* (x), \qquad A = \frac{1}{\lambda} \, S_\lambda(x).$$

Open problem 3.1.35. *Show that equation (3.31) is bi-Hamiltonian with Hamiltonian operators*

$$\mathcal{H}_1 = \mathrm{ad}_x, \qquad \mathcal{H}_2 = \mathrm{ad}_x^1,$$

where ad_x^1 corresponds to the multiplication (3.14).

Example 3.1.36. In the case of Example 3.1.29, we obtain the Manakov equation (see Example 1.2.2 and Appendix 3.1.15)

$$\frac{dx}{dt} = [x, \, xr + rx] = x^2 \, r - r \, x^2$$

for a matrix $x(t)$.

Example 3.1.37. In the case of Remark 3.1.32, we have equation [92]

$$x_t = [x, \; bxa + axb + xba + bax], \qquad x, a, b \in \mathrm{Mat}_m, \tag{3.32}$$

where $a^2 = b^2 = \mathbf{1}_m$. Equation (3.32) admits the following skew-symmetric reduction

$$x^T = -x, \qquad b = a^T.$$

The various integrable $\mathfrak{so}_m$-models generated by this reduction are described by the canonical form

$$a = \begin{pmatrix} \mathbf{1}_p & T \\ 0 & -\mathbf{1}_{m-p} \end{pmatrix} \tag{3.33}$$

of the matrix a. Here, $\mathbf{1}_s$ denotes the unity matrix of dimension $s \times s$, and $T = \{t_{ij}\}$, where $t_{ij} = \delta_{ij}\alpha_i$. This canonical form is determined by the natural parameter p and the continuous parameters $\alpha_1, \ldots, \alpha_r$, where $p \le m/2$, $r = \min(p, m - p)$.

For $m = 4$, the cases of $p = 2$ and $p = 1$ lead to the integrable Steklov and Poincaré [23] $\mathfrak{so}_4$-models, respectively. For arbitrary m, the choice of $p = [m/2]$ and $p = 1$ gives $\mathfrak{so}_m$-generalizations of these models.

In the cases of Example 3.1.29 and Remark 3.1.32, we have $R(x) = r\,x$ and $R(x) = axb + bax$. Both of these R-operators are written in the form (3.30). In the next sections, we construct generalizations of these operators associated with the affine Dynkin diagrams of types $A, D,$ and E.

$\mathcal{M}$-structures

As already noted, if a multiplication (3.14) is associative, then it is compatible with the matrix one. Consider the operators R of the form (3.30) for which the multiplication (3.14) is associative. Our goal is to investigate [91, 93] the structure of associative algebra $\mathcal{M}$ generated by the matrices $\mathbf{a}_i, \mathbf{b}^i, \mathbf{c}$ and $\mathbf{1}$. Denote by $\mathcal{L}$ the vector space of dimension $2p + 2$ spanned by these matrices.

Lemma 3.1.38. *If a multiplication* (3.14) *is associative, then*

$$\mathbf{a}_i \mathbf{a}_j = \sum_k \phi_{i,j}^k \mathbf{a}_k + \mu_{i,j}\mathbf{1}, \qquad \mathbf{b}^i \mathbf{b}^j = \sum_k \psi_k^{i,j} \mathbf{b}^k + \lambda^{i,j}\mathbf{1}$$

for some tensors $\phi_{i,j}^k, \mu_{i,j}, \psi_k^{i,j}, \lambda^{i,j}$.

Lemma 3.1.38 means that the vector spaces spanned by $\mathbf{1}, \mathbf{a}_1, \ldots \mathbf{a}_p$ and $\mathbf{1}, \mathbf{b}^1, \ldots \mathbf{b}^p$ are associative algebras. We denote them by $\mathcal{A}$ and $\mathcal{B}$, respectively. It turns out that the vector space $\mathcal{L}$ is a left module over $\mathcal{B}$ and a right one over $\mathcal{A}$.

The simplest example of such a structure is the finite-dimensional associative bialgebras [97].

Example 3.1.39. Let $\mathcal{A}$ and $\mathcal{B}$ be associative algebras with bases $A_1, \ldots, A_p$ and $B^1, \ldots, B^p$ and the structural constants $\phi^i_{j,k}$ and $\psi^{\alpha,\beta}_\gamma$, respectively. Suppose that the structural constants are connected by the relations

$$\phi^s_{j,k}\psi^{l,i}_s = \phi^l_{s,k}\psi^{s,i}_j + \phi^i_{j,s}\psi^{l,s}_k, \qquad 1 \le i, j, k, l \le p.$$

Then, $\mathcal{M}$ is an algebra of dimension $2p + p^2$ with basis $A_i, B^j, A_i B^j$ and the relations

$$B^i A_j = \psi^{k,i}_j A_k + \phi^i_{j,k} B^k$$

and it is associative. It is clear that the vector space $\mathcal{L}$ spanned by $A_1, \ldots, A_p$, $B^1, \ldots, B^p$ is a left module over $\mathcal{B}$ and a right module over $\mathcal{A}$.

Our situation is different from Example 3.1.39 because of the presence of an additional element $\mathbf{c}$ and the fact that the intersection of the algebras $\mathcal{A}$ and $\mathcal{B}$ is nontrivial (contains the unity $\mathbf{1}$). As a result, the algebra $\mathcal{M}$ turns out to be infinite-dimensional.

We describe the algebraic structure arising here, first at the level of relations between the structural constants of the algebras $\mathcal{A}$ and $\mathcal{B}$, and then in invariant terms.

Lemma 3.1.40. *If the operation* (3.14) *is associative, then*

$$\phi^s_{j,k}\psi^{l,i}_s = \phi^l_{s,k}\psi^{s,i}_j + \phi^i_{j,s}\psi^{l,s}_k + \delta^l_k t^i_j - \delta^i_j t^l_k - \delta^i_j \phi^l_{s,r} \psi^{r,s}_k,$$

and

$$\mathbf{b}^i \mathbf{a}_j = \psi^{k,i}_j \mathbf{a}_k + \phi^i_{j,k} \mathbf{b}^k + t^i_j \mathbf{1} + \delta^i_j \mathbf{c} \qquad (3.34)$$

for some tensor t^i_j.

The matrix $\mathbf{c}$ satisfies the following relations:

Lemma 3.1.41. *If the multiplication* (3.14) *is associative, then*

$$\mathbf{b}^i \mathbf{c} = \lambda^{k,i} \mathbf{a}_k - t^i_k \mathbf{b}^k - \phi^i_{k,l} \psi^{l,k}_s \mathbf{b}^s - \phi^i_{k,l} \lambda^{l,k} \mathbf{1}, \qquad (3.35)$$

$$\mathbf{c}\, \mathbf{a}_j = \mu_{j,k} \mathbf{b}^k t^k_j \mathbf{a}_k - \phi^s_{k,l} \psi^{l,k}_j \mathbf{a}_s - \mu_{k,l} \psi^{l,k}_j \mathbf{1}, \qquad (3.36)$$

where

$$\phi^s_{j,k} t^i_s = \psi^{s,i}_j \mu_{s,k} + \phi^i_{j,s} t^s_k - \delta^i_j \psi^{s,r}_k \mu_{r,s}, \qquad \psi^{k,i}_s t^s_j = \phi^i_{j,s} \lambda^{k,s} + \psi^{s,i}_j t^k_s - \delta^i_j \phi^k_{s,r} \lambda^{r,s}.$$

The relations (3.34)–(3.36) mean that the vector space $\mathcal{L}$ is a left $\mathcal{B}$-module and a right $\mathcal{A}$-module.

Invariant description

In this section, we forget that $\mathcal{L}$ consists of matrices and give a purely algebraic description of the structure that has arisen above.

Definition 3.1.42. By a weak $\mathcal{M}$-structure on a linear space $\mathcal{L}$ we mean the following data set:

- Two subspaces $\mathcal{A} \subset \mathcal{L}$ and $\mathcal{B} \subset \mathcal{L}$ and a distinguished element $1 \in \mathcal{A} \cap \mathcal{B}$;

- A nondegenerate symmetric scalar product $(\cdot, \cdot)$ on the vector space $\mathcal{L}$.

- Two associative multiplications $\mathcal{A} \times \mathcal{A} \to \mathcal{A}$ and $\mathcal{B} \times \mathcal{B} \to \mathcal{B}$ with unity 1;

- A left action $\mathcal{B} \times \mathcal{L} \to \mathcal{L}$ of the algebra $\mathcal{B}$ and a right action $\mathcal{L} \times \mathcal{A} \to \mathcal{L}$ of the algebra $\mathcal{A}$ on the space $\mathcal{L}$ that commute with each other.

This data supposed to satisfy the following conditions:

1. The intersection of the spaces $\mathcal{A}$ and $\mathcal{B}$ is one-dimensional and is generated by the unity 1;

2. The restriction of the action of $\mathcal{B} \times \mathcal{L} \to \mathcal{L}$ on the subspace $\mathcal{B} \subset \mathcal{L}$ coincides with the multiplication in $\mathcal{B}$. The restriction of the action of $\mathcal{L} \times \mathcal{A} \to \mathcal{L}$ to $\mathcal{A} \subset \mathcal{L}$ coincides with the multiplication in $\mathcal{A}$;

3. The scalar product has the following properties:

$$(a_1, a_2) = (b^1, b^2) = 0, \qquad (b^1 b^2, v) = (b^1, b^2 v), \qquad (v, a_1 a_2) = (v a_1, a_2),$$

for any $a_1, a_2 \in \mathcal{A}$, $b^1, b^2 \in \mathcal{B}$ and $v \in \mathcal{L}$.

Remark 3.1.43. *From these properties it follows that* $(\cdot, \cdot)$ *defines a nondegenerate pairing between the quotient spaces* $\mathcal{A}/\mathbb{C}$ *and* $\mathcal{B}/\mathbb{C}$, *so* $\dim \mathcal{A} = \dim \mathcal{B}$ *and* $\dim \mathcal{L} = 2 \dim \mathcal{A}$.

Definition 3.1.44. An associative algebra $U(\mathcal{L})$ is called the weak $\mathcal{M}$-algebra associated with a weak $\mathcal{M}$-structure on $\mathcal{L}$ if the following conditions hold:

1. $\mathcal{L} \subset \mathcal{U}(\mathcal{L})$ and the actions $\mathcal{B} \times \mathcal{L} \to \mathcal{L}$, $\mathcal{L} \times \mathcal{A} \to \mathcal{L}$ are the restrictions of the product in $U(\mathcal{L})$;

2. For any algebra X satisfying Property 1, there exists a unique homomorphism of algebras $X \to U(\mathcal{L})$ identical on $\mathcal{L}$.

Explicit formulas for $U(\mathcal{L})$

Let $\{\bar{A}_1, \ldots, \bar{A}_p\}$ be a basis in $\mathcal{A}/\mathbb{C}$ and $\{\bar{B}^1, \ldots, \bar{B}^p\}$ be the dual basis in $\mathcal{B}/\mathbb{C}$. This means that $(\bar{A}_i, \bar{B}^j) = \delta_i^j$. It is clear that $1, A_1, \ldots, A_p$ and $1, B^1, \ldots B^p$, where $A_i \in \bar{A}_i$ and $B_i \in \bar{B}_i$, are bases in $\mathcal{A}$ and $\mathcal{B}$, respectively. The element $C \in \mathcal{L}$ does not belong to the sum of $\mathcal{A}$ and $\mathcal{B}$. Since the scalar product $(\cdot, \cdot)$ is nondegenerate, $(1, C) \neq 0$. Without loss of generality, we assume that $(1, C) = 1$, $(C, C) = (C, A_i) = (C, B^j) = 0$. For given bases in $\mathcal{A}$ and in $\mathcal{B}$, such an element C is uniquely defined.

Proposition 3.1.3. *The algebra $U(\mathcal{L})$ is defined by the following relations:*

$$A_i A_j = \phi_{i,j}^k A_k + \mu_{i,j} \, 1, \qquad B^i B^j = \psi_k^{i,j} B^k + \lambda^{i,j} \, 1,$$

$$B^i A_j = \psi_j^{k,i} A_k + \phi_{j,k}^i B^k + t_j^i \, 1 + \delta_j^i C,$$

$$B^i C = \lambda^{k,i} A_k + u_k^i B^k + p^i \, 1, \qquad C A_j = \mu_{j,k} B^k + u_j^k A_k + q_i \, 1$$

for some tensors $\phi_{i,j}^k, \psi_k^{i,j}, \mu_{i,j}, \lambda^{i,j}, u_k^i, p^i, q_i$.

Let us define the element $K \in U(\mathcal{L})$ by the formula

$$K = A_i B^i + C. \tag{3.37}$$

Definition 3.1.45. A weak $\mathcal{M}$-structure on $\mathcal{L}$ is called the $\mathcal{M}$-structure, if $K \in U(\mathcal{L})$ is the central element of the algebra $U(\mathcal{L})$.

Theorem 3.1.46. (cf. Example 3.1.39) *For any $\mathcal{M}$-structure the elements K^s, $A_i K^s$, $B_j K^s$, $A_i B^j K^s$, where $i, j = 1, \ldots, p$, and $s = 0, 1, 2, \ldots$, form a basis of the algebra $U(\mathcal{L})$.*

Theorem 3.1.47. *For any representation $U(\mathcal{L}) \to \mathrm{Mat}_m$, where*

$$A_1 \mapsto \mathbf{a}_1, \ldots, A_p \mapsto \mathbf{a}_p, \quad B^1 \mapsto \mathbf{b}^1, \ldots, B^p \mapsto \mathbf{b}^p, \quad C \mapsto \mathbf{c},$$

the formulas

$$R(x) = \mathbf{a}_1 \, x \, \mathbf{b}^1 + \cdots + \mathbf{a}_p \, x \, \mathbf{b}^p + \mathbf{c} \, x$$

and

$$X \circ Y = R(X) \, Y + X \, R(Y) - R(X \, Y)$$

define an associative multiplication on Mat_m, compatible with ordinary matrix multiplication.

Example 3.1.48. Suppose algebras $\mathcal{A}$ and $\mathcal{B}$ are generated by elements $A \in \mathcal{A}$ and $B \in \mathcal{B}$, such that $A^p = B^p = 1$. Assume that $(B^i, A^{-i}) = \epsilon^i - 1$, $(1, C) = 1$, and other scalar products are equal to zero. Here, ϵ is a primitive root of unity of degree p. Let

$$B^i A^j = \frac{\epsilon^{-j} - 1}{\epsilon^{-i-j} - 1} A^{i+j} + \frac{\epsilon^i - 1}{\epsilon^{i+j} - 1} B^{i+j}$$

for $i + j \neq 0$ modulo p and

$$B^i A^{-i} = 1 + (\epsilon^i - 1)C, \qquad C A^i = \frac{1}{1 - \epsilon^i} A^i + \frac{1}{\epsilon^i - 1} B^i,$$

$$B^i C = \frac{1}{\epsilon^{-i} - 1} A^i + \frac{1}{1 - \epsilon^{-i}} B^i,$$

for $i \neq 0$ modulo p. Then, these formulas define an $\mathcal{M}$-structure.

Since the scalar product is defined differently than it was done above (namely, $(B^i, A^{-i}) \neq 1$), then the central element K has the form (cf. (3.37))

$$K = C + \sum_{0 < i < p} \frac{1}{\epsilon^i - 1} A^{-i} B^i.$$

Any representation of $U(\mathcal{L})$ defines an associative multiplication (3.14), where

$$R(x) = \sum_{0 < i < p} \frac{1}{\epsilon^i - 1} \mathbf{a}^{p-i} x \, \mathbf{b}^i + \mathbf{c} \, x.$$

There is the following convenient way to satisfy all the algebraic relations between A, B and C. Let $\mathbf{a}$, $\mathbf{t}$ be linear operators on some vector space. Suppose that[4] $\mathbf{a}^p = 1$, $\mathbf{a} \mathbf{t} = \epsilon \mathbf{t} \mathbf{a}$ and that $\mathbf{t} - 1$ is invertible. It is easy to verify that the formulas

$$A \mapsto \mathbf{a}, \qquad B \mapsto \frac{\epsilon \mathbf{t} - 1}{\mathbf{t} - 1} \mathbf{a}, \qquad C \mapsto \frac{\mathbf{t}}{\mathbf{t} - 1}$$

define a representation of algebra $U(\mathcal{L})$.

Exercise 3.1.49. Verify that for $p = 2$ the multiplication from Example 3.1.48 coincides with the multiplication from Remark 3.1.32.

The case of semisimple $\mathcal{A}$ and $\mathcal{B}$

Proposition 3.1.4. *Suppose that in a weak $\mathcal{M}$-structure the algebra $\mathcal{A}$ is semisimple:*

$$\mathcal{A} = \oplus_{1 \leq i \leq r} \operatorname{End}(V_i), \qquad \dim V_i = m_i.$$

Then, the space $\mathcal{L}$, as the right $\mathcal{A}$-module, is isomorphic to $\oplus_{1 \leq i \leq r} (V_i^*)^{2m_i}$.

---

[4]In contrast with the formula (3.19), we do not assume that $\mathbf{t}^p = 1$.

Proof. As any right $\mathcal{A}$-module, $\mathcal{L}$ has the form

$$\mathcal{L} = \oplus_{1 \le i \le r} \mathcal{L}_i, \quad \text{where} \quad \mathcal{L}_i = (V_i^*)^{l_i}$$

for some $l_1, \ldots, l_r \ge 0$. Note that $\mathcal{A} \subset \mathcal{L}$ and $\text{End}(V_i) \subset \mathcal{L}_i$ for $i = 1, \ldots, r$. In addition, $\text{End}(V_i) \perp \mathcal{L}_j$ for $i \ne j$. Indeed, for any $v \in \mathcal{L}_j$, $a \in \text{End}(V_i)$ we have

$$(v, a) = (v, \mathbf{I}_i a) = (v \, \mathbf{I}_i, a) = 0,$$

where $\mathbf{I}_i$ is the unity of the subalgebra $\text{End}(V_i)$. Since $(\cdot, \cdot)$ is nondegenerate and, according to Property 3 of a weak $\mathcal{M}$-structure, $\text{End}(V_i) \perp \text{End}(V_i)$, we obtain that $\dim \mathcal{L}_i \ge 2 \dim \text{End}(V_i)$. But,

$$\sum_i \dim \mathcal{L}_i = \dim \mathcal{L} = 2 \dim \mathcal{A} = \sum_i 2 \dim \text{End}(V_i)$$

and, therefore, $\dim \mathcal{L}_i = 2 \dim \text{End}(V_i)$ for $i = 1, \ldots, r$, which is equivalent to the statement of the proposition. $\square$

Theorem 3.1.50. *Suppose that the vector space $\mathcal{L}$ is endowed with a weak $\mathcal{M}$-structure such that the associative algebras $\mathcal{A}$ and $\mathcal{B}$ are semisimple:*

$$\mathcal{A} = \oplus_{1 \le i \le r} \text{End}(V_i), \qquad \mathcal{B} = \oplus_{1 \le j \le s} \text{End}(W_j), \tag{3.38}$$

$$\dim V_i = m_i, \qquad \dim W_j = n_j.$$

Then, $\mathcal{L}$, as $\mathcal{A} \otimes \mathcal{B}$-module, is given by

$$\mathcal{L} = \oplus_{1 \le i \le r, 1 \le j \le s} (V_i^* \otimes W_j)^{a_{i,j}} \tag{3.39}$$

for some nonnegative integers $a_{i,j}$ such that

$$\sum_{j=1}^s a_{i,j} \, n_j = 2 \, m_i, \qquad \sum_{i=1}^r a_{i,j} \, m_i = 2 \, n_j. \tag{3.40}$$

Proof. It is known that any $\mathcal{A} \otimes \mathcal{B}$-module has the form (3.39). Applying Proposition 3.1.4, we get that $\dim \mathcal{L}_i = 2 \, m_i^2$, where

$$\mathcal{L}_i = \oplus_{1 \le j \le s} (V_i^* \otimes W_j)^{a_{i,j}}.$$

This gives the first of the relations (3.40). The second is obtained similarly. $\square$

Remark 3.1.51. *Since the dimensions of $\mathcal{A}$ and $\mathcal{B}$ are the same, we have*

$$\sum_{i=1}^r m_i^2 = \sum_{i=1}^s n_i^2.$$

Definition 3.1.52. Matrix $A = \{a_{i,j}\}$ of dimension $r \times s$ from Theorem 3.1.50 is called the *matrix of multiplicities* of the weak $\mathcal{M}$-structures.

Definition 3.1.53. The $r \times s$ matrix A is called *decomposable* if there are partitions $\{1, \ldots, r\} = I \cup I'$ and $\{1, \ldots, s\} = J \cup J'$ such that $a_{i,j} = 0$ for $(i, j) \in I \times J'$ or for $(i, j) \in I' \times J$.

Lemma 3.1.54. [91] *The matrix of multiplicities of the weak $\mathcal{M}$-structure is indecomposable.*

Consider (3.40) as a system of linear equations with respect to the vector $(m_1, \ldots, m_r, n_1, \ldots, n_s)$. The matrix of the system has the form

$$Q = \begin{pmatrix} 2 & -A \\ -A^T & 2 \end{pmatrix}.$$

According to E. Vinberg [98], if such a matrix is indecomposable and its kernel contains a positive integer vector, then Q is the Cartan matrix of some affine Dynkin diagram. In addition, the structure of the matrix Q (two diagonal scalar blocks) implies that this is a so-called "simply-laced" affine Dynkin diagram for which there is a partition of the vertices into two subsets (white and black vertices in the diagrams below) such that vertices from one subset are not connected.

Theorem 3.1.55. *Let A be the $r \times s$ matrix of multiplicities of the weak $\mathcal{M}$-structure. Then, after permutations of rows and columns, the matrix A coincides with one of the following list:*

1. $A = (2)$. *Here, $r = s = 1$, $n_1 = m_1 = m$. The corresponding Dynkin diagram is of type $\tilde{A}_1$.*

2. $a_{i,i} = a_{i,i+1} = 1$ *and* $a_{i,j} = 0$ *for other pairs i, j. Here, $r = s = k \geq 2$, indices are taken modulo k and $n_i = m_i = m$. The corresponding Dynkin diagram is $\tilde{A}_{2k-1}$.*

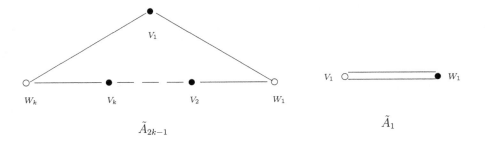

3. $A = \begin{pmatrix} 1 & 1 & 0 & 0 \\ 1 & 0 & 1 & 0 \\ 1 & 0 & 0 & 1 \end{pmatrix}$. *Here,* $r = 3$, $s = 4$ *and* $n_1 = 3m$, $n_2 = n_3 = n_4 = m$, $m_1 = m_2 = m_3 = 2m$. *The Dynkin diagram is* $\tilde{E}_6$:

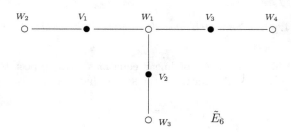

4. $A = \begin{pmatrix} 1 & 1 & 0 & 0 & 0 \\ 0 & 1 & 1 & 1 & 0 \\ 0 & 0 & 0 & 1 & 1 \end{pmatrix}$. *Here,* $r = 3$, $s = 5$ *and* $n_1 = m$, $n_2 = 3m$, $n_3 = 2m$, $n_4 = 3m$, $n_5 = m$, $m_1 = 2m$, $m_2 = 4m$, $m_3 = 2m$. *The Dynkin diagram is* $\tilde{E}_7$:

5. $A = \begin{pmatrix} 1 & 0 & 0 & 0 & 0 \\ 1 & 1 & 1 & 0 & 0 \\ 0 & 0 & 1 & 1 & 0 \\ 0 & 0 & 0 & 1 & 1 \end{pmatrix}$. *Here,* $r = 4$, $s = 5$ *and* $n_1 = 4m$, $n_2 = 3m$, $n_3 = 5m$, $n_4 = 3m$, $n_5 = m$, $m_1 = 2m$, $m_2 = 6m$, $m_3 = 4m$, $m_4 = 2m$. *The Dynkin diagram is* $\tilde{E}_8$:

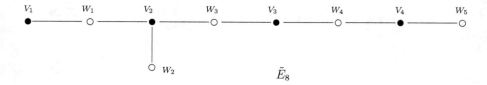

6. $A = (1, 1, 1, 1)$. *Here,* $r = 1, s = 4$ *and* $n_1 = n_2 = n_3 = n_4 = m$, $m_1 = 2m$. *The corresponding Dynkin diagram is* $\tilde{D}_4$.

7. $a_{1,1} = a_{1,2} = a_{1,3} = 1$, $a_{2,3} = a_{2,4} = a_{3,4} = a_{3,5} = \cdots = a_{k-2,k-1} = a_{k-2,k} = 1$, $a_{k-1,k} = a_{k-1,k+1} = a_{k-1,k+2} = 1$, *and* $a_{i,j} = 0$ *for other pairs* (i, j).

In this case, $r = k - 1$, $s = k + 2$ *and* $n_1 = n_2 = n_{k+1} = n_{k+2} = m$, $n_3 = \cdots = n_k = 2m$, $m_1 = \cdots = m_l = 2m$. *The corresponding Dynkin diagram is* $\tilde{D}_{2k}$, *where* $k \geq 3$.

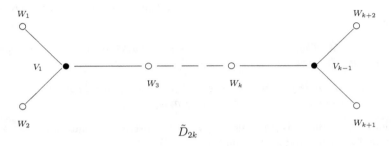

$$\tilde{D}_{2k}$$

8. $a_{1,1} = a_{1,2} = a_{1,3} = 1$, $a_{2,3} = a_{2,4} = a_{3,4} = a_{3,5} = \cdots = a_{k-2,k-1} = a_{k-2,k} = 1$, $a_{k-1,k} = a_{k,k} = 1$, *and* $a_{i,j} = 0$ *for other pairs* (i, j).

Here, $r = s = k \geq 3$, $n_1 = n_2 = m$, $n_3 = \cdots = n_k = 2m$, $m_1 = \cdots = m_{k-2} = 2m$, $m_{k-1} = m_k = m$. *The Dynkin diagram is* $\tilde{D}_{2k-1}$:

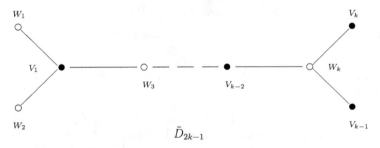

$$\tilde{D}_{2k-1}$$

Note that if $k = 3$, *then* $a_{1,1} = a_{1,2} = a_{1,3} = 1$, $a_{2,3} = a_{3,3} = 1$.

Summary

Let $\mathcal{L}$ be a $\mathcal{M}$-structure with semisimple algebras (3.38). Then, there is an affine Dynkin diagram of type A, D, or E such that:

1. There is a one-to-one correspondence between the set of vertices and the set of vector spaces $\{V_1, \ldots, V_r, W_1, \ldots, W_s\}$.

2. For any i, j the spaces V_i, V_j, as well as the spaces W_i, W_j, are not connected by edges.

3. The vector

$$(\dim V_1, \ldots, \dim V_r, \dim W_1, \ldots, \dim W_s)$$

equals mJ, $m \in \mathbb{N}$, where J is the minimal imaginary positive root of the Dynkin diagram.

Question 3.1.56. *Can we construct the corresponding $\mathcal{M}$-structure for any Dynkin diagram from Theorem 3.1.55? Does this $\mathcal{M}$-structure exist for any m?*

Answer 3.1.57. *It is possible to prove that for any Dynkin diagram of A, D, E-type the corresponding $\mathcal{M}$-structure exists if $m = 1$.*

In order to define this $\mathcal{M}$-structure, we will construct an embedding $\mathcal{A} \to \mathcal{L}$, $\mathcal{B} \to \mathcal{L}$ and a scalar product $(\cdot, \cdot)$ on the vector space $\mathcal{L}$.

Fixing a generic element $\mathbf{1} \in \mathcal{L}$, we can define the embedding $\mathcal{A} \to \mathcal{L}$, $\mathcal{B} \to \mathcal{L}$ by the formula $a \mapsto \mathbf{1}a$, $b \mapsto b\mathbf{1}$ for $a \in \mathcal{A}$, $b \in \mathcal{B}$.

Thus, in order to construct the $\mathcal{M}$-structure corresponding to the Dynkin diagram, we have to

- choose an element in $\mathcal{L} = \oplus_{1 \le i \le r, 1 \le j \le s}(V_i^\star \otimes W_j)^{a_{i,j}}$;

- find its simplest canonical form, choosing bases in the vector spaces $V_1, \ldots, V_r$, $W_1, \ldots, W_s$;

- calculate the embedding $\mathcal{A} \to \mathcal{L}$, $\mathcal{B} \to \mathcal{L}$ and the scalar product $(\cdot, \cdot)$ in the vector space $\mathcal{L}$.

The classification of elements $\mathbf{1} \in \mathcal{L}$ of a general form up to a choice of bases in $V_1, \ldots, V_r, W_1, \ldots, W_s$ is equivalent to the classification of irreducible representations of quivers corresponding to our affine Dynkin diagram. Therefore, we can apply known results about such representations.

The $\mathcal{M}$-structure corresponding to the Dynkin diagram of type $\tilde{A}_{2p-1}$ is described in Example 3.1.48. Explicit formulas for $\mathcal{M}$-structures corresponding to the diagrams $\tilde{D}_{2p-1}$ and $\tilde{D}_{2p}$ can be found in the appendix to the paper [91]. In all these cases, the operator (3.13) has the form

$$S_\lambda = \mathbf{1} + R\,\lambda.$$

3.2 Polynomial forms of Calogero–Moser elliptic systems

In this section, we discuss polynomial forms of both quantum and classical elliptic Calogero–Moser Hamiltonians. In the quantum case, integrable (see Remark 1.3.4) differential operators with polynomial coefficients appear. They belong to a certain class of operators that deserve detailed study. In particular, in the case of ordinary differential operators, the polynomial form of the Darboux–Treibich–Verdier operator [99] belongs to this class.

For a small number of particles, it is shown that the classical limit of these integrable polynomial differential operators has a bi-Hamiltonian nature associated with the elliptic Odesskii–Feigin algebras (see Theorem 3.0.4).

3.2.1 Calogero–Moser Hamiltonians

A wide class of quantum integrable Hamiltonians of the form

$$H = -\Delta + U(x_1, \ldots, x_n), \qquad \text{where} \qquad \Delta = \sum_{i=1}^{n} \frac{\partial^2}{\partial x_i^2}, \tag{3.41}$$

is associated with simple Lie algebras [100]. For these Hamiltonians, the potential U is a rational, trigonometric, or elliptic function.

Observation 3.2.1. (A. Turbiner) *For most of these integrable Hamiltonians, there is a composition of a change of variables and a gauge transformation, reducing the Hamiltonian to a differential operator with polynomial coefficients.*

Such polynomial forms are convenient for further generalizations to the difference and q-difference case. The limiting process from a polynomial form to a classical integrable Hamiltonian is usually almost obvious. In addition, one can try to "lift" a set of commuting differential operators with polynomial coefficients to a commutative subalgebra in the universal enveloping algebra of a proper Lie algebra. The study of maximal commutative subalgebras in universal enveloping algebras and the construction of examples of such subalgebras is one of the main algebraic problems in the theory of quantum integrable systems.

The elliptical quantum Calogero–Moser Hamiltonian is given by the formula

$$H_N = -\Delta + \beta(\beta - 1) \sum_{i \neq j}^{N+1} \wp(x_i - x_j). \tag{3.42}$$

Here, β is a parameter, and $\wp(x)$ is the Weierstrass $\wp$-function with the invariants g_2, g_3. The classic counterpart of (3.42) is

$$h_N = -\sum_{i=1}^{N+1} p_i^2 + \beta(\beta-1)\sum_{i\neq j}^{N+1} \wp(q_i - q_j),\qquad (3.43)$$

where p_i, q_i are the symplectic coordinates associated with the constant Poisson bracket (1.24).

In the new coordinates

$$X = \frac{1}{N+1}\sum_{i=1}^{N+1} x_i, \qquad y_i = x_i - X$$

operator (3.42) takes the form

$$H_N = -\frac{1}{N+1}\frac{\partial^2}{\partial X^2} + \mathcal{H}_N\left(y_1, y_2, \ldots y_N\right),$$

where

$$\mathcal{H}_N = -\frac{N}{N+1}\sum_{i=1}^{N}\frac{\partial^2}{\partial y_i^2} + \frac{1}{N+1}\sum_{i\neq j}^{N}\frac{\partial^2}{\partial y_i \partial y_j} + \beta(\beta-1)\sum_{i\neq j}^{N+1}\wp(y_i - y_j). \quad (3.44)$$

We have to replace y_{N+1} by $-\sum_{i=1}^{N} y_i$ in the last term.

In the paper [101] the transformation $(y_1, \ldots, y_N) \mapsto (u_1, \ldots, u_N)$ defined by

$$\begin{pmatrix} \wp(y_1) & \wp'(y_1) & \cdots & \wp^{(N-2)}(y_1) & \wp^{(N-1)}(y_1) \\ \wp(y_2) & \wp'(y_2) & \cdots & \wp^{(N-2)}(y_2) & \wp^{(N-1)}(y_2) & \vdots & \vdots \\ \wp(y_N) & \wp'(y_N) & \cdots & \wp^{(N-2)}(y_N) & \wp^{(N-1)}(y_N) \end{pmatrix}\begin{pmatrix} u_1 \\ u_2 \\ \vdots \\ u_N \end{pmatrix} = \begin{pmatrix} 1 \\ 1 \\ \vdots \\ 1 \end{pmatrix} \quad (3.45)$$

was considered. Denote by $D_N(y_1, \ldots, y_N)$, its Jacobian.

Conjecture 3.2.2. *The gauge transform $\mathcal{H}_N \to D_N^{-\frac{\beta}{2}}\mathcal{H}_N D_N^{\frac{\beta}{2}}$ and the succeeding change of variables (3.45) bring (3.44) to a differential operator P_N with polynomial coefficients.*

If $N = 2$, then (3.45) coincides with the transformation

$$u_1 = \frac{\wp'(y_2) - \wp'(y_1)}{\wp(y_1)\wp'(y_2) - \wp(y_2)\wp'(y_1)}, \qquad u_2 = \frac{\wp(y_1) - \wp(y_2)}{\wp(y_1)\wp'(y_2) - \wp(y_2)\wp'(y_1)}, \quad (3.46)$$

found in [102]. In addition to the explicit form of P_2, in this paper a polynomial form for the elliptic G_2-model was discovered. Polynomial forms for rational and trigonometric Calogero–Moser Hamiltonians have been described in [103].

Remark 3.2.3. *Obviously, for any polynomial form P of Hamiltonian of the form*
(3.41):

1. *The contravariant metric defined by the symbol of P is flat;*

2. *The operator P can be reduced to a self-adjoint operator by a gauge transformation of the form $P \mapsto fPf^{-1}$, where f is some function.*

In addition to the obvious properties 1,2, we will take into account the following

Observation 3.2.4. (A. Turbiner) *For most of the known integrable cases, the
polynomial form P preserves some nontrivial finite-dimensional vector space of
polynomials.*

For Hamiltonians (3.44), the polynomial forms seem to have the following structure. Consider differential operators $e_{i,j} = E_{i-1,j-1}$, where

$$E_{ij} = y_i \frac{\partial}{\partial y_j}, \qquad E_{0i} = \frac{\partial}{\partial y_i},$$

$$E_{00} = -\sum_{j=1}^{N} y_j \frac{\partial}{\partial y_j} + \beta\,(N+1), \qquad E_{i0} = y_i E_{00}. \tag{3.47}$$

It is easy to verify that they satisfy the commutation relations

$$e_{ij}e_{kl} - e_{kl}e_{ij} = \delta_{j,k}e_{il} - \delta_{i,l}e_{kj}, \qquad i,j = 1,\ldots,N+1, \tag{3.48}$$

and, therefore, define a representation of the Lie algebra $\mathfrak{gl}_{N+1}$ and its universal
enveloping algebra $U(\mathfrak{gl}_{N+1})$. The latter representation is not exact.

Conjecture 3.2.5. *The differential operator P_N from Conjecture 3.2.2 can be
written as a linear combination of anti-commutators of the operators E_{ij}.*

Conjectures 3.2.2 and 3.2.5 were verified in [101], for $N = 2, 3$. In addition,
differential operators with polynomial coefficients were found, commuting with P_2
and P_3. These operators can also be written as noncommutative polynomials in
E_{ij}.

Remark 3.2.6. *If $k = -\beta(N+1)$ is a positive integer, then operators (3.47)
preserve the vector space of polynomials in the variables $y_1, \ldots, y_N$ of degree not
higher than k [103].*

3.2.2   Quasi-exactly solvable differential operators

Definition 3.2.7. A linear differential operator

$$Q = \sum_{i_1+\cdots+i_N\leq m} a_{i_1,\ldots,i_N}\partial_{y_1}^{i_1}\cdots\partial_{y_N}^{i_N} \tag{3.49}$$

of order m with polynomial coefficients is called *quasi-exactly solvable*, if it preserves the vector space of all polynomials in $y_1,\ldots,y_N$ of degree not higher than k for some $k \geq m$.

Theorem 3.2.8. [104] *For any quasi-exactly solvable differential operator* (3.49), *the estimate*

$$\deg(a_{i_1,\ldots,i_N}) \leq m + i_1 + \cdots + i_N$$

is valid.

Open problem 3.2.9. *Prove that any quasi-exactly solvable operator can be represented as a (noncommutative) polynomial in variables* (3.47), *where* $k = -\beta(N+1)$.

Remark 3.2.10. *Such a representation is not unique.*

Ordinary quasi-exactly solvable operators

Consider the case $N = 1$.

Lemma 3.2.11. *Any quasi-exactly solvable operator P of second order has the following structure:*

$$P = (a_4x^4+a_3x^3+a_2x^2+a_1x+a_0)\frac{d^2}{dx^2} + (b_3x^3+b_2x^2+b_1x+b_0)\frac{d}{dx}+c_2x^2+c_1x+c_0,$$

where the coefficients are related by

$$b_3 = 2(1-k)\,a_4, \qquad c_2 = k(k-1)\,a_4, \qquad c_1 = k(a_3 - ka_3 - b_2).$$

The transformation group

$$x \mapsto \frac{s_1x+s_2}{s_3x+s_4}, \qquad P \mapsto (s_3x+s_4)^{-k}P(s_3x+s_4)^k \tag{3.50}$$

acts on the 9-dimensional vector space of such operators. The coefficient $a(x)$ of the second derivative is a fourth degree polynomial which is transformed by the rule

$$a(x) \mapsto (s_3x+s_4)^4 a\!\left(\frac{s_1x+s_2}{s_3x+s_4}\right).$$

If $a(x)$ has four distinct roots, we call the operator P *elliptic*. In the elliptic case, using a transformation (3.50), we can convert a to

$$a(x) = 4\,x(x-1)(x-\kappa),$$

where κ is an essential "elliptic" parameter.

We define the parameters $n_1, \ldots, n_5$ by means of the following equalities:

$$b_0 = 2(1 + 2n_1), \quad b_1 = -4\Big((\kappa+1)(n_1+1) + \kappa n_2 + n_3\Big),$$

$$b_2 = -2\,(3 + 2n_1 + 2n_2 + 2n_3),$$

$$k = -\frac{1}{2}(n_1 + n_2 + n_3 + n_4),$$

$$n_5 = c_0 + n_2(1 - n_2) + \kappa n_3(1 - n_3) + (n_1 + n_3)^2 + \kappa(n_1 + n_2)^2.$$

Then, the operator $H = hPh^{-1}$, where

$$h = x^{\frac{n_1}{2}} (x-1)^{\frac{n_2}{2}} (x-\kappa)^{\frac{n_3}{2}},$$

has the form

$$H = a(x)\frac{d^2}{dx^2} + \frac{a'(x)}{2}\frac{d}{dx} + n_5 + n_4(1 - n_4)\,x + \frac{n_1(1 - n_1)\kappa}{x} +$$

$$\frac{n_2(1 - n_2)(1 - \kappa)}{x - 1} + \frac{n_3(1 - n_3)\kappa(\kappa - 1)}{x - \kappa}.$$

Now, after the transformation $y = f(x)$, where

$$f'^2 = 4f(f - 1)(f - \kappa),$$

we arrive at the operator

$$H = \frac{d^2}{dy^2} + n_5 + n_4(1 - n_4)\,f + \frac{n_1(1 - n_1)\kappa}{f} + \frac{n_2(1 - n_2)(1 - \kappa)}{f - 1} + \frac{n_3(1 - n_3)\kappa(\kappa - 1)}{f - \kappa}.$$

Here, n_i are arbitrary parameters.

Another form of this Hamiltonian (up to a constant term) is given by the formula

$$H = \frac{d^2}{dy^2} + n_4(1 - n_4)\,\wp(y) + n_1(1 - n_1)\,\wp(y + \omega_1) +$$

$$n_2(1 - n_2)\,\wp(y + \omega_2) + n_3(1 - n_3)\,\wp(y + \omega_1 + \omega_2),$$

where ω_i is the half-periods of the Weierstrass function $\wp(x)$. For $n_1 = n_2 = n_3 = 0$, we get the Lamé operator. In general, this is the Darboux–Trebisch–Verdier operator [99].

When

$$k = -\frac{1}{2}(n_1 + n_2 + n_3 + n_4)$$

is a natural number, this operator H preserves a finite-dimensional space of elliptic functions that corresponds to the space of polynomials which is preserved by the original operator P.

Two-dimensional operators

Consider second order differential operators of the form

$$P = a(x,y)\frac{\partial^2}{\partial x^2} + 2b(x,y)\frac{\partial^2}{\partial x \partial y} + c(x,y)\frac{\partial^2}{\partial y^2} + d(x,y)\frac{\partial}{\partial x} + e(x,y)\frac{\partial}{\partial y} + f(x,y) \quad (3.51)$$

with polynomial coefficients. Denote by $D(x,y)$ the determinant $a(x,y)c(x,y) - b(x,y)^2$. We assume that $D \neq 0$.

Lemma 3.2.12. *The operator* (3.51) *is quasi-exactly solvable iff its coefficients have the following structure*

$$a = q_1 x^4 + q_2 x^3 y + q_3 x^2 y^2 + z_1 x^3 + z_2 x^2 y + z_3 xy^2 + a_1 x^2 + a_2 xy + a_3 y^2 + a_4 x + a_5 y + a_6;$$

$$b = q_1 x^3 y + q_2 x^2 y^2 + q_3 xy^3 + \frac{1}{2}\left(z_4 x^3 + (z_1 + z_5)x^2 y + (z_2 + z_6)xy^2 + z_3 y^3\right)$$
$$+ b_1 x^2 + b_2 xy + b_3 y^2 + b_4 x + b_5 y + b_6;$$

$$c = q_1 x^2 y^2 + q_2 xy^3 + q_3 y^4 + z_4 x^2 y + z_5 xy^2 + z_6 y^3 + c_1 x^2 + c_2 xy + c_3 y^2 + c_4 x + c_5 y + c_6;$$

$$d = (1-k)\left(2(q_1 x^3 + q_2 x^2 y + q_3 xy^2) + z_7 x^2 + (z_2 + z_8 - z_6)xy + z_3 y^2\right) + d_1 x + d_2 y + d_3;$$

$$e = (1-k)\left(2(q_1 x^2 y + q_2 xy^2 + q_3 y^3) + z_4 x^2 + (z_5 + z_7 - z_1)xy + z_8 y^2\right) + e_1 x + e_2 y + e_3;$$

$$f = k(k-1)\left(q_1 x^2 + q_2 xy + q_3 y^2 + (z_7 - z_1)x + (z_8 - z_6)y\right) + f_1.$$

The group GL_3 acts on the space of these operators in a projective manner according to the formula

$$\tilde{x} = \frac{a_1 x + a_2 y + a_3}{c_1 x + c_2 y + c_3}, \qquad \tilde{y} = \frac{b_1 x + b_2 y + b_3}{c_1 x + c_2 y + c_3}, \tag{3.52}$$
$$\tilde{P} = (c_1 x + c_2 y + c_3)^{-k} P \circ (c_1 x + c_2 y + c_3)^k.$$

This transformation corresponds to the matrix

$$\begin{pmatrix} a_1 & a_2 & a_3 \\ b_1 & b_2 & b_3 \\ c_1 & c_2 & c_3 \end{pmatrix} \in GL_3.$$

The space of the corresponding GL_3-representation is the 36-dimensional space of the coefficients of operators (3.51). This representation is the sum of the irreducible representations on vector spaces W_1, W_2, and W_3 of dimensions 27, 8 and 1, respectively. The polynomials

$$x_1 = 5z_7 - z_5 - 7z_1, \qquad x_2 = 5z_8 - z_2 - 7z_6, \qquad x_3 = 5d_1 + 2(k-1)(2a_1 + b_2),$$

$$x_4 = 5e_1 + 2(k-1)(2b_1 + c_2), \qquad x_5 = 5d_2 + 2(k-1)(2b_3 + a_2),$$

$$x_6 = 5e_2 + 2(k-1)(2c_3 + b_2), \qquad x_7 = 5d_3 + 2(k-1)(a_4 + b_5),$$

$$x_8 = 5e_3 + 2(k-1)(b_4 + c_5)$$

form a basis in W_2. The maximal orbit of the group action on W_2 is of dimension 6. There are two polynomial invariants of this action:

$$I_1 = x_3^2 - x_3 x_6 + x_6^2 + 3x_4 x_5 + 3(k-1)(x_1 x_7 + x_2 x_8),$$

and

$$I_2 = 2x_3^3 - 3x_3^2 x_6 - 3x_3 x_6^2 + 2x_6^3 + 9x_4 x_5(x_3 + x_6) +$$
$$9(k-1)(x_1 x_3 x_7 + x_2 x_6 x_8 - 2x_1 x_6 x_7 - 2x_2 x_3 x_8 + 3x_2 x_4 x_7 + 3x_1 x_5 x_8).$$

Flat polynomial metrics

According to Remark 3.2.3, the contravariant metric

$$g^{1,1} = a \qquad g^{1,2} = g^{2,1} = b \qquad g^{2,2} = c,$$

determined by the coefficients a, b, c of the operator (3.51), has to be flat (i.e. $R_{1,2,1,2} = 0$) for any polynomial form P of an operator (3.41).

Open problem 3.2.13. *Describe, up to transformations (3.52), all flat contravariant metrics defined by the polynomials a, b, c from Lemma 3.2.12.*

Some partial results on this subject can be found in [104].

Example 3.2.14. For any constant κ, the metric g with the coefficients

$$a = (x^2 - 1)(x^2 - \kappa) + (x^2 + \kappa) y^2, \qquad b = xy (x^2 + y^2 + 1 - 2\kappa),$$

$$c = (\kappa - 1)(x^2 - 1) + (x^2 + 2 - \kappa) y^2 + y^4$$

is flat. This metric corresponds to the polynomial form [105] for an integrable elliptic Inozemtsev's BC_2-Hamiltonian

$$H = \Delta + 2m(m-1)(\wp(x+y) + \wp(xy)) + \sum_{i=0}^{3} n_i(n_i - 1)(\wp(x + \omega_i) + \wp(y + \omega_i)),$$

where $\omega_0 = 0, \omega_3 = \omega_1 + \omega_2$ and ω_1, ω_2 are the half-periods of the Weierstrass function $\wp(x)$. In fact, we have a pencil (with respect to the parameter κ) of compatible polynomial flat contravariant metrics [106].

3.2.3   Commutative subalgebras in $U(\mathfrak{gl}_{N+1})$ and Calogero–Moser quantum Hamiltonians

Commutative subalgebras in $U(\mathfrak{gl}_n)$ is one of algebraic objects related to quantum integrability. A class of commutative subalgebras was constructed in [107]. These subalgebras are quantizations of commutative Poisson subalgebras generated by compatible constant and linear $\mathfrak{gl}_n$-Poisson brackets. The quantization procedure is very simple: each product $\prod_1^k x_i$ in the commuting generators should be replaced by $\dfrac{1}{k!} \sum\limits_{\sigma \in S_k} \prod y_{\sigma(i)}$, where y_i are noncommutative generators.

The universal enveloping algebra $U(\mathfrak{gl}_{N+1})$ is an associative algebra generated by the elements e_{ij} and the relations (3.48). Consider the case $N = 2$. It turns out that the element of the universal enveloping algebra $U(\mathfrak{gl}_3)$

$$H = H_0 + H_1 g_2 + H_2 g_2^2 + H_3 g_3, \qquad (3.53)$$

where

$$H_0 = 12e_{12}e_{11} - 12e_{32}e_{13} - 12e_{33}e_{12} - e_{23}^2,$$

$$H_1 = -e_{21} + 2e_{21}e_{11} - e_{22}e_{21} - e_{31}e_{23} - 12e_{32}^2 - e_{33}e_{21},$$

$$H_2 = -e_{31}^2, \qquad H_3 = 36e_{32}e_{31} + 3e_{21}^2,$$

commutes with the following two third order elements

$$K = K_0 + K_1 g_2 + K_2 g_3,$$

$$M = M_0 + M_1 g_2 + M_2 g_3 + M_3 g_2^2 + M_4 g_2 g_3 + M_5 g_3^2 + M_6 g_2^3.$$

Here, g_2 and g_3 are arbitrary parameters[5] and

$$K_0 = -e_{23} + 2e_{21}e_{13} - e_{23}e_{22} - 36e_{32}e_{12} + e_{33}e_{23} - e_{21}e_{13}e_{11} - e_{22}e_{21}e_{13} + e_{23}e_{11}^2 +$$
$$2e_{23}e_{21}e_{12} - e_{23}e_{22}e_{11} + 12e_{31}e_{12}^2 - e_{31}e_{23}e_{13} - 12e_{32}e_{12}e_{11} - e_{32}e_{23}^2 -$$
$$12e_{32}^2 e_{13} + 2e_{33}e_{21}e_{13} - e_{33}e_{23}e_{11} + e_{33}e_{23}e_{22} + 12e_{33}e_{32}e_{12},$$

$$K_1 = 3e_{31}e_{11} - 3e_{31}e_{22} - 2e_{32}e_{21} + e_{31}e_{21}e_{12} + e_{31}e_{22}e_{11} - e_{31}e_{22}^2 + e_{31}^2 e_{13} -$$
$$2e_{32}e_{21}e_{11} + e_{32}e_{22}e_{21} - 2e_{32}e_{31}e_{23} - e_{33}e_{31}e_{11} + e_{33}e_{31}e_{22} + e_{33}e_{32}e_{21},$$

$$K_2 = 3\left(2e_{31}e_{21} + e_{31}e_{22}e_{21} + e_{31}^2 e_{23} - e_{32}e_{21}^2 - e_{33}e_{31}e_{21}\right);$$

$$M_0 = 2\Big(12e_{13}e_{11} - 6e_{22}e_{13} - 6e_{33}e_{13} - 12e_{13}e_{11}^2 - 6e_{22}e_{13}e_{11} + 6e_{22}^2 e_{13} + 18e_{23}e_{12}e_{11} -$$
$$18e_{23}e_{22}e_{12} + e_{23}^3 - 216e_{32}e_{12}^2 + 18e_{32}e_{23}e_{13} + 30e_{33}e_{13}e_{11} - 6e_{33}e_{22}e_{13} - 12e_{33}^2 e_{13}\Big),$$

---

[5]By their origin, they are the invariants of the elliptic curve.

$$M_1 = -3\Big(2e_{23}e_{21} - 36e_{31}e_{12} + 20e_{32}e_{11} - 28e_{32}e_{22} + 8e_{33}e_{32} - 4e_{23}e_{21}e_{11} + 2e_{23}e_{22}e_{21} - $$
$$12e_{31}e_{12}e_{11} - e_{31}e_{23}^2 + 8e_{32}e_{11}^2 + 36e_{32}e_{21}e_{12} + 4e_{32}e_{22}e_{11} - 4e_{32}e_{22}^2 - 24e_{32}e_{31}e_{13} - $$
$$12e_{32}^2e_{23} + 2e_{33}e_{23}e_{21} + 12e_{33}e_{31}e_{12} - 20e_{33}e_{32}e_{11} + 4e_{33}e_{32}e_{22} + 8e_{33}^2e_{32}\Big),$$

$$M_2 = -18\Big(4e_{31}e_{11} - 2e_{31}e_{22} - 2e_{33}e_{31} - e_{23}e_{21}^2 - 2e_{31}e_{11}^2 - 6e_{31}e_{21}e_{12} + 2e_{31}e_{22}e_{11} - 2e_{31}e_{22}^2 + $$
$$6e_{31}^2e_{13} + 6e_{32}e_{22}e_{21} + 24e_{32}^3 + 2e_{33}e_{31}e_{11} + 2e_{33}e_{31}e_{22} - 6e_{33}e_{32}e_{21} - 2e_{33}^2e_{31}\Big),$$

$$M_3 = -3\Big(2e_{31}e_{21} - 2e_{31}e_{21}e_{11} + e_{31}e_{22}e_{21} + e_{31}^2e_{23} - 24e_{32}^2e_{31} + e_{33}e_{31}e_{21}\Big),$$

$$M_4 = 9\Big(e_{31}e_{21}^2 - 12e_{32}e_{31}^2\Big), \qquad M_5 = 108e_{31}^3, \qquad M_6 = -2e_{31}^3.$$

It is easy to check that $[K, M] = 0$. Thus, we have a maximal commutative subalgebra in $U(\mathfrak{gl}_3)$ generated by the elements H, K, M and by three central elements of $U(\mathfrak{gl}_3)$ of orders 1, 2, and 3.

This subalgebra generates "integrable" (see Remark 1.3.4) operators through various representations of $U(\mathfrak{gl}_3)$ by differential, difference, and q-difference operators.

In particular, the substitution of differential operators (3.47) with two independent variables for e_{ij} maps the element H to a polynomial form P_2 for the elliptic Calogero–Moser Hamiltonian (3.44) with $N = 2$, the element M to a third order differential operator that commutes with P_2, and the element K to zero. The parameters g_2 and g_3 coincide with the invariants of the Weierstrass function $\wp(x)$ from (3.44).

Remark 3.2.15. *The representation of $U(\mathfrak{gl}_3)$ by the matrix unities in* Mat_3 *maps H, K and M to zero.*

The representation

$$e_{ij} \to z_i\frac{\partial}{\partial z_j}$$

maps H to a homogeneous differential operator $\mathcal{H} = \sum_{i \geq j} a_{ij}\frac{\partial^2}{\partial z_i \partial z_j}$ with three independent variables, where the coefficients are given by

$$a_{11} = -2g_2z_1z_2 - 3g_3z_3^2 + g_2^2z_3^2, \qquad a_{22} = 12g_2z_3^2, \qquad a_{33} = z_3^2,$$

$$a_{21} = -12z_1^2 + g_2z_2^2 - 36g_3z_3^2, \qquad a_{31} = 2\,g_2z_2z_3, \qquad a_{32} = 24\,z_1z_3.$$

The element M becomes a differential operator of the form $\mathcal{M} = \sum_{i \geq j \geq k} b_{ijk}\frac{\partial^3}{\partial z_i \partial z_j \partial z_k}$ which commutes with $\mathcal{H}$. Interestingly, the terms of lower order are absent in both $\mathcal{H}$ and $\mathcal{M}$.

A similar commutative subalgebra in $U(\mathfrak{gl}_4)$ [101] leads to a polynomial form of the elliptic Calogero–Moser Hamiltonian (3.44) with $N = 3$.

3.2.4 Bi-Hamiltonian origin of the classical elliptic Calogero–Moser model

Consider the following standard limit procedure. Every element $f \in U(\mathfrak{gl}_n)$ is a polynomial in noncommutative variables e_{ij} that satisfy the commutator relations (3.48). Choosing all terms of the highest degree in f and replacing e_{ij} in them with commutative variables x_{ij}, we get a polynomial which is called *symbol* of f and is denoted by $\mathrm{symb}(f)$.

It is well known that for any elements $f, g \in U(\mathfrak{gl}_n)$

$$\mathrm{symb}([f, g]) = \{\mathrm{symb}(f), \mathrm{symb}(g)\},$$

where $\{\,,\}$ is a linear Poisson bracket defined by

$$\{x_{ij}, x_{kl}\} = \delta_{j,k}\, x_{il} - \delta_{i,l}\, x_{kj}, \qquad i, j = 1, \ldots, n. \tag{3.54}$$

This bracket corresponds to the Lie algebra $\mathfrak{gl}_n$. In particular, if $[f, g] = 0$, then

$$\{\mathrm{symb}(f), \ \mathrm{symb}(g)\} = 0.$$

Consider polynomials in the commutative variables x_{ij}. We will consider x_{ij} as entries of a matrix X. Applying the limit procedure to the generators of the commutative subalgebra in $U(\mathfrak{gl}_3)$ described in Section 3.2.3, we get the polynomials

$$c_1 = \mathrm{tr}\, X, \qquad c_2 = \mathrm{tr}\, X^2, \qquad c_3 = \mathrm{tr}\, X^3,$$

$$h = h_0 + h_1 g_2 + h_2 g_2^2 + h_3 g_3, \qquad k = k_0 + k_1 g_2 + k_2 g_3, \tag{3.55}$$

$$m = m_0 + m_1 g_2 + m_2 g_3 + m_3 g_2^2 + m_4 g_2 g_3 + m_5 g_3^2 + m_6 g_2^3,$$

where

$$h_0 = 12 x_{12} x_{11} - 12 x_{32} x_{13} - 12 x_{33} x_{12} - x_{23}^2,$$

$$h_1 = 2 x_{21} x_{11} - x_{22} x_{21} - x_{31} x_{23} - 12 x_{32}^2 - x_{33} x_{21},$$

$$h_2 = -x_{31}^2, \qquad h_3 = 36 x_{32} x_{31} + 3 x_{21}^2,$$

$$k_0 = -x_{21}x_{13}x_{11} - x_{22}x_{21}x_{13} + x_{23}x_{11}^2 + 2x_{23}x_{21}x_{12} -$$
$$x_{23}x_{22}x_{11} + 12x_{31}x_{12}^2 - x_{31}x_{23}x_{13} - 12x_{32}x_{12}x_{11} - x_{32}x_{23}^2 -$$
$$12x_{32}^2x_{13} + 2x_{33}x_{21}x_{13} - x_{33}x_{23}x_{11} + x_{33}x_{23}x_{22} + 12x_{33}x_{32}x_{12},$$

$$k_1 = x_{31}x_{21}x_{12} + x_{31}x_{22}x_{11} - x_{31}x_{22}^2 + x_{31}^2x_{13} - 2x_{32}x_{21}x_{11} +$$
$$x_{32}x_{22}x_{21} - 2x_{32}x_{31}x_{23} - x_{33}x_{31}x_{11} + x_{33}x_{31}x_{22} + x_{33}x_{32}x_{21},$$

$$k_2 = 3\left(x_{31}x_{22}x_{21} + x_{31}^2x_{23} - x_{32}x_{21}^2 - x_{33}x_{31}x_{21}\right);$$

$$m_0 = 2\Big(-12x_{13}x_{11}^2 - 6x_{22}x_{13}x_{11} + 6x_{22}^2x_{13} + 18x_{23}x_{12}x_{11} - 18x_{23}x_{22}x_{12} +$$
$$x_{23}^3 - 216x_{32}x_{12}^2 + 18x_{32}x_{23}x_{13} + 30x_{33}x_{13}x_{11} - 6x_{33}x_{22}x_{13} - 12x_{33}^2x_{13}\Big),$$

$$m_1 = -3\Big(-4x_{23}x_{21}x_{11} + 2x_{23}x_{22}x_{21} - 12x_{31}x_{12}x_{11} - x_{31}x_{23}^2 + 8x_{32}x_{11}^2 +$$
$$36x_{32}x_{21}x_{12} + 4x_{32}x_{22}x_{11} - 4x_{32}x_{22}^2 - 24x_{32}x_{31}x_{13} - 12x_{32}^2x_{23} +$$
$$2x_{33}x_{23}x_{21} + 12x_{33}x_{31}x_{12} - 20x_{33}x_{32}x_{11} + 4x_{33}x_{32}x_{22} + 8x_{33}^2x_{32}\Big),$$

$$m_2 = -18\Big(-x_{23}x_{21}^2 - 2x_{31}x_{11}^2 - 6x_{31}x_{21}x_{12} + 2x_{31}x_{22}x_{11} - 2x_{31}x_{22}^2 + 6x_{31}^2x_{13} +$$
$$6x_{32}x_{22}x_{21} + 24x_{32}^3 + 2x_{33}x_{31}x_{11} + 2x_{33}x_{31}x_{22} - 6x_{33}x_{32}x_{21} - 2x_{33}^2x_{31}\Big),$$

$$m_3 = -3\Big(-2x_{31}x_{21}x_{11} + x_{31}x_{22}x_{21} + x_{31}^2x_{23} - 24x_{32}^2x_{31} + x_{33}x_{31}x_{21}\Big),$$

$$m_4 = 9\Big(x_{31}x_{21}^2 - 12x_{32}x_{31}^2\Big), \qquad m_5 = 108x_{31}^3, \qquad m_6 = -2x_{31}^3.$$

These six polynomials commute with each other with respect to the linear $\mathfrak{gl}_3$-Poisson bracket (3.54).

It can be verified that the corresponding elements of the universal enveloping algebra can be reconstructed from polynomials (3.55) by the quantization procedure described at the beginning of Section 3.2.3.

Quadratic Poisson bracket

Consider the following quadratic bracket

$$\{f,g\}_2 = \{f,g\}_a + \kappa\{f,g\}_b + \kappa^2\{f,g\}_c, \tag{3.56}$$

where κ is an arbitrary parameter,

$$\{f,g\}_a = -3\left(x_{11} + x_{22} + x_{33}\right)\{f,g\}_1, \qquad \{f,g\}_c = Z_1(f)Z_2(g) - Z_1(g)Z_2(f),$$

$$\{f,g\}_b = Z_3(\{f,g\}_1) - \{Z_3(f),g\}_1 - \{f,Z_3(g)\}_1.$$

Here, $\{\cdot,\cdot\}_1$ is the linear Poisson bracket (3.54), and the vector fields Z_i appearing in the bracket are defined as

$$Z_1(f) = \sum_{i=1}^{3} \frac{\partial f}{\partial x_{ii}}, \qquad Z_2(f) = \{h, f\}_1,$$

where the Hamiltonian h is given by the formula (3.55), and

$$Z_3(f) = \sum_{i,j=1}^{3} G_{i,j} \frac{\partial f}{\partial x_{ij}}.$$

The coefficients of the vector field Z_3 are

$G_{1,1} = (-2x_{11}x_{23} + x_{22}x_{23} + 36x_{12}x_{32} + x_{23}x_{33}) + x_{31}(x_{11} - 2x_{22} + x_{33})\,g_2 + 9\,x_{21}x_{31}\,g_3,$

$G_{2,2} = -G_{1,1}, \qquad G_{3,3} = 0,$

$G_{1,2} = (x_{11}x_{13} + x_{13}x_{22} - 3\,x_{12}x_{23} - 2x_{13}x_{33}) + (3x_{12}x_{31} + 5x_{11}x_{32} - 4x_{22}x_{32} - x_{32}x_{33})\,g_2 -$
$\qquad 3(2x_{11}x_{31} - x_{22}x_{31} - 3x_{21}x_{32} - x_{31}x_{33})\,g_3,$

$G_{1,3} = 3x_{13}x_{23} - (x_{11} - x_{22})(x_{11} + x_{22} - 2x_{33})\,g_2 - 3x_{21}(x_{11} + x_{22} - 2x_{33})\,g_3,$

$G_{2,1} = -3(x_{21}x_{23} + 12x_{12}x_{31} + 4x_{11}x_{32} - 8x_{22}x_{32} + 4x_{32}x_{33}) - 6x_{21}x_{31}\,g_2,$

$G_{2,3} = 3(4x_{11}x_{12} + 4x_{12}x_{22} + x_{23}^2 - 8x_{12}x_{33}) + x_{21}(x_{11} + x_{22} - 2x_{33})\,g_2,$

$G_{3,1} = 2(x_{11}x_{21} + x_{21}x_{22} + 18x_{32}^2 - 2x_{21}x_{33}) - 3x_{31}^2\,g_2,$

$G_{3,2} = -(x_{11} - x_{22})(x_{11} + x_{22} - 2x_{33}) + 6x_{31}x_{32}\,g_2 - 9x_{31}^2\,g_3.$

Theorem 3.2.16. (i) *Formula (3.56) defines a Poisson bracket;*

(ii) *This quadratic bracket is compatible with the linear $\mathfrak{gl}_3$-Poisson bracket (3.54);*

(iii) *The Casimir functions of the pencil of these brackets generate (see Theorem 3.0.2) the commutative Poisson subalgebra described in Section 3.2.4.*

Conjecture 3.2.17. *The Poisson bracket (3.56) is an elliptic Poisson bracket of type $q_{9,2}$ (see [89]), written in an unusual basis.*

Remark 3.2.18. *In the case $N = 3$, there exists a similar quadratic Poisson bracket compatible with the linear bracket $\mathfrak{gl}_4$. By means of Theorem 3.0.2 the pencil of these brackets generates a commutative Poisson subalgebra containing a quadratic element h.*

In order to get the classical elliptic Calogero–Moser Hamiltonian from the quadratic element h in cases $N = 2, 3$, the following reduction has to be done:

$$x_{i+1,j+1} = q_i\,p_j, \qquad x_{1,i+1} = p_i, \qquad x_{i+1,1} = q_i\,x_{1,1},$$

$$x_{1,1} = -\sum_{j=1}^{N} q_j p_j + \beta(N+1), \tag{3.57}$$

where p_i and q_i are canonical variables in the constant Poisson bracket (1.24). It is easy to verify that, for arbitrary N, the functions p_i, q_i are Darboux coordinates on the minimal symplectic leaf of the linear bracket $\mathfrak{gl}_{N+1}$ which is the orbit of the diagonal matrix $\mathrm{diag}((N+1)\,\beta, 0, 0, \ldots, 0)$.

In the case of $N = 2$, after substituting (3.57) into (3.55), we obtain polynomials in the canonical variables p_i, q_i commuting with respect to the bracket (1.24). The element h becomes the polynomial form of the classical Calogero–Moser Hamiltonian, the element m generates a first integral, cubic in momenta p_i, the element k becomes zero, and the Casimir functions c_i turn to constants.

In order to reduce the obtained Hamiltonian and its cubic integral to the standard Calogero–Moser form (3.43), one has to apply the canonical transformation, the change of the coordinates q_i in which is defined by the formula (3.45).

The situation with $N = 3$ is similar.

Remark 3.2.19. *The parameter β in the minimal symplectic leaf of the linear bracket coincides with the parameter in formula (3.43).*

Remark 3.2.20. *The quadratic bracket cannot be restricted to the minimal symplectic leaf and, therefore, upon reduction (3.57) the bi-Hamiltonian property of the model is destroyed.*

Conjecture 3.2.21. *For any N, the classical elliptic Calogero–Moser Hamiltonian (3.43) can be obtained from the elliptic quadratic Poisson bracket of type $q_{(N+1)^2,N}$ by the procedure described above. Namely, the quadratic bracket is compatible with the linear bracket isomorphic to $\mathfrak{gl}_{N+1}$ (see Conjecture 3.0.5). Theorem 3.0.2 generates an integrable quadratic Hamiltonian. After its restriction to the minimal symplectic leaf of the linear bracket, a polynomial Hamiltonian arises which can be reduced to the Hamiltonian (3.43) by a canonical transformation.*

The main obstacle in the proof of this conjecture is that, for known explicit expressions for the $q_{(N+1)^2,N}$-bracket, the linear bracket compatible with it has coefficients expressed in terms of θ-constants. Because of algebraic relations between the θ-constants, it is almost impossible to deal with such a form of the linear bracket.

Open problem 3.2.22. *For the elliptic bracket $\{,\}$ of type $q_{(N+1)^2,N}$ find a basis in which the linear bracket $\{,\}_1$ compatible with $\{,\}$ has the canonical form (3.54).*

Part 3. Symmetry approach
to integrability

Chapter 4

Basic concepts of symmetry approach

The symmetry approach to the classification of integrable partial differential equations with two independent variables is based on the existence of local higher symmetries and/or conservation laws for the equation (see Sections 1.4 and 1.5 of the Introduction).

4.1 Description of some classification results

4.1.1 Hyperbolic equations

The first classification result obtained in the framework of the symmetry approach in 1979 is formulated in the Introduction (see Theorem 2).

Open problem 4.1.1. *A complete classification of integrable hyperbolic equations of the form*

$$u_{xy} = \Psi(u, u_x, u_y) \tag{4.1}$$

is still an open problem. Moreover, even the possible form of the dependence of the function Ψ on the variables u_x and u_y is not determined. Some partial results were obtained in [108].

Example 4.1.2. The following equation [109, 110]

$$u_{xy} = S(u)\sqrt{u_x^2 + 1}\sqrt{u_y^2 + 1}, \qquad \text{where} \qquad S'' - 2S^3 + c\,S = 0,$$

possesses infinite series of higher symmetries and conservation laws. The simplest symmetry is of third order.

In paper [112] integrable hyperbolic systems of the form

$$u_x = p(u, v), \qquad v_y = q(u, v)$$

were investigated.

The following problem was solved in [111]. In the case of hyperbolic equations (4.1), the symmetry approach requires the existence of both x-symmetries of the form

$$u_t = A(u, u_x, u_{xx}, \ldots,),$$

and y-symmetries of the form

$$u_\tau = B(u, u_y, u_{yy}, \ldots,).$$

For example, the famous integrable sine-Gordon equation

$$u_{xy} = \sin u$$

has symmetries

$$u_t = u_{xxx} + \frac{1}{2}u_x^3, \qquad u_\tau = u_{yyy} + \frac{1}{2}u_y^3.$$

It was assumed in [111] that both the x and y-symmetries of equation (4.1) are third order *integrable* evolution equations. All such hyperbolic equations (4.1) were found. Lists are given in Appendix 9.1.

Section 5 is devoted to a special class of integrable hyperbolic equations (4.1), the so-called equations of *Liouville type* or *Darboux integrable* equations [110].

4.1.2   Evolution equations

For evolution equations of the form (1.26), necessary conditions for the existence of higher symmetries, independent of the orders of symmetries, were obtained in [113, 3] (see Section 4.2.5). These conditions lead to an overdetermined system of partial differential equations with respect to the right-hand side of equation (1.26). The solutions of this system are not always rational (and even not always algebraic [114]).

Example 4.1.3. It turns out that the dependence of the right-hand side of any integrable third order equation

$$u_t = F(u, u_x, u_{xx}, u_{xxx}) \tag{4.2}$$

on u_{xxx} is determined by the ordinary differential equation[1]

$$9\,(F')^2 F'''' - 45\,F'F''F''' + 40\,(F'')^3 = 0,$$

where $'$ means the derivative with respect to u_{xxx}. Solving this equation, we find that there are three different types of integrable equations (4.2) [5]:

---

[1] This fact follows from the formula (4.39).

(1) $$u_t = a\,u_{xxx} + b;$$

(2) $$u_t = \frac{a}{(u_{xxx} + b)^2} + c;$$

and

(3) $$u_t = \frac{2a\,u_{xxx} + b}{\sqrt{a\,u_{xxx}^2 + b\,u_{xxx} + c}} + d,$$

where a, b, c and d are some functions of the variables u, u_x, u_{xx}.

Remark 4.1.4. *For integrable quasilinear equations of Case* (1) *of the previous example, one can show that the possible dependence of the right-hand side on the variable u_{xx} (up to contact and point transformations) is as follows:*

(1) $$u_t = u_{xxx} + \alpha_2 u_{xx}^2 + \alpha_1 u_{xx} + \alpha_0;$$

(2) $$u_t = \frac{u_{xxx}}{u^3} + \alpha_2 u_{xx}^2 + \alpha_1 u_{xx} + \alpha_0;$$

(3) $$u_t = \frac{u_{xxx}}{\alpha_3^3} + \alpha_2 u_{xx}^2 + \alpha_1 u_{xx} + \alpha_0, \qquad \frac{\partial^2 \alpha_3}{\partial u_x^2} \neq 0;$$

and

(4) $$u_t = (\alpha_1 u_{xx} + \alpha_0)^{-\frac{3}{2}} (u_{xxx} + \alpha_4 u_{xx}^2 + \alpha_3 u_{xx} + \alpha_2) + \alpha_5, \qquad \alpha_1 \neq 0,$$

where α_i are some functions of the variables u, u_x.

In [115], the necessary integrability conditions were generalized to equations with infinite series of local conservation laws. These conditions turned out to be stronger than the conditions for the existence of symmetries. This is not surprising, since there exist equations that have higher symmetries but have no higher conservation laws. The simplest example[2] is the heat equation $u_t = u_{xx}$.

Definition 4.1.5. An equation (1.26) is called *S-integrable* (in the terminology of F. Calogero) if it has infinitely many symmetries and conservation laws. An equation is called *C-integrable* if it has an infinite series of symmetries, but does not have an infinite series of conservation laws.

The most famous example of a nonlinear *S*-integrable equation is the KdV equation (1.12), while the Burgers equation (1.34) is *C*-integrable.

---

[2]Examples of evolution equations with infinite series of conservation laws, but without symmetries, are not known.

Remark 4.1.6. *Informally speaking, S-integrable are equations that have a Lax pair and to which the inverse scattering method is applicable, and C-integrable are equations that can be linearized by differential substitutions. However, to bring Definition 4.1.5 in accordance with this insight and to eliminate obvious exceptions, we need to refine the definition (see Definition 4.2.34). Otherwise, it turns out (see Example 1.5.1) that the linear equation $u_t = u_{xxx}$ is S-integrable.*

Proposition 4.1.1. (see Proposition 4.2.3 or Theorem 29 in [8]) *A scalar evolution equation* (1.26) *of even order cannot have an infinite series of conservation laws.*

There are two types of classification results within the framework of the symmetry approach: the "weak" version when equations with conservation laws are sought, and the "strong" version associated with the existence of symmetries. In the first case, the list of integrable equations contains only S-integrable equations, while in the second case both S and C-integrable equations are to be found.

Second order equations

All nonlinear equations of the form

$$u_t = F(x, t, u, u_1, u_2), \tag{4.3}$$

having symmetries were found in [116] and in [117]. The following list of such equations

$$u_t = u_2 + 2uu_x + h(x),$$

$$u_t = u^2 u_2 - \lambda x u_1 + \lambda u,$$

$$u_t = u^2 u_2 + \lambda u^2,$$

$$u_t = u^2 u_2 - \lambda x^2 u_1 + 3\lambda x u$$

is complete up to contact transformations (1.55), (1.53). According to Proposition 4.1.1 all these equations are C-integrable. They are connected [29] to the heat equation $v_t = v_{xx}$ by group differential substitutions (see Section 1.6.2).

The first three equations of the list have local higher symmetries. All such equations were obtained in [116]. In this paper, it was assumed that the right-hand side of the equation was independent of t.

The latter equation has the so-called *quasi-local symmetries* (see [117]). All such equations (4.3) were found in [117].

Third order equations

The classification of S-integrable KdV type equations was done in [115].

Theorem 4.1.7. *The complete — up to point and quasi-local (see Definition 1.6.20) transformations — list of equations*

$$u_t = u_{xxx} + f(u, u_x, u_{xx}), \tag{4.4}$$

possessing an infinite series of conservation laws, can be written as:

$$u_t = u_{xxx} + u\, u_x,$$

$$u_t = u_{xxx} + u^2\, u_x,$$

$$u_t = u_{xxx} - \frac{1}{2}u_x^3 + (\alpha e^{2u} + \beta e^{-2u})u_x, \tag{4.5}$$

$$u_t = u_{xxx} - \frac{1}{2}Q''\, u_x + \frac{3}{8}\frac{(Q - u_x^2)_x^2}{u_x\,(Q - u_x^2)}, \tag{4.6}$$

$$u_t = u_{xxx} - \frac{3}{2}\frac{u_{xx}^2 + Q(u)}{u_x}, \tag{4.7}$$

where $Q'''''(u) = 0$.

For the classification of C-integrable equations (4.4), see [118] and Appendix 9.3. The proof of the corresponding statement can be found in the survey [14].

As for the integrable equations (4.2) from the more general classes described in Example 4.1.3, their classification is not yet complete. Some partial results were obtained in [119, 120].

Conjecture 4.1.8. *Any S-integrable third order evolution equation is connected with the Krichever–Novikov equation* (4.7)[3] *by a sequence of differential substitutions (see Section 1.6.2).*

Some results in this direction can be found in [49, 114].

Fifth order equations

All equations of the form

$$u_t = u_5 + F(u, u_x, u_2, u_3, u_4), \tag{4.8}$$

possessing an infinite series of conservation laws, were found in [121]. We give an intermediate result of the classification.

---

[3]Mostly with a degenarate polynomial Q.

Lemma 4.1.9. *The possible dependence of function F on u_4, u_3, u_2 for such equations is given by the following formula:*

$$u_t = u_5 + (A_1 u_2 + A_2) u_4 + A_3 u_3^2 + (A_4 u_2^2 + A_5 u_2 + A_6) u_3 +$$
$$A_7 u_2^4 + A_8 u_2^3 + A_9 u_2^2 + A_{10} u_2 + A_{11},$$

where $A_i = A_i(u, u_x)$.

The list of integrable cases contains both well-known equations such as (1.42), (1.43) and

$$u_t = u_5 + 5(u_1 - u^2) u_3 + 5u_2^2 - 20 \, uu_1u_2 - 5u_1^3 + 5u^4u_1, \tag{4.9}$$

(see [45]) and several new equations. One of them (lost in [122]) has the form

$$u_t = u_5 + 5(u_2 - u_1^2 + \lambda_1 e^{2u} - \lambda_2^2 e^{-4u}) u_3 - 5u_1 u_2^2 +$$
$$15(\lambda_1 e^{2u} u_3 + 4\lambda_2^2 e^{-4u}) u_1 u_2 + u_1^5 - 90\lambda_2^2 e^{-4u} u_1^3 + 5(\lambda_1 e^{2u} - \lambda_2^2 e^{-4u})^2 u_1$$

and resembles equation (4.5). A "stronger" version of this classification result concerning both S-integrable and C-integrable equations (see Appendix 9.3) was published in [14].

At first glance, the problem of classifying integrable equations

$$u_t = u_n + F(u, u_x, u_{xx}, \ldots, u_{n-1}), \qquad u_i = \frac{\partial^i u}{\partial x^i} \tag{4.10}$$

for arbitrary n seems to be far from a conclusive solution. This is not quite so. Each integrable equation together with all its symmetries forms the so-called hierarchy of integrable equations. For S-integrable equations, all equations of the hierarchy have the same L-operator. This fact underlies (see Theorem 2.1.16 and Proposition 2.2.2) the commutativity of hierarchies (each equation of the hierarchy is a symmetry for all others). Another explanation of commutativity is contained in Remark 1.4.8. For homogeneous equations the commutativity can be easily proved. A general rigorous statement about "almost" commutativity of symmetries for an equation of the form (4.10) can be found in [123].

Under the assumption that the right-hand side of equation (4.10) is polynomial and homogeneous, it was proved in the papers [30, 124] that the hierarchy of any such integrable equation contains an equation of second, third, or fifth order.

Some additional references concerning the classification of integrable scalar evolution equations are contained in the surveys [3, 4, 5, 8, 14]. Here I would like to mention the earlier articles [125]–[128].

In some classification papers, an explicit dependence of the right-hand side of equation (1.26) and its symmetries on x was allowed. This assumption sometimes turns out to be essential, in particular, because even if the right-hand side of the equation does not depend on x, its symmetry may depend on x.

Example 4.1.10. The equation

$$u_t = D\left(\frac{u_{xx}}{u^3} - 3\frac{u_x^2}{u^4}\right) + 1$$

possesses an infinite series of symmetries that depend on x polynomially.

The assumption of the explicit dependence of the right-hand side of equation (1.26) on t makes it very difficult to apply the symmetry approach to the classification of integrable cases. The main reason is that in this case the kernel of the total derivative D does not consist of constants, but of functions of t. An attempt to overcome this difficulty was made in [117]. Examples of integrable equations, where the explicit dependence on t could not be eliminated by an invertible transformation, are unknown to me.

4.1.3 Systems of two equations

In [129, 130], the necessary integrability conditions were generalized to the case of systems of evolution equations. However, componentwise calculations in this case are much more tedious than in the case of scalar equations. The only serious classification problem was completely solved in [129, 130, 4], resulting in the list of all systems of the form

$$u_t = u_2 + F(u, v, u_1, v_1), \qquad v_t = -v_2 + G(u, v, u_1, v_1) \qquad (4.11)$$

possessing higher conservation laws. In other words, the authors found all S-integrable systems (4.11).

In addition to the NLS equation, written in the form of a system of two equations (1.13), the following basic models from the long list of integrable systems can be noted:

- A version of the Boussinesq equation

$$u_t = u_2 + (u + v)^2, \qquad v_t = -v_2 + (u + v)^2; \qquad (4.12)$$

- The two-component form of the Landau–Lifschitz equation[4]

$$\begin{cases} u_t = u_2 - \dfrac{2u_1^2}{u + v} - \dfrac{4\left(p(u, v)\,u_1 + r(u)\,v_1\right)}{(u + v)^2}, \\[3mm] v_t = -v_2 + \dfrac{2v_1^2}{u + v} - \dfrac{4\left(p(u, v)\,v_1 + r(-v)\,u_1\right)}{(u + v)^2}, \end{cases} \qquad (4.13)$$

where $\quad r(y) = c_4 y^4 + c_3 y^3 + c_2 y^2 + c_1 y + c_0 \quad$ and

$$p(u, v) = 2c_4 u^2 v^2 + c_3 (uv^2 - vu^2) - 2c_2 uv + c_1 (u - v) + 2c_0.$$

---

[4]In fact, it can be derived from (2.45) using the stereographic projection.

The complete list of integrable systems (4.11), distinct up to point transformations of the form

$$u \mapsto \Phi(u), \qquad v \mapsto \Psi(v), \tag{4.14}$$

contains more than 100 systems. Such a list has never been published. Instead, in [4] a list of canonical forms of integrable systems with respect to "almost invertible" transformations [131] was presented. In principle, the text of this paper contains an algorithm for reducing any S-integrable system (4.11) to one of these canonical forms.

Remark 4.1.11. *It turned out that all S-integrable systems (4.11) have fourth order symmetry of the form*

$$\begin{cases} u_\tau = u_{xxxx} + f(u, v, u_x, v_x, u_{xx}, v_{xx}, u_{xxx}, v_{xxx}), \\ v_\tau = -v_{xxxx} + g(u, v, u_x, v_x, u_{xx}, v_{xx}, u_{xxx}, v_{xxx}). \end{cases} \tag{4.15}$$

Most of them also have a third order symmetry. The latter systems (systems of NLS type) have a Lax pair in the Lie algebra $\mathfrak{sl}_2$, while coefficients of Lax operators for systems of the Boussinesq equation, having no symmetry of order 3, belong to $\mathfrak{sl}_3$.

Conjecture 4.1.12. *Any S-integrable system of NLS type is connected with the Landau–Lifschitz equation (4.13) by a sequence of differential substitutions.*

In the paper [132], already after the publication of articles [129, 130, 4], integrable quasilinear systems of the form

$$\begin{cases} u_t = u_{xx} + A_1(u, v)\, u_x + A_2(u, v)\, v_x + A_0(u, v), \\ v_t = -v_{xx} + B_1(u, v)\, v_x + B_2(u, v)\, u_x + B_0(u, v) \end{cases} \tag{4.16}$$

were considered. There are at least three reasons why this paper was written.

Reason 1. There exist C-integrable cases that are absent in the Mikhailov–Shabat–Yamilov classification. One of C-integrable systems is given by

$$\begin{cases} u_t = u_{xx} - 2uu_x - 2vu_x - 2uv_x + 2u^2v + 2uv^2, \\ v_t = -v_{xx} + 2vu_x + 2uv_x + 2vv_x - 2u^2v - 2uv^2. \end{cases} \tag{4.17}$$

The system (4.17) was discovered in [133]. It can be reduced to the linear system

$$U_t = U_{xx}, \qquad V_t = -V_{xx}$$

by the following Cole–Hopf type substitution:

$$u = \frac{U_x}{(U+V)}, \qquad v = \frac{V_x}{(U+V)}.$$

Open problem 4.1.13. *Find all C-integrable systems of the form* (4.11).

Reason 2. Integrable systems (4.15) can be easily classified without using any quasi-local transformations (see Definition 1.6.20). In particular, such systems allow only linear transformations of the form (4.14). It turned out that the right-hand sides of integrable systems (4.15) are polynomial.

Reason 3. The results of any serious classification should be verified independently. Only after that one can be sure that nothing is lost in the lists.

When classifying systems (4.15), a naive version of the symmetry approach was used (cf. Section 1.4).

We formulate an intermediate classification result.

Lemma 4.1.14. *If a system* (4.16) *has a fourth order symmetry* (4.15) *(see Remark 4.1.11), then this system has the following structure:*

$$\begin{cases} u_t = u_{xx} + (a_{12}uv + a_1u + a_2v + a_0)u_x + (p_2v + p_{11}u^2 + p_1u + p_0)v_x + A_0(u,v), \\ v_t = -v_{xx} + (b_{12}uv + b_1v + b_2u + b_0)v_x + (q_2u + q_{11}v^2 + q_1v + q_0)u_x + B_0(u,v), \end{cases}$$

where A_0 and B_0 are polynomials of degree at most five.

The coefficients of the system have to satisfy algebraic equations, the simplest and most essential of which are

$$p_2(b_{12} - q_{11}) = 0, \qquad p_2(a_{12} - p_{11}) = 0, \qquad p_2(a_{12} + 2b_{12}) = 0,$$

$$q_2(b_{12} - q_{11}) = 0, \qquad q_2(a_{12} - p_{11}) = 0, \qquad q_2(b_{12} + 2a_{12}) = 0,$$

$$a_{12}(a_{12} - b_{12} + q_{11} - p_{11}) = 0, \qquad b_{12}(a_{12} - b_{12} + q_{11} - p_{11}) = 0,$$

$$(a_{12} - p_{11})(p_{11} - q_{11}) = 0, \qquad (b_{12} - q_{11})(p_{11} - q_{11}) = 0,$$

$$(a_{12} - p_{11})(a_{12} - b_{12}) = 0, \qquad (b_{12} - q_{11})(a_{12} - b_{12}) = 0.$$

As usual, factorized equations lead to a branching tree.

When solving this overdetermined algebraic system, we do not consider so-called triangular systems like the following:

$$u_t = u_{xx} + 2uv_x, \qquad v_t = -v_{xx} - 2vv_x.$$

Here, we have an equation for u and v and an equation solely for v. That is, the first is a linear equation with variable coefficients for any given solution v of the second equation.

The classification statement is formulated in Appendix 9.4.

4.2   Necessary integrability conditions

Below, for a rigor of exposition, we will use the language of differential algebra. For the notation see Section 1.1.3 and the beginning of Section 2.1.

4.2.1   Evolutionary vector fields, recursion operator and variational derivative

The main local object associated with the dynamical system (1.29) is the finite-dimensional vector field (1.31). In the case of the evolution equations (1.26), the infinite-dimensional vector field (2.2) plays a similar role. This vector field commutes with the total x-derivative D.

The set of all evolutionary vector fields (1.31) forms a Lie algebra over $\mathbb{C}$: $[D_G, D_H] = D_K$, where K is defined by the formula (2.3). An integrable hierarchy of evolutionary equations is nothing more than a commutative infinite-dimensional subalgebra in this Lie algebra.

Proposition 4.2.1. *Suppose that an operator*[5] *$\mathcal{R}$ satisfies the operator equation*

$$D_t(\mathcal{R}) = F_* \, \mathcal{R} - \mathcal{R} \, F_*. \qquad (4.18)$$

Then, for any symmetry (1.32) of equation (1.26), the function $\mathcal{R}(G)$ is a generator of symmetry for (1.26).

In what follows, D_t is the total t-derivative operator with respect to equation (1.26) (see Remark 2.0.3), and

$$D_t\Big(\sum s_i D^i\Big) \overset{def}{=} \sum D_t(s_i)\, D^i.$$

Proof. Let us rewrite (4.18) as

$$[D_t - F_*, \mathcal{R}] = 0. \qquad (4.19)$$

Now, the statement follows from the formula (2.4). □

Definition 4.2.1. An operator $\mathcal{R} : \mathcal{F} \to \mathcal{F}$ satisfying relation (4.18) is called a *recursion operator* for equation (1.26).

---

[5]Usually $\mathcal{R}$ is a ratio of differential operators (see Section 4.3.1).

Variational derivative

Definition 4.2.2. The variational derivative of a function $a \in \mathcal{F}$ is the function

$$\frac{\delta a}{\delta u} \overset{def}{=} \sum_k (-1)^k D^k \left(\frac{\partial a}{\partial u_k} \right) = a_*^+(1).$$

The operator $\dfrac{\delta}{\delta u}$ is called the Euler operator.

If the function a is a total x-derivative: $a = D(b)$, $b \in \mathcal{F}$ (in this case, we say that $a \in \operatorname{Im} D$), then its variational derivative is equal to zero. Moreover, the vanishing of the variational derivative is "almost" a criterion of the fact that the function belongs to $\operatorname{Im} D$ [134]:

Theorem 4.2.3. *The variational derivative of a function $a \in \mathcal{F}$ equals zero iff $a \in \operatorname{Im} D + \mathbb{C}$.*

Lemma 4.2.4. *The following identities*

$$(ab)_* = ab_* + ba_*, \qquad (D(a))_* = D\,a_* = D(a_*) + a_*\,D,$$

$$(D_t(a))_* = D_t(a_*) + a_*\,F_*, \qquad (a_*(b))_* = D_b(a_*) + a_*\,b_*,$$

$$\left(\frac{\delta a}{\delta u} \right)_* = \left(\frac{\delta a}{\delta u} \right)_*^+, \qquad \frac{\delta}{\delta u}(D_t(a)) = D_t \left(\frac{\delta a}{\delta u} \right) + F_*^+ \left(\frac{\delta a}{\delta u} \right)$$

hold for any $a, b, F \in \mathcal{F}$.

4.2.2 Formal symmetries

In this section, we use the notation and results from Section 2.1.1.

Definition 4.2.5. A pseudo-differential series

$$\Lambda = l_1 D + l_0 + l_{-1} D^{-1} + \cdots, \tag{4.20}$$

where $l_k = l_k(u, \ldots, u_{s_k}) \in \mathcal{F}$, is called *a formal symmetry* (or *a formal recursion operator*)[6] for (1.26) if $R = \Lambda$ satisfies the equation

$$D_t(R) = [F_*, R], \qquad \text{where} \qquad F_* = \sum_{i=0}^{n} \frac{\partial F}{\partial u_i} D^i. \tag{4.21}$$

---

[6]According to Proposition 4.2.1, any "genuine" operator satisfying the relation (4.21) maps symmetries of equation (1.26) to symmetries.

Definition 4.2.6. An equation (1.26) is called *formally integrable* if it has a formal symmetry Λ of the form (4.20).

Proposition 4.2.2. [3] *Suppose a pseudo-differential series R of order k satisfies equation (4.21). Then,*

(1) *The series $R^{\frac{1}{k}}$ is a formal symmetry;*

(2) *If R_1 and R_2 satisfy equation (4.21), then the product $R_1 \circ R_2$ satisfies (4.21);*

(3) *The series $R^{\frac{i}{k}}$ satisfies this equation for any $i \in \mathbb{Z}$;*

(4) *If Λ is a formal symmetry, then the series R can be written as*

$$R = \sum_{-\infty}^{k} a_i \Lambda^i, \qquad k = \mathrm{ord}\, R, \quad a_i \in \mathbb{C};$$

(5) *In particular, any formal symmetry $\bar{\Lambda}$ has the form*

$$\bar{\Lambda} = \sum_{-\infty}^{1} c_i \Lambda^i, \qquad c_i \in \mathbb{C}. \tag{4.22}$$

The coefficients of the formal symmetry are to be found from relation (4.21).

Example 4.2.7. Let us consider equations of the form

$$u_t = u_3 + f(u, u_1) \tag{4.23}$$

and find several coefficients $l_1, l_0, \ldots$ of the formal symmetry Λ for such equations. We substitute

$$F_* = D^3 + \frac{\partial f}{\partial u_1} D + \frac{\partial f}{\partial u}, \qquad \Lambda = l_1 D + l_0 + l_{-1} D^{-1} + \cdots$$

into (4.21) and equate the coefficients of $D^3, D^2, \ldots$ in this relation. As a result, we get

$$D^3: \quad 3D(l_1) = 0; \qquad\qquad D^2: \quad 3D^2(l_1) + 3D(l_0) = 0;$$

$$D: \quad D^3(l_1) + 3D^2(l_0) + 3D(l_{-1}) + \frac{\partial f}{\partial u_1} D(l_1) = D_t(l_1) + l_1 D\left(\frac{\partial f}{\partial u_1}\right).$$

It follows from the first equation (see Remark 2.0.1) that l_1 is a constant. Let us set $l_1 = 1$. Now the second equation means that l_0 is a constant. Subtracting the trivial constant solution of equation (4.21) from Λ, we assume that $l_0 = 0$. From the third equation we find that

$$D(l_{-1}) = D\left(\frac{1}{3}\frac{\partial f}{\partial u_1}\right)$$

and therefore

$$l_{-1} = \frac{1}{3}\frac{\partial f}{\partial u_1} + c_{-1}, \qquad c_{-1} \in \mathbb{C}.$$

The integration constant c_{-1} can be set equal to zero without loss of generality (see the formula (4.22)). So

$$\Lambda = D + \frac{1}{3}\frac{\partial f}{\partial u_1}D^{-1} + \cdots. \tag{4.24}$$

Notice, that when finding the first three coefficients of the formal symmetry, we had to solve at each step an equation of the form

$$D(l_k) = S_k, \qquad \text{where} \qquad S_k \in \mathcal{F}. \tag{4.25}$$

Moreover, it turned out that $S_1, S_0, S_{-1} \in \operatorname{Im} D$ for any function $f(u, u_1)$. The same is true for S_{-2}. A first obstacle to the existence of formal symmetry will arise when we look for the coefficient l_{-3}. Namely, $S_{-3} \in \operatorname{Im} D$ only for some special functions f (see Section 4.2.7).

Remark 4.2.8. *For any evolution equation* (1.26) *we determine the coefficients Λ from the relation* (4.21) *step by step, solving equations of the form* (4.25). *This equation is solvable only if $S_k \in \operatorname{Im} D$ (see Theorem 4.2.3). Thus, there are infinitely many obstacles to the existence of formal symmetry. In Section 4.2.5, we consider in more detail the question of what these obstacles are.*

Theorem 4.2.9. [113] *If an equation $u_t = F$ has an infinite series of higher symmetries*

$$u_{\tau_i} = G_i(u, \ldots, u_{m_i}), \qquad m_i \to \infty, \tag{4.26}$$

then this equation has a formal symmetry.

Proof. A simple proof [3] of Theorem 4.2.9 is based on the fact that coefficients of the formal symmetry are "similar" to coefficients of the Fréchet derivative of the higher symmetry generator [3]. Suppose that equation (1.26) has a symmetry with generator G. Then, G satisfies the relation (2.4). We calculate the Fréchet derivative of the left side of this relation. Using the identities of Lemma 4.2.4, we obtain the equality

$$D_t(G_*) + G_*F_* = D_G(F_*) + F_*G_*,$$

which can be rewritten as

$$D_t(G_*) - [F_*, G_*] = D_G(F_*). \tag{4.27}$$

Remark 4.2.10. *A less rigorous but more explanatory proof of identity (4.27) is as follows. If the equations $u_t = F$ and $u_\tau = G$ are compatible, then the corresponding linearized equations $v_t = F_*(v)$ and $v_\tau = G_*(v)$ are compatible also (see Section 2.0.1). This implies that the linearization operators $D_t - F_*$ and $D_\tau - G_*$ commute with each other, which is equivalent to (4.27).*

If function G has a sufficiently high order m, then the order of the left-hand side of (4.27) is much higher than the order of the right-hand side. Therefore, the equations for determining the first few coefficients of the differential operator G_* are exactly the same (cf. (4.27) with (4.21)) as for determining the coefficients of the series Λ^m. Therefore, we can take the coefficients of the series $G_*^{\frac{1}{m}}$ as the first coefficients of Λ. The number of "correct" coefficients is indicated in [3], where a procedure for gluing "approximate" solutions of (4.21) generated by the symmetries (4.26) into a single infinite series Λ with coefficients from $\mathcal{F}$ is also described.   □

4.2.3   Conservation laws

The concept of the first integral, in contrast to infinitesimal symmetry, cannot be directly generalized to the case of partial differential equations. It is replaced by the concept of a local conservation law (see Section 1.5).

Definition 4.2.11. A function $\rho \in \mathcal{F}$ is called the *density* of a local conservation law for equation (1.26) if there exists a function $\sigma \in \mathcal{F}$ such that

$$D_t(\rho) = D(\sigma). \tag{4.28}$$

Obviously, the relation (4.28) is satisfied if $\rho = D(h)$, $\sigma = D_t(h)$, where $h \in \mathcal{F}$ is any function. Such "conservation laws" are called *trivial*.

Definition 4.2.12. The two densities ρ_1, ρ_2 are called equivalent ($\rho_1 \sim \rho_2$) if the difference $\rho_1 - \rho_2$ is a trivial density (that is, $\rho_1 - \rho_2 \in \operatorname{Im} D$).

Remark 4.2.13. *Actually, it is natural to regard a conserved density as the equivalence class of densities with respect to the relation $\sim$. The Euler operator is well defined on such equivalence classes.*

Lemma 4.2.14. *Any nontrivial density ρ is equivalent to a density $\bar{\rho}(u, \dots, u_k)$ such that*

$$\frac{\partial^2 \bar{\rho}}{\partial u_k^2} \neq 0.$$

Proof. If the function ρ is linear in the highest derivative, then we can reduce the order of the function ρ (see Definition 1.1.2) by subtracting the appropriate element from $\operatorname{Im} D$.   □

Definition 4.2.15. The number k is called order of conserved density ρ. We denote k by ord (ρ).

Remark 4.2.16. *It is easy to verify that the order of the differential operator $\left(\dfrac{\delta\rho}{\delta u}\right)_*$ is equal to $2\operatorname{ord}(\rho)$.*

Question 4.2.17. *Suppose that an equation (1.26) and a function $\rho \in \mathcal{F}$ are given. How to find out whether there is a function $\sigma \in \mathcal{F}$ satisfying (4.28)?*

The first method is associated with the Euler operator. If we apply it to both sides of the relation (4.28), then, according to Theorem 4.2.3, we obtain

$$\frac{\delta}{\delta u}\Big(D_t(\rho)\Big) = 0. \qquad (4.29)$$

This identity can be verified straightforwardly.

Unfortunately, we cannot find the function σ in this way. However, there is a well-known inductive algorithm for that. The left-hand side of (4.28) is known and the problem reduces to the following:

Question 4.2.18. *How to solve an equation of the form $D(X) = S$ (cf. (4.25)) for a given function $S(u,\ldots,u_m) \in \mathcal{F}$?*

From (2.1) it follows that the function S has to be linear in the highest derivative:

$$S = A(u,u_1,\ldots,u_{m-1})\,u_m + B(u,u_1,\ldots,u_{m-1}), \qquad m \geq 1.$$

If this is not the case, then our equation has no solutions. In the linear case, we can reduce the order of the function S by subtracting an expression of the form $D(r(u,u_1,\ldots,u_{m-1}))$ from both sides of the equation, where r is any solution of the equation $\dfrac{\partial r}{\partial u_{m-1}} = A$, solvable in quadratures. Continuing this procedure, we either come to the case of a nonlinear function S, or reduce it to the case $S = c$, where c is some constant. If $c \neq 0$, then the original equation has no solutions.[7] If $c = 0$, then we have solved the original equation using quadratures. According to Remark 2.0.1, the solution is unique up to a constant term.

4.2.4 Formal symplectic operator

According to Lemma 4.2.4, function $X = \dfrac{\delta\rho}{\delta u}$ satisfies the conjugate equation of (2.4):

$$D_t(X) + F_*^+(X) = 0. \qquad (4.30)$$

---

[7]However, it can be solved, if we add the independent variable x to the variables (1.1).

Definition 4.2.19. *Any solution* $X \in \mathcal{F}$ *of equation* (4.30) *is called* cosymmetry.

Definition 4.2.20. *A pseudo-differential series*

$$S = s_1 D + s_0 + s_{-1} D^{-1} + \cdots, \qquad s_1 \neq 0 \qquad (4.31)$$

is called a formal symplectic operator[8] *if it satisfies the relation*

$$D_t(S) + S\, F_* + F_*^+ \, S = 0\,. \qquad (4.32)$$

Remark 4.2.21. *If* S *is a formal symplectic operator, then* S^+ *and* $S - S^+$ *are also formal symplectic operators. Therefore, if there exists any formal symplectic operator of the form* (4.31), *then there is a skew-symmetric formal symplectic operator.*

Lemma 4.2.22. *The ratio* $R = S_1^{-1} S_2$ *of two solutions* S_1 *and* S_2 *of equation* (4.32) *satisfies equation* (4.21). *The product* $S_2 = S_1 R$ *of series* S_1 *and* R *satisfying* (4.32) *and* (4.21) *satisfies equation* (4.32).

Theorem 4.2.23. [115] *If an equation* $u_t = F$ *possesses an infinite series of local conservation laws of the form*

$$D_t\Big(\rho_i(u, \ldots, u_{m_i})\Big) = D(\sigma_i), \qquad \frac{\partial^2 \rho_i}{\partial u_{m_i}^2} \neq 0, \quad m_i \to \infty,$$

then this equation has a formal symmetry Λ *and a formal symplectic operator* S *such that*

$$S^+ = -S$$

and

$$\Lambda^+ = -S\Lambda S^{-1}. \qquad (4.33)$$

Proof. A proof is contained in the survey [3]. Here, we will only establish a connection between the coefficients of the series satisfying (4.32) and the coefficients of the differential operator

$$T = \left(\frac{\delta\rho}{\delta u}\right)_*,$$

where ρ is the density of a conservation law of order k. Using the identities of Lemma 4.2.4, let us find the Fréchet derivative of the left-hand side of equation (4.30), where $X = \dfrac{\delta\rho}{\delta u}$. As a result, we come to the relation

$$D_t(T) + T\, F_* + F_*^+ \, T = Q, \qquad (4.34)$$

---

[8]The relation (4.32) can be rewritten as $(D_t + F_*^+) \circ S = S(D_t - F_*)$. Consequently, if the S-operator acts from $\mathcal{F}$ to $\mathcal{F}$, then it maps symmetries in cosymmetries. If equation (1.26) is Hamiltonian, then the inverse of the Hamiltonian operator is a symplectic operator S satisfying (4.32) [135].

where

$$Q = -\sum_{k=0}^{n} D^k \left(\frac{\delta\rho}{\delta u}\right)(H_k)_*,$$

and H_k are the coefficients of the differential operator $F_*^+ = \sum H_k D^k$. The order of the differential operator Q does not exceed $2n$, and the order of the operator T is $2k$ (see Remark 4.2.16). Therefore, if the order k of the conservation law is sufficiently large, then the first few coefficients of the operator T coincide with the coefficients of the series S of order $2k$ satisfying (4.32). □

Proposition 4.2.3. *No scalar evolution equation*

$$u_t = F(u, u_1, \ldots, u_{2m}) \tag{4.35}$$

of even order can have a conservation law of order greater than m.

Proof. Let ρ be the density of a conservation law of order k, where $k > m$. Then, $\operatorname{ord} T = 2k > 2m$. Since $\operatorname{ord} Q \le 4m$, comparing the coefficients of D^{2k+2m} in (4.34), we get $\dfrac{\partial F}{\partial u_{2m}} = 0$. □

Remark 4.2.24. *This statement is not true in the case of systems of evolution equations.*

Resume

Any evolution equation with an infinite series of local symmetries and/or conservation laws is formally integrable (see Definition 4.2.6).

4.2.5 Canonical densities and integrability conditions

In this section, we study the necessary integrability conditions for equations of the form (1.26). According to Theorems 4.2.9 and 4.2.23, the existence of higher symmetries and/or higher conservation laws for the equation entails the existence of a formal symmetry. In Remark 4.2.8, we discussed where the obstacles to the existence of formal symmetry come from. Here we show that these obstacles can be written in the form of conservation laws.

For equations (1.26) having a formal symmetry Λ, we define an infinite sequence of *canonical conserved densities*.

Definition 4.2.25. The functions

$$\rho_i = \operatorname{res}(\Lambda^i), \qquad i = -1, 1, 2, \ldots, \quad \text{and} \quad \rho_0 = \operatorname{res}\log(\Lambda) \tag{4.36}$$

are called *canonical densities* for equation (1.26).

Remark 4.2.26. *Despite the fact that the formal symmetry of Λ is not unique, the sequence of canonical densities is well defined in the following sense. From the formula (4.22) it follows that the canonical density $\bar{\rho}_i$, corresponding to the formal symmetry $\bar{\Lambda}$ is a linear combination of the densities ρ_j, $j \leq i$, defined by the formal symmetry Λ.*

Theorem 4.2.27. *Equation* (1.26) *possesses a formal symmetry Λ iff the canonical densities define local conservation laws*

$$D_t(\rho_i) = D(\sigma_i), \qquad \sigma_i \in \mathcal{F}, \qquad i = -1, 0, 1, 2, \ldots \qquad (4.37)$$

for equation (1.26).

Proof. Suppose that a formal symmetry Λ exists. Since it satisfies equation (4.21), it follows from the Proposition 4.2.2 that the series Λ^k, $k = -1, 1, 2, 3 \ldots$ also satisfy (4.21). Using Adler's Theorem 2.1.5, we get

$$D_t(\rho_k) = D_t(\mathrm{res}\,\Lambda^k) = \mathrm{res}\,([F_*, \Lambda^k]) \in \mathrm{Im}\,D, \quad k = -1, 1, 2, 3 \ldots.$$

In addition,

$$D_t(\rho_0) = \mathrm{res}\,\left(D_t(\Lambda)\,\Lambda^{-1}\right) = \mathrm{res}\,\left([F_*, \Lambda]\,\Lambda^{-1}\right) = \mathrm{res}\,\left([F_*\Lambda^{-1}, \Lambda]\right) \in \mathrm{Im}\,D.$$

A proof of the converse statement can be found in [3] (see also discussion of Question 4.2.36). □

Remark 4.2.28. *Since the formal symmetry is not applicable to functions from $\mathcal{F}$, it cannot be regarded as a "genuine" recursion operator (see Proposition 4.2.1). However, we see that it can be used to construct local conservation laws for equation* (1.26). *The problem is that some (or even all) of these conservation laws may be trivial.*

Example 4.2.29. For any n, the differential operator $\Lambda = D$ is a formal symmetry for the linear equation $u_t = u_n$. Therefore, all canonical densities for such an equation are equal to zero.

Theorem 4.2.30. *Under the assumptions of Theorem 4.2.23 there exists a formal symmetry Λ such that all even canonical densities ρ_{2j} are trivial (that is, they belong to $\mathrm{Im}\,D$).*

Proof. It follows from (4.33) that

$$(\Lambda^{2j})^+ = S\,\Lambda^{2j}\,S^{-1}.$$

Since $\mathrm{res}\,(\Lambda^{2j}) = -\mathrm{res}\,\left((\Lambda^{2j})^+\right)$, the statement of the theorem follows from Corollary 2.1.6. □

Remark 4.2.31. *Theorems 4.2.23 and 4.2.30 imply that, when classifying equations* (1.26) *with the higher conservation laws, the conditions* (4.37) *can be strengthened by the assumption that* $\rho_{2j} = D(\theta_j)$, *where* $\theta_j \in \mathcal{F}$.

Example 4.2.32. The KdV equation (1.12) has the recursion operator (see Section 2.1.2)

$$\hat{\Lambda} = D^2 + 4u + 2u_1 D^{-1}, \tag{4.38}$$

which satisfies equation (4.21). One can take $\Lambda = \hat{\Lambda}^{1/2}$ for a formal symmetry of the KdV equation. The first five canonical densities have the form (cf. Example 1.5.3)

$$\rho_{-1} = 1, \qquad \rho_0 = 0, \qquad \rho_1 = 2u, \qquad \rho_2 = 2u_1, \qquad \rho_3 = 2u_2 + u^2.$$

We see that even canonical densities are trivial.

Example 4.2.33. The Burgers equation has a recursion operator

$$\Lambda = D + u + u_1 D^{-1},$$

which can be chosen as a formal symmetry. The canonical densities for the Burgers equation are

$$\rho_{-1} = 1, \qquad \rho_0 = u, \qquad \rho_1 = u_1, \qquad \rho_2 = u_2 + 2uu_1, \ldots.$$

Although the density ρ_0 defines a nontrivial conservation law, it can be verified that all other densities are trivial. This is consistent with the Proposition 4.2.3.

Now we are able to refine (see Remark 4.1.6) Definition 4.1.5.

Definition 4.2.34. An equation of the form (1.26) is called *S-integrable* if it has a formal symmetry generating infinitely many linearly independent nontrivial canonical densities. An equation is called *C-integrable* if it has a formal symmetry that generates only a finite number of linearly independent nontrivial canonical densities.

The concepts of S and C-integrability are well defined by Remark 4.2.26.

Remark 4.2.35. *According to Definition 4.2.34, the equation* $u_t = u_{xxx}$, *which has* (*see Example 1.5.1*) *an infinite series of conservation laws, is C-integrable because all its canonical densities are trivial.*

Suppose equation (1.26) has a formal symmetry Λ.

Question 4.2.36. *How are the functions* ρ_i, σ_i *in the canonical series of conservation laws* (4.37) *related to the right-hand side F of equation* (1.26)?

The first $n-1$ coefficients $l_1, l_0, \ldots, l_{3-n}$ of the formal symmetry Λ can be chosen equal to the first $n-1$ coefficients of the series $(F_*)^{1/n}$. Indeed $u_\tau = F$ is a symmetry of equation (1.26) and we can use the main idea of the proof of Theorem 4.2.9. Carefully calculating (see [3]) the number of the "correct" coefficients, we arrive at the ansatz

$$\Lambda = (F_*)^{1/n} + \tilde{l}_{2-n}D^{2-n} + \tilde{l}_{1-n}D^{1-n} + \cdots .$$

Knowing the $n-1$ coefficient of Λ, we can find the first $n-1$ canonical densities $\rho_{-1}, \rho_0, \ldots, \rho_{n-3}$ explicitly in terms of the coefficients

$$F_i = \frac{\partial F}{\partial u_i}$$

of the Fréchet derivative

$$F_* = F_n D^n + F_{n-1}D^{n-1} + \cdots + F_0$$

of the right-hand side of the equation. For example,

$$\rho_{-1} = F_n^{-\frac{1}{n}}. \tag{4.39}$$

Since $[(F_*)^{1/n}, F_*] = 0$, the coefficients of D^k for $k > 1$ in (4.21) are automatically canceled out. Comparing the coefficients of D, we find that the first of the unknown coefficients $\tilde{l}_{2-n} \in \mathcal{F}$ of the series Λ exists iff (4.39) is a density of a local conservation law for the equation (1.26), that is, there exists a function $\sigma_{-1} \in \mathcal{F}$ such that $D_t(\rho_{-1}) = D(\sigma_{-1})$.

If $D_t(\rho_{-1}) \notin \operatorname{Im} D$, then equation (1.26) cannot have an infinite sequence of higher symmetries or conservation laws. If the function $\sigma_{-1} \in \mathcal{F}$ exists, then the coefficient $\tilde{l}_{2-n}$ can be explicitly expressed in terms of the coefficients $F_n, \ldots, F_0$ and σ_{-1}. Similarly, the next coefficient $\tilde{l}_{1-n}$ can be found (as an element of $\mathcal{F}$) iff $D_t(\rho_0) = D(\sigma_0)$ for some function $\sigma_0 \in \mathcal{F}$. In this case, $\tilde{l}_{1-n}$ is expressed through $F_n, \ldots, F_0, \sigma_{-1}, \sigma_0$, and so on.

Example 4.2.37. Let us consider evolution equations of second order

$$u_t = F(u, u_1, u_2). \tag{4.40}$$

The computations described above show that for such equations the first three canonical densities can be written as

$$\rho_{-1} = F_2^{-1/2}, \qquad \rho_0 = F_2^{-1/2}\sigma_{-1} - F_2^{-1}F_1,$$

$$\rho_1 = \rho_{-1}F_0 - \frac{\rho_0^2}{4\rho_{-1}} + \frac{\rho_0\sigma_{-1}}{2} - \frac{\rho_{-1}\sigma_0}{2}.$$

Using the approach of [125, 136], it is possible to construct recurrent relations for the whole infinite chain of canonical densities ρ_i. For equations of the form (4.4), these relations were obtained in [14]. They have the form

$$\rho_{n+2} = \frac{1}{3}\left[\sigma_n - \delta_{n,0}f_0 - f_1\rho_n - f_2\left(D(\rho_n) + 2\rho_{n+1} + \sum_{s=0}^{n}\rho_s\,\rho_{ns}\right)\right] - \sum_{s=0}^{n+1}\rho_s\,\rho_{n+1-s}$$

$$- \frac{1}{3}\sum_{0\leq s+k\leq n}\rho_s\,\rho_k\,\rho_{n-s-k} - D\left[\rho_{n+1} + \frac{1}{2}\sum_{s=0}^{n}\rho_s\,\rho_{n-s} + \frac{1}{3}D(\rho_n)\right], \qquad n\geq 0,$$

$$(4.41)$$

where the first two terms of the sequence ρ_i are given by

$$\rho_0 = -\frac{1}{3}f_2, \qquad \rho_1 = \frac{1}{9}f_2^2 - \frac{1}{3}f_1 + \frac{1}{3}D(f_2). \qquad (4.42)$$

In the relation (4.41), $\delta_{i,j}$ is the Kronecker delta, $f_i = \dfrac{\partial f}{\partial u_i}$, where $i = 0, 1, 2$.

For a given equation of the form (4.4), one can check on the computer as many conditions (4.37) of the formal integrability as the computer resources allow. The corresponding algorithm is based on the relation (4.41) and on the procedure described in Section 4.2.3.

4.2.6 Invariance of integrability conditions with respect to changes of variables

Let us explore how the formal symmetry changes and what happens with the canonical series of conservation laws under differential substitutions.

Generally speaking, the existence of local symmetries and conservation laws is not invariant with respect to differential substitutions. Let equations (1.60) and (1.61) be connected by a differential substitution (1.59).

Proposition 4.2.4. *Let $\bar{D}_t(\bar{\rho}) = \bar{D}(\bar{\sigma})$ be a conservation law for equation (1.61). Then, $D_t(\rho) = D(\sigma)$, where*

$$\rho = \bar{\rho}\,D(\phi), \qquad \sigma = \bar{\sigma} + \bar{\rho}\,D_t(\phi), \qquad (4.43)$$

is a conservation law for equation (1.60).

If, on the contrary, equation (1.60) has a local conservation law $D_t(\rho) = D(\sigma)$, then it cannot be claimed that the corresponding conservation law of equation (1.61) is local. For its locality, it is necessary that the functions $\bar{\rho}$ and $\bar{\sigma}$ from the formula (4.43) can be written as functions of variables $\bar{x}, \bar{u}, \dots$.

The situation is even worse in the case of local symmetries. If equation (1.60) has a local symmetry $g(x, u, \dots)$, then the corresponding symmetry of equation (1.61) is determined from the relation

$$\bar{g} = Q(g),$$

where the differential operator Q is defined by (1.58). For the locality of the symmetry $\bar{g}$ it is necessary that it could be written as a function of variables $\bar{x}, \bar{u}, \dots$. Moreover, if equation (1.61) possesses a local symmetry, one cannot state that the corresponding symmetry of equation (1.60) is local.

Exercise 4.2.38. Verify that equations[9]

$$u_t = u_3 - \frac{3}{4}\frac{u_2^2}{u_1} \tag{4.44}$$

and

$$\bar{u}_t = \bar{u}_3 \tag{4.45}$$

are connected by the differential substitution

$$\bar{u} = \sqrt{u_1}.$$

Show that the symmetries of the form $\bar{u}_\tau = \bar{u}_{2i}$, $i \in \mathbb{N}$ of the linear equation correspond to nonlocal symmetries of equation (4.44).

Obviously, if the differential substitution is invertible (i.e. in the case of point or contact transformations), there is a one-to-one correspondence between both local symmetries and the local conservation laws for equations (1.60) and (1.61).

A formal symmetry as well as a formal symplectic operator can always be "lifted" from equation (1.61) to equation (1.60).[10]

Proposition 4.2.5. *Let $\bar{L}$ and $\bar{S}$ be a formal symmetry and a formal symplectic operator of equation (1.61). Then, the series*

$$L = Q^{-1}\bar{L}\,Q \tag{4.46}$$

and

$$S = Q^+ \bar{S}\, Q,$$

where Q is the differential operator (1.58), are a formal symmetry and a formal symplectic operator for equation (1.60). It is assumed that the coefficients of the series L and S are expressed in terms of the variables $x, u, u_1, \dots$ using the formula (1.59), and the operator $\bar{D}$ is replaced by D by the rule (1.56).

---

[9]Equation (4.44) differs from the S-integrable equation (1.71) only by a numerical coefficient, but it turns out to be C-integrable.

[10]This is an advantage of the formal symmetry compared to the usual higher symmetries.

According to formula (4.46), the series $\bar{L}$ and L are conjugated.

Proposition 4.2.6. *The pseudo-differential series*

$$A = p\,D + \sum_{i=0}^{\infty} a_{-i} D^{-i} \quad and \quad B = p\,D + \sum_{i=0}^{\infty} b_{-i} D^{-i}$$

are conjugated (i.e. $B = TAT^{-1}$ for some series T with the coefficients from $\mathcal{F}$) iff (see Corollary 2.1.6)

$$\frac{a_0 - b_0}{p} \in \text{Im } D, \quad \text{res } (A^i - B^i) \in \text{Im } D, \quad i \in \mathbb{N}.$$

From this statement and the formula (4.46) it follows that the canonical densities (4.36) of equation (1.61) transform (up to the total x-derivatives) to the canonical densities of equation (1.60). This fact can be effectively used when searching for a differential substitution that relates two given evolution equations (1.60) and (1.61).

If equation (1.61) is linear, then $\bar{L} = \bar{D}$ and all the canonical densities of equation (1.60) are trivial. Therefore, if equation (1.60) has only a few nontrivial canonical densities, for linearization of this equation it is necessary first to "kill" nontrivial canonical densities by the potentiation.

Example 4.2.39. Equation (1.39) possesses the only nontrivial canonical density ρ_0 proportional to u^2. By introducing a new nonlocal variable $v = D^{-1}(u^2)$, we obtain the evolution equation

$$v_t = v_{xxx} - \frac{3}{4}\frac{v_{xx}^2}{v_x} + 3v_x v_{xx} + v_x^3$$

with trivial canonical densities. After this, the differential substitution

$$\bar{u} = \sqrt{v_x}\,\exp{(v)},$$

which reduces the equation for v to (4.45), can be easily found.

4.2.7   Classification of integrable equations of KdV type

The first few integrability conditions allow us to find all equations of a prescribed type that may have a formal symmetry. After that, for each equation from the resulting list, we need to prove its integrability. To do this, one can try to find:

(1) A transformation that reduces this equation to a known integrable equation;

(2) A Lax representation;

(3) A recursion operator;

(4) An auto-Bäcklund transformation depending on an arbitrary parameter (see [137]).

In order to demonstrate how efficient the necessary integrability conditions are, we solve in this section a simple classification problem [138, 126].

Let us consider evolution equations of the form (4.23). According to (4.42), for such equations we have $\rho_0 = \sigma_0 = 0$ and

$$D_t \left(\frac{\partial f}{\partial u_1} \right) = D(\sigma_1), \tag{4.47}$$

where σ_1 is a function depending on $u, u_1, \ldots, u_3$.

Example 4.2.40. For the mKdV equation

$$u_t = u_3 + u^2 u_1 \tag{4.48}$$

the conservation law (4.47) has the form:

$$D_t(u^2) = D \left(2uu_2 - u_1^2 + \frac{1}{2}u^4 \right).$$

Applying the Euler operator to both sides of (4.47) and using the condition (4.29), we obtain

$$0 = \frac{\delta}{\delta u} D_t \left(\frac{\partial f}{\partial u_1} \right) = 3u_4 \left(u_2 \frac{\partial^4 f}{\partial u_1^4} + u_1 \frac{\partial^4 f}{\partial u_1^3 \partial u} \right) + O(3), \tag{4.49}$$

where $O(3)$ denotes terms of order not higher than 3. Relation (4.49) has to be satisfied identically with respect to the variables $u, u_1, \ldots, u_4$. Equating to zero the coefficient of u_4 and using the fact that f does not depend on u_2, we obtain

$$f(u, u_1) = \mu u_1^3 + A(u)u_1^2 + B(u)u_1 + C(u),$$

where μ is some constant. It is easy to verify that for such a function f the condition (4.49) is equivalent to the following system of ordinary differential equations:

$$\mu A' = 0, \qquad B''' + 8\mu B' = 0, \qquad (B'C)' = 0, \qquad AB' + 6\mu C' = 0.$$

The next integrability condition (4.41) has the form

$$D_t \left(\frac{\partial f}{\partial u} \right) = D(\sigma_2),$$

which, according to (4.29), can be written as

$$\frac{\delta}{\delta u} D_t \left(\frac{\partial f}{\partial u} \right) = 0. \tag{4.50}$$

The latter condition leads to additional equations

$$A' = 0, \qquad AC'' = 0, \qquad (C''' + 2\mu C')' = 0, \qquad (CC'')' = 0.$$

In the case of $\mu \neq 0$, solving the system of ODE obtained above, we find the functions A, B and C. As a result, up to a scaling of the form $u \mapsto \text{const } u$, we arrive at the equations

$$u_t = u_{xxx} - \frac{1}{2} u_x^3 + \left(c_1 e^{2u} + c_2 e^{-2u} + c_3 \right) u_x \tag{4.51}$$

and

$$u_t = u_{xxx} + c_1 u_x^3 + c_2 u_x^2 + c_3 u_x + c_4, \tag{4.52}$$

where c_i are arbitrary constants.

If $\mu = 0$, then solving the system of ODE for functions A, B, C, we obtain that the equation has the form

$$u_t = u_{xxx} + c_0 u_x^2 + (c_1 u^2 + c_2 u + c_3) u_x + c_4 u + c_5,$$

where

$$c_0 c_1 = 0, \qquad c_0 c_2 = 0, \qquad c_4 c_1 = 0, \qquad c_4 c_2 = 0, \qquad c_1 c_5 = 0.$$

From the third integrability condition (see formula (4.41) for ρ_3), we find additional algebraic relations

$$c_0 c_4 = 0, \qquad c_2 c_5 = 0.$$

In the case of $c_0 \neq 0$, we arrive at a special case of equation (4.52). If $c_0 = 0$, two cases are possible: (a) $c_4 = c_5 = 0$ and (b) $c_1 = c_2 = 0$. In case (a) we obtain the equation

$$u_t = u_{xxx} + (c_1 u^2 + c_2 u + c_3) u_x. \tag{4.53}$$

Case (b) leads to a linear equation, which has an infinite set of higher symmetries and a formal symmetry $\Lambda = D$. Equation (4.53) can be reduced to the KdV equation or to the mKdV equation by scalings and shift of u. Equation (4.52) is related to (4.53) by the potentiation (see Definition 1.6.18). Equation (4.51) was found in [139]. It is connected with the mKdV equation by a differential substitution of the Miura type [49] (see Example 4.2.54).

4.2.8   Integrable equations of Harry–Dym type

The Harry-Dym equation

$$u_t = u^3 u_{xxx} \tag{4.54}$$

is one of the well-known integrable evolution equations. Equations of the form

$$u_t = f(u) u_3 + Q(u, u_1, u_2), \qquad f'(u) \neq 0 \tag{4.55}$$

are called *Harry–Dym type equations*.

The first of the integrability conditions (4.37) for equations (1.26) is given by $D_t(\rho_{-1}) = D(\sigma)$, where (see formula (4.39))

$$\rho_{-1} = \left(\frac{\partial F}{\partial u_n} \right)^{-\frac{1}{n}}.$$

Let us show that the use of only this condition allows us to reduce the function f in any integrable equation of the form (4.55) to 1 by some quasi-local transformation (see Definition 1.6.20).

Theorem 4.2.41. [3] *Any equation (4.55) for which the first integrability condition is satisfied can be reduced to the form*

$$v_t = v_3 + G(v_1, v_2) \tag{4.56}$$

by the potentiation and point transformations.

Proof. First we make the point transformation $\tilde{u} = f(u)^{-1/3}$ in order to bring equation (4.55) to the form

$$\tilde{u}_t = \frac{\tilde{u}_3}{\tilde{u}^3} + \tilde{Q}(\tilde{u}, \tilde{u}_1, \tilde{u}_2).$$

For such an equation, we have $\rho_{-1} = \tilde{u}$. Since function $\tilde{u}$ is the density of the conservation law, the equation has the form

$$\tilde{u}_t = D \left(\frac{\tilde{u}_2}{\tilde{u}^3} + \Psi(\tilde{u}, \tilde{u}_1) \right).$$

The second step is the potentiation $\hat{u} = D^{-1}\tilde{u}$. As a result, we obtain

$$\hat{u}_t = \frac{\hat{u}_3}{\hat{u}_1^3} + \Psi(\hat{u}_1, \hat{u}_2). \tag{4.57}$$

The last step is the point transformation

$$\hat{t} = t, \qquad \hat{x} = v, \qquad \hat{u} = x. \tag{4.58}$$

For this transformation we have (see Section 1.6)

$$\hat{u}_1 = \frac{1}{v_1}, \qquad \hat{u}_2 = -\frac{v_2}{v_1^3}, \qquad \hat{u}_3 = -\frac{v_3}{v_1^4} + \frac{3v_2^2}{v_1^5}, \qquad \hat{u}_t = -\frac{v_t}{v_1}.$$

Using these formulas, it is easy to verify that any equation (4.57) transforms into some equation of the form (4.56). □

Example 4.2.42. For the Harry–Dym equation, after the transformation $\tilde{u} = \dfrac{1}{u}$ and the potentiation, the following equation appears:

$$\hat{u}_t = \frac{\hat{u}_3}{\hat{u}_1^3} - \frac{3\,\hat{u}_2^2}{2\,\hat{u}_1^4}.$$

Using the transformation (4.58), we arrive at the Schwartz–KdV equation (see Exercise 1.6.22)

$$v_t = v_3 - \frac{3\,v_2^2}{2\,v_1},$$

which is a special case of the Krichever–Novikov equation (4.7).

4.2.9 Nonlocal variables, evolution equations with constraints and inversion of differential substitutions

In this section, we generalize the nonlocal transformation $v = D^{-1}(u)$ (see Definition 1.6.18, Remark 1.6.19 and Example 4.4.17). This transformation means that we add the new variable v to variables (1.1). Since u is a conserved density, both expressions $D(v)$ and $D_t(v)$ can be expressed in terms of variables (1.1).

Let us add to local variables (1.1) a finite set of (nonlocal) variables $v^1, \ldots, v^m$, that satisfy a system of ordinary differential equations of the form

$$D(v^i) = \Psi_i(\mathbf{v}, u, \ldots u_k), \qquad k \geq 0, \quad i = 1, \ldots m, \tag{4.59}$$

where $\mathbf{v} = (v^1, \ldots, v^m)$. The relations (4.59) mean that we extend the total x-derivative (1.2) to the vector field

$$D = \sum_{j=1}^m \Psi_i \frac{\partial}{\partial v^j} + \sum_{i=0}^{\infty} u_{i+1} \frac{\partial}{\partial u_i}. \tag{4.60}$$

Let $\bar{\mathcal{F}}$ be a differential field of functions of variables $v^1, \ldots, v^m, u, u_1, \ldots$.

Increasing the number m with the help of the redefining $v^{m+1} = u, \ldots, v^{m+k} = u_k$, $\bar{u} = u_{k+1}, \bar{u}_1 = u_{k+2}, \ldots$, we can assume, without loss of generality, that

$$D(v^i) = a_i(\mathbf{v}) + b_i(\mathbf{v})\, u, \qquad i = 1, \ldots m, \tag{4.61}$$

and, therefore,

$$D = A + uB + \sum_{i=0}^{\infty} u_{i+1} \frac{\partial}{\partial u_i}, \tag{4.62}$$

where

$$A = \sum_{j=1}^{m} a_i \frac{\partial}{\partial v^j}, \qquad B = \sum_{j=1}^{m} b_i \frac{\partial}{\partial v^j}.$$

One of the main properties of the total x-derivative D defined by (1.2) is the triviality of the kernel: $\operatorname{Ker} D = \mathbb{C}$. We assume that the same holds for the vector field (4.62). It is easy to verify that this is equivalent to the assumption that the Lie algebra generated by the vector fields A and B is of dimension m over the field $\mathcal{F}_0$ of functions in the variables $v^1, \dots, v^m$.

Evolutionary vector fields

A vector field ∇ is called *evolutionary* if $[D, \nabla] = 0$. In the case of local variables (1.1), evolutionary vector fields (see Section 4.2.1) are described by formula (2.2), where the generator F is an arbitrary function from $\mathcal{F}$. In the presence of nonlocal variables, every evolutionary vector field has the following structure:

$$D_t = \sum_{j=1}^{m} G_j \frac{\partial}{\partial v^j} + \sum_{i=0}^{\infty} D^i(H) \frac{\partial}{\partial u_i}, \qquad H, G_i \in \bar{\mathcal{F}}. \tag{4.63}$$

Definition 4.2.43. In the case when the operators of the total derivatives with respect to x and t are given by formulas of the form (4.60) and (4.63), we will call the variables v^i *quasi-local*.

The vector field (4.63) corresponds to the system of evolution equations

$$u_t = H(\mathbf{v}, u, \dots, u_n), \qquad (v^i)_t = G_i(\mathbf{v}, u, \dots, u_{n-1}) \tag{4.64}$$

with differential constraints (4.61).[11] The number n is called the *order* of equation (4.64). The condition $[D, D_t] = 0$ imposes restrictions on the functions H, G_i.

Question 4.2.44. *How to describe all evolutionary vector fields (4.63) for a given system of constraints (4.61)?*

Consider a $\bar{\mathcal{F}}$-module $\mathcal{M}$ consisting of 1-forms spanned by dx, dv^i, $i = 1, \dots, m$, du_i, $i = 0, 1, 2, \dots$ and its submodule $\bar{\mathcal{M}}$ of the Cartan forms [11] with a basis consisting of

$$\tau_i = dv^i - D(v^i)dx, \qquad i = 1, \dots, m \tag{4.65}$$

---

[11]The fact that the order of functions G_i is not higher than $n - 1$ follows from the condition $[D, D_t] = 0$.

and of
$$\sigma_i = du_i - u_{i+1}dx.$$

The structure of a $\mathcal{F}[D]$-module on $\mathcal{M}$ and $\bar{\mathcal{M}}$ is defined by the formula

$$D(a\,db) = D(a)\,db + a\,d(D(b)).$$

It is known [140] that $\bar{\mathcal{M}}$ is a free one-dimensional $\mathcal{F}[D]$-module. Its generator ω can be found as follows. Denote by Ω the vector space over $\bar{\mathcal{F}}$ generated by the forms (4.65). The required form

$$\omega = \sum_{i=1}^{m} s_i \tau_i \tag{4.66}$$

is uniquely (up to a multiplier from $\bar{\mathcal{F}}$) defined by the conditions

$$D^i(\omega) \in \Omega, \qquad i = 0, \ldots, m-1.$$

If the nonlocal variables are absent, then $\omega = du - u_1 dx$.

Open problem 4.2.45. *Find the form ω in the case of constraints* (4.78).

For any function $a \in \bar{\mathcal{F}}$ we define the differential operator a_* by the formula

$$d\,a - D(a)\,dx = a_*(\omega).$$

If the constraints (4.61) are absent, then $\omega = du - u_1 dx$ and a_* coincides with the Fréchet derivative (1.4).

For any element $F \in \bar{\mathcal{F}}$, we define an operator $\nabla_F : \bar{\mathcal{F}} \to \bar{\mathcal{F}}$ by the formula

$$\nabla_F(f) = f_*(F). \tag{4.67}$$

Lemma 4.2.46. *The operator ∇_F defines an evolutionary vector field. Any evolutionary vector field D_t is defined by the formula* (4.67), *where* $F = \sum s_i D_t(v^i)$. *Here, the functions s_i are the coefficients in the formula* (4.66).

We will call the system (4.64) *integrable* if it has infinitely many symmetries of the form
$$u_\tau = X(\mathbf{v}, u, \ldots, u_k), \qquad (v^i)_\tau = Y_i(\mathbf{v}, u, \ldots, u_{k-1}). \tag{4.68}$$

In other words, the algebra of all evolutionary vector fields that commute with the vector field (4.63), is infinite-dimensional over $\mathbb{C}$.

Denote by R a differential operator such that the vector field (4.63) and the form (4.66) satisfy the relation

$$D_t(\omega) = R(\omega). \tag{4.69}$$

The following statement was proved in [141] (cf. Theorems 4.2.9 and 4.2.23).

Theorem 4.2.47. *If an evolution equation with the constraints (4.64) has an infinite series of symmetries (4.68) with $k \to \infty$, then*

(i) *There exists a formal Lax pair*

$$L_t = [R, L],$$

where

$$L = a_1 D + a_0 + a_{-1} D^{-1} + \cdots, \qquad a_i \in \bar{\mathcal{F}}. \qquad (4.70)$$

Here R is a differential operator defined by the formula (4.69);

(ii) *The following functions*

$$\rho_{-1} = \frac{1}{a_1}, \qquad \rho_0 = \frac{a_0}{a_1}, \qquad \rho_i = \operatorname{res} L^i, \qquad i \in \mathbb{N} \qquad (4.71)$$

are densities of local conservation laws[12] for equation (4.64).

Theorem 4.2.48. *If equation (4.64) possesses an infinite series of local conservation laws, then*

(i) *There exist a formal Lax pair L, R from Theorem 4.2.47 and a series S of the form*

$$S = s_1 D + s_0 + s_{-1} D^{-1} + s_{-2} D^{-2} + \cdots, \qquad s_i \in \bar{\mathcal{F}}$$

such that

$$S_t + R^+ S + S R = 0, \qquad S^+ = -S, \qquad L^+ = -S^{-1} L S;$$

(ii) *The densities (4.71) with $i = 2k$ are trivial, i.e. they have the form $\rho_{2k} = D(\sigma_k)$ for some functions $\sigma_k \in \bar{\mathcal{F}}$.*

Quasi-generators and differential substitutions

Definition 4.2.49. The function $Z(\mathbf{v}, u, u_1, \dots)$ is called a *quasi-generator* for the system of constraints (4.61) if any of the variables $v^1, v^2, \dots, v^m, u, u_1, \dots$ is a function of the variables

$$Z_0 = Z, \quad Z_1 = D(Z), \quad \dots, \quad Z_i = D^i(Z), \quad \dots. \qquad (4.72)$$

Lemma 4.2.50. *A function Z is a quasi-generator iff the functions $Z_0, Z_1, \dots, Z_{m-1}$*

---

[12]The local conservation law is defined by the formula (4.28), where $\rho, \sigma \in \bar{\mathcal{F}}$.

(1) *Depend only on variables* $\mathbf{v}$;

(2) *Are functionally independent.*

If the conditions of the lemma are satisfied, then

$$Z = \Psi(v^1, \ldots, v^m) \tag{4.73}$$

and

$$v^i = \xi_i(Z_0, Z_1, \ldots, Z_{m-1}) \tag{4.74}$$

for some functions Ψ and ξ_i.

According to Lemma 4.2.50, the function $Z_1 = A(Z) + uB(Z)$ does not depend on u. Therefore, $B(Z) = 0$ and $Z_1 = A(Z)$. Similarly, $B(Z_1) = BA(Z) = 0$. Therefore, $[B, A](Z) = 0$. Using similar arguments, one can show that the quasi-generator $Z(\mathbf{v})$ satisfies the system of linear PDEs

$$B_i(Z) = 0, \qquad i = 0, 1, \ldots, m-2, \qquad \text{where} \qquad B_i = (\text{ad } A)^i(B). \tag{4.75}$$

The existence of a nonconstant solution Z is equivalent to the condition that the dimension over $\mathcal{F}_0$ of the Lie algebra generated by the vector fields B_i, $i = 0, 1, \ldots, m-2$, is equal to $m-1$.[13]

The condition $\text{Ker } D = \mathbb{C}$ implies that $B_{m-1}(Z) \neq 0$. Therefore, the variables $u, u_1, \ldots$ can be expressed in terms of variables (4.72). In particular, there exists a relation of the form

$$u = \psi(Z_0, Z_1, \ldots, Z_m). \tag{4.76}$$

Proposition 4.2.7. *Suppose that the system of constraints* (4.61) *has a quasi-generator* $Z(\mathbf{v})$. *Then, any evolution system* (4.64) *is equivalent to a single evolution equation of the form*

$$Z_t = S(Z, Z_1, \ldots, Z_n) \tag{4.77}$$

for the quasi-generator.

Proof. It follows from Lemma 4.2.50 and formula (4.64) that $D_t(Z_{m-1})$ depends on the variables $\mathbf{v}, u, \ldots, u_{n-1}$ only. Therefore, $D_t(Z)$ is a function of the variables $\mathbf{v}, u, \ldots, u_{n-m}$. The formulas (4.74) and (4.76) entail (4.77). $\qquad\square$

Remark 4.2.51. *If the function H in equation* (4.64) *does not depend on* $\mathbf{v}$, *we have an evolution equation* $u_t = H$, *extended to nonlocal variables* $v^1, \ldots, v^m$. *If the system* (4.61) *has a quasi-generator Z, then* (4.76) *is a differential substitution from equation* (4.77) *to the equation* $u_t = H$, *and* (4.73) *defines the inverse map from the extension* (4.64) *of the equation* $u_t = H$ *into equation* (4.77).

---

[13]If this dimension is less than $m-1$, then $\text{Ker } D \neq \mathbb{C}$.

Extensions of the KdV equation

In the following important example, the nonlocal variables v^i are the logarithmic derivatives of several eigenfunctions of the Schrödinger operator $D^2 + u$. Let

$$D^2(\varphi_i) + u\,\varphi_i = \lambda_i(\varphi_i), \qquad v^i = \frac{D(\varphi_i)}{\varphi_i}, \qquad i = 1, \ldots, p.$$

Then, system (4.61) has the form

$$D(v^i) = -(v^i)^2 - u + \lambda_i, \qquad i = 1, \ldots, p. \tag{4.78}$$

Exercise 4.2.52. Prove that for any p the kernel of the total derivative (4.60) is trivial.

The corresponding evolution equation with constraints is an extension of the KdV equation by means of nonlocal variables v^i. The t-dynamics of these variables is determined by the A-operator (2.6). Equation (4.64) has the form

$$
\begin{aligned}
u_t &= u_3 + 6uu_1, \\
(v^i)_t &= -u_2 + 2v^i u_1 - 2u^2 - 2(v^i)^2 u - 2\lambda_i u - 4\lambda_i(v^i)^2 + 4\lambda_i^2,
\end{aligned}
\tag{4.79}
$$

where $i = 1, \ldots, p$.

Example 4.2.53. In the case $p = 1$, the function $Z = v^1$ is a quasi-generator. It is well known that Z satisfies the mKdV equation

$$Z_t = Z_3 - 6Z^2 Z_1 + 6\lambda_1 Z$$

and the substitution (4.76) is exactly the Miura transformation

$$u = -Z_1 - Z^2 + \lambda_1.$$

Example 4.2.54. In the case $p = 2$, the vector field B in (4.62) has the form $B = -\dfrac{\partial}{\partial v^1} - \dfrac{\partial}{\partial v^2}$. The conditions (4.75) for determining Z consist of one equation $B(Z) = 0$ and therefore the function $Z = v^1 - v^2$ is a quasi-generator. A simple calculation shows that

$$Z_t = Z_3 - \frac{3Z_1 Z_2}{Z} + \frac{3}{2}\frac{Z_1^3}{Z^2} - \frac{3}{2}\left(Z^2 - 2(\lambda_1 + \lambda_2) + \frac{(\lambda_1 - \lambda_2)^2}{Z^2}\right)Z_1$$

and the differential substitution (4.76) from this equation into the KdV equation is given by

$$u = \frac{Z_2}{2Z} - \frac{3}{4}\frac{Z_1^2}{Z^2} + \frac{(\lambda_1 - \lambda_2)Z_1}{Z^2} - \frac{Z^2}{4} + \frac{\lambda_1 + \lambda_2}{2} - \frac{(\lambda_1 - \lambda_2)^2}{4Z^2}. \tag{4.80}$$

Exercise 4.2.55. Find a point transformation that reduces the equation for Z to equation (4.5).

Exercise 4.2.56. Check that in the case $p = 3$ a quasi-generator has the form

$$\bar{Z} = \frac{v^1 - v^2}{v^1 - v^3}.$$

Show that the evolution equation for the quasi-generator is

$$\bar{Z}_t = \bar{Z}_3 - \frac{3}{2}\bar{Z}_1 \frac{\left(\bar{Z}_2 + \frac{1}{2}P'(\bar{Z})\right)^2}{\bar{Z}_1^2 + P(\bar{Z})} + 12(\lambda_1 - \lambda_2)\,\bar{Z}\,\bar{Z}_1 + 6(\lambda_3 + \lambda_2 - \lambda_1)\,\bar{Z}_1,$$

where $P(u) = 4u(u-1)\big((\lambda_1 - \lambda_3)\,u + \lambda_2 - \lambda_1\big)$. This equation is equivalent to equation (4.6) with a generic polynomial Q. The differential substitution (4.76) into the KdV equation can be written as a composition of (4.80) and the substitutions

$$Z = \frac{\bar{Z}_1 + \sqrt{\bar{Z}_1^2 + P(\bar{Z})}}{2(\bar{Z} - 1)}.$$

From the results of [35], it follows that for $p > 3$ quasi-generators do not exist.

Remark 4.2.57. *For exact solutions $u(x, t)$ of the KdV equations (see Section 1.2.3), the eigenfunctions of the Lax operator $D^2 + u$ are known. Therefore, using the formulas found above, we can find explicitly solutions of the mKdV equation as well as equations (4.5) and (4.6).*

The paper [35] developed a general approach to the description of differential substitutions that relate the KdV equation to other integrable evolution equations. This problem was reduced to a description of Lie subgroups of codimension ≤ 3 in the Lie group SL_2^p. A similar approach was used in [121] for fifth order evolution equations.

4.3 Weakly nonlocal recursion and Hamiltonian operators

Recursion and Hamiltonian operators relate symmetries and cosymmetries (see Definition 4.2.19) of evolution equations (1.26).

Recall that the recursion operator is an operator $\mathcal{R}$ which satisfies relation (4.18) and, therefore, maps the symmetries of equation (1.26) into symmetries. The usual way to obtain all symmetries of equation (1.26) is to apply a recursion operator to the simplest symmetry u_x.

Lemma 4.3.1. *Let $\mathcal{R}$ be a recursion operator. Then, the operator $\mathcal{R}^+$ maps cosymmetries to cosymmetries.*

Proof. From (4.19) it follows that $[D_t + F_*^+, \mathcal{R}^+] = 0$. Using Definition 4.2.19, we arrive at the statement of the lemma. □

The analogue of the operator identity (4.19) for Hamiltonian operators is the relation

$$(D_t - F_*)\mathcal{H} = \mathcal{H}(D_t + F_*^+), \tag{4.81}$$

which means that $\mathcal{H}$ maps cosymmetries (and, in particular, the variational derivatives of the conserved densities) to symmetries. It is easy to verify that if an operator S satisfies relation (4.32), then the operator $\mathcal{H} = S^{-1}$ satisfies (4.81).

According to Theorems 4.2.9 and 4.2.23, for any S-integrable equation there exists pseudo-differential series $\mathcal{R} = \Lambda$ and $\mathcal{H} = S^{-1}$ that satisfy (4.18) and (4.81) respectively. The orders of these series can be chosen arbitrarily (see Proposition 4.2.2 and Lemma 4.2.22).

Question 4.3.2. *How to "collapse" these series (i.e. how to rewrite them as operators that can be applied to some functions from $\mathcal{F}$)?*

The main difficulty here is that, for almost all integrable models, the recursion operator is nonlocal, that is, it contains D^{-1} (see, for example, (4.38)). Therefore, if we want to obtain a function from $\mathcal{F}$ as a result, then we can only apply it to very special functions from $\mathcal{F}$.

The following section provides a general weakly nonlocal ansatz for recursion operators — this helps to answer Question 4.3.2. Another approach was developed in [144].

4.3.1 Weakly nonlocal recursion operators

Observation 4.3.3. *The known recursion operators for scalar evolution equations (1.26) have the following nonlocal structure:*

$$\mathcal{R} = R + \sum_{i=1}^{k} G_i\, D^{-1} \circ g_i, \qquad g_i, G_i \in \mathcal{F}, \tag{4.82}$$

where R is a differential operator. Without loss of generality, we can assume that the functions $G_1, \ldots, G_k$ (as well as the functions $g_1, \ldots, g_k$) are linearly independent over $\mathbb{C}$.

Definition 4.3.4. Operators of the form (4.82) are called *weakly nonlocal* or *quasi-local*.

Operators of the form (4.82) belong to the skew field of fractions of differential operators [43]. It is natural to draw the following analogy. The partial fraction expansion of a rational function $\dfrac{P(x)}{Q(z)}$ is possible if the roots of the polynomial Q are known. The formula (4.82) is an analogue of the partial fraction expansion for a ratio of differential operators.

Exercise 4.3.5. Show that a ratio AB^{-1} of two differential operators can be represented in the form (4.82) if $\operatorname{Ker} B \subset \mathcal{F}$.

Lemma 4.3.6. *Let* $\mathcal{R}$ *be defined by the formula* (4.82) *and*

$$\mathcal{R} = \bar{R} + \sum_{i=1}^{\bar{k}} \bar{G}_i \, D^{-1} \circ \bar{g}_i. \tag{4.83}$$

Then, the product $\mathcal{R} \circ \bar{\mathcal{R}}$ *is weakly nonlocal if* $g_i \bar{G}_j = D(a_{ij})$, $a_{ij} \in \mathcal{F}$ *for all* i, j.

Proof. It suffices to verify that the expression

$$G_i D^{-1} \circ g_i \, \bar{G}_j D^{-1} \circ \bar{g}_j$$

is weakly nonlocal. This follows from the chain of identities

$$G_i D^{-1} \circ g_i \, \bar{G}_j D^{-1} \circ \bar{g}_j = G_i D^{-1} (D \circ a_{ij} - a_{ij} D) D^{-1} \circ \bar{g}_j =$$

$$G_i a_{ij} D^{-1} \circ \bar{g}_j - G_i D^{-1} \circ a_{ij} \bar{g}_j.$$

$\square$

Definition 4.3.7. An operator $\mathcal{R}$ of the form (4.82) is called a *weakly nonlocal recursion operator* for equation (1.26) if

(1) Operator $\mathcal{R}$, considered as a pseudo-differential series, satisfies relation (4.18);

(2) Functions G_i are symmetry generators for equation (1.26);

(3) Functions g_i are the variational derivatives of conserved densities for equation (1.26).

Probably, the weakly nonlocal ansatz was first used to find the recursion operator for the Krichever–Novikov equation in [41]. It is a quite effective way for finding the recursion operator: we calculate the simplest symmetries and conserved densities, from the ansatz (4.82) and substitute it into (4.18) to find the coefficients of R. The order of R and the number k have to be determined experimentally. A hint there could give the orders of symmetries and a homogeneity of the equations of the hierarchy.

Another, more systematic, method of finding the recursion operator based on the Lax representation was proposed in [42] (see Sections "Recursion operator for the KvD and the NLS equations").

Example 4.3.8. It is easy to see that the recursion operator (4.38) for the KdV equation is weakly nonlocal with $k = 1$, $G_1 = \dfrac{u_x}{2}$ and $g_1 = \dfrac{\delta u}{\delta u} = 1$.

Remark 4.3.9. *It may be reasonable to add* [142] *the hereditary property* [143] *of operator $\mathcal{R}$ to the requirements 1–3 in Definition 4.3.7.*

Remark 4.3.10. *Substituting* (4.82) *into* (4.18) *and equating the nonlocal terms, we get*

$$\sum_{i=1}^{k}(D_t - F_*)(G_i)\, D^{-1} \circ g_i + \sum_{i=1}^{k} G_i\, D^{-1} \circ (D_t + F_*^{+})(g_i) = 0.$$

This "almost" means that the functions G_i are symmetries, and the functions g_i are cosymmetries.

Question 4.3.11. *Why do we get local symmetries when applying a weakly nonlocal recursion operator $\mathcal{R}$ to local symmetries?*

Question 4.3.12. *Why the product of weakly nonlocal recursion operators is a weakly nonlocal?*

Suppose that equation (1.26) belongs to the infinite-dimensional commutative hierarchy of S-integrable evolution equations (see Remark 1.4.8 and Example 1.5.1). By definition this means that

(i) Every equation from the hierarchy is a symmetry for all others;

(ii) Every conserved density for any equation is a conserved density for all equations from the hierarchy.

In this case, the expression $\mathcal{R}(g)$, where g is any symmetry of equation (1.26), belongs to $\mathcal{F}$. This fact follows from the condition (ii) and the following statement.

Lemma 4.3.13. [12] *The product of the right-hand side F of equation (1.26) and the variational derivative $\dfrac{\delta\rho}{\delta u}$ of any conserved density for this equation belongs to* Im D.

Proof. It is sufficient to show that

$$\left(F\frac{\delta\rho}{\delta u}\right)_*^{+}(1) = 0.$$

Using the identities of Lemma 4.2.4, we find that the left-hand side is equal to

$$F_*^{+}\left(\frac{\delta\rho}{\delta u}\right) + \left(\frac{\delta\rho}{\delta u}\right)_*(F).$$

This expression is zero since $\dfrac{\delta\rho}{\delta u}$ is a cosymmetry. □

Remark 4.3.14. *A different choice of the integration constants when applying the operator $\mathcal{R}$ results in the addition of a linear combination of symmetries $G_1, \ldots, G_k$.*

Let $\mathcal{R}$ and $\bar{\mathcal{R}}$, given by formulas (4.82) and (4.83), be weakly nonlocal recursion operators. Then, the nonlocal terms in the product $\mathcal{R} \circ \bar{\mathcal{R}}$ are given by

$$\sum_{i=1}^{\bar{k}} \mathcal{R}(\bar{G}_i)\, D^{-1} \circ \bar{g}_i + \sum_{i=1}^{k} G_i\, D^{-1} \circ \bar{\mathcal{R}}^+(g_i).$$

To prove that the other terms are local, we need to use Lemma 4.3.6 and the fact[14] that for any differential operator S and functions $p, q \in \mathcal{F}$ the expression $pS(q) - qS^+(p)$ belongs to Im D.

We see that the operator $\mathcal{R} \circ \bar{\mathcal{R}}$ satisfies the requirements (1) (see Proposition 4.2.2) and (2) of Definition 4.3.7. A problem arises with the requirement (3). It follows from Lemma 4.3.1 that the expression $\bar{\mathcal{R}}^+(g_i)$ is a cosymmetry (but not necessarily the variational derivative of a conserved density).

However, for a particular equation, one can try to prove that any cosymmetry is the variational derivative of some conserved density. Perhaps this is the case for all S-integrable equations. We briefly outline the proof of this fact for the KdV equation.

1. Use the fact that for the KdV equation, the operator $\dfrac{\partial}{\partial u}$ converts any cosymmetry to a cosymmetry while reducing its order by two, to prove that every cosymmetry has an even order;

2. Derive from (4.30) that any cosymmetry S of order $2k$ has the form $S = c\, u_{2k} + O(2k - 1)$;

3. Use the fact that for any k the KdV equation has a conserved density of the form $\rho_k = u_k^2 + O(k - 1)$ to reduce the order of the cosymmetry: it is clear that $S_1 = S + (-1)^{k+1} \dfrac{c}{2} \dfrac{\delta \rho_k}{\delta u}$ has a smaller order than S;

4. Use the induction on k.

Open problem 4.3.15. *Prove a similar statement for the Krichever–Novikov equation* (4.7).

Remark 4.3.16. *The function u_1 is a cosymmetry of the linear equation $u_t = u_{xxx}$, which is not the variational derivative of a conserved density.*

---

[14] It can be chosen as a definition of the adjoint operator.

So, the set of all weakly nonlocal recursion operators for the KdV equation forms a commutative (see Proposition 4.2.2) associative algebra A_{rec} over $\mathbb{C}$. It can be proved that this algebra is generated by operator (4.38). In other words, A_{rec} is isomorphic to the algebra of polynomials in one variable.

It turns out that [145] this is not true for the Krichever–Novikov equation and the two-component Landau–Lifschitz system, which are fundamental (see Conjectures 4.1.8 and 4.1.12) integrable models within their classes of integrable equations. In these cases, A_{rec} is isomorphic to the coordinate ring of an elliptic curve.

4.3.2 Weakly nonlocal Hamiltonian operators

The weakly nonlocal ansatz for a Hamiltonian operator, defining a Hamiltonian structure (see Section 1.3) for an equation of the form (1.26), can be written as

$$\mathcal{H} = H + \sum_{i=1}^{m} G_i D^{-1} \bar{G}_i + \bar{G}_i D^{-1} G_i, \qquad (4.84)$$

where H is a skew-symmetric differential operator.

Remark 4.3.17. *Clearly,* $\mathcal{H}^+ = -\mathcal{H}$.

Definition 4.3.18. An operator $\mathcal{H}$ of the form (4.84) is called a *weakly nonlocal Hamiltonian operator* for (1.26) if

(1) Operator $\mathcal{H}$, regarded as a pseudo-differential series, satisfies relation (4.81);

(2) The functions G_i and $\bar{G}_i$ are symmetry generators [146, 147] for equation (1.26).

The fact that by applying the operator $\mathcal{H}$ to the variational derivatives of conserved densities, we obtain local symmetries, follows from Lemma 4.3.13.

Remark 4.3.19. *We did not include the Jacobi identity*

$$\{\{f, g\}, h\} + \{\{g, h\}, f\} + \{\{h, f\}, g\} \in \operatorname{Im} D \qquad (4.85)$$

for the bracket

$$\{f, g\} = \frac{\delta f}{\delta u} \mathcal{H}\left(\frac{\delta g}{\delta u}\right) \qquad (4.86)$$

as a condition in Definition 4.3.18.[15] *For nonlocal operators, checking the Jacobi identity is rather difficult and requires a special technique (see, for example, [12,*

---

[15]In this sense, the name "Hamiltonian operator" is somewhat misleading, and it might be worth changing the terminology.

section 7.1] *or* [28]), *which we do not discuss here. As it was shown in* [147], *the Item* (2) *of the definition 4.3.18 follows from the Jacobi identity.*

As noted in Section 3.0.2, the KdV equation has two local Hamiltonian operators (3.4). The first example

$$\mathcal{H}_0 = u_x D^{-1} u_x \qquad (4.87)$$

of a weakly nonlocal Hamiltonian operator for the Krichever–Novikov equation was found in [41] (see also [145]).

4.3.3 Recursion operators for Krichever–Novikov equation

The Krichever–Novikov equation has the form

$$u_{t_1} = u_{xxx} - \frac{3}{2} \frac{u_{xx}^2}{u_x} + \frac{P(u)}{u_x}, \qquad P^{(V)} = 0. \qquad (4.88)$$

Denote by G_1 the right-hand side of (4.88). The fifth order symmetry for equation (4.88) is given by

$$G_2 = u_5 - 5\frac{u_4 u_2}{u_1} - \frac{5}{2}\frac{u_3^2}{u_1} + \frac{25}{2}\frac{u_3 u_2^2}{u_1^2} - \frac{45}{8}\frac{u_2^4}{u_1^3} -$$

$$\frac{5}{3}P\frac{u_3}{u_1^2} + \frac{25}{6}P\frac{u_2^2}{u_1^3} - \frac{5}{3}P'\frac{u_2}{u_1} - \frac{5}{18}\frac{P^2}{u_1^3} + \frac{5}{9}u_1 P''.$$

The three simplest conserved densities of equation (4.88) are

$$\rho_1 = -\frac{1}{2}\frac{u_2^2}{u_1^2} - \frac{1}{3}\frac{P}{u_1^2}, \qquad \rho_2 = \frac{1}{2}\frac{u_3^2}{u_1^2} - \frac{3}{8}\frac{u_2^4}{u_1^4} + \frac{5}{6}P\frac{u_2^2}{u_1^4} + \frac{1}{18}\frac{P^2}{u_1^4} - \frac{5}{9}P'',$$

$$\rho_3 = \frac{u_4^2}{u_1^2} + 3\frac{u_3^3}{u_1^3} - \frac{19}{2}\frac{u_3^2 u_2^2}{u_1^4} + \frac{7}{3}P\frac{u_3^2}{u_1^4} + \frac{35}{9}P'\frac{u_2^3}{u_1^4} + \frac{45}{8}\frac{u_2^6}{u_1^6} - \frac{259}{36}\frac{u_2^4 P}{u_1^6} + \frac{35}{18}P^2\frac{u_2^2}{u_1^6}$$

$$-\frac{14}{9}P''\frac{u_2^2}{u_1^2} + \frac{1}{27}\frac{P^3}{u_1^6} - \frac{14}{27}\frac{P''P}{u_1^2} - \frac{7}{27}\frac{P'^2}{u_1^2} - \frac{14}{9}P''''u_1^2.$$

Since equation (4.88) has symmetries of all odd orders, one could assume that, as in the case of the KdV equation, the weakly nonlocal recursion operator is of order 2. However, it is easy to verify that such an operator exists only if the polynomial P has at least one multiple root.

The following weakly nonlocal fourth order recursion operator was found in [41]:

$$\mathcal{R}_1 = D^4 + a_1 D^3 + a_2 D^2 + a_3 D + a_4 + u_x D^{-1} \circ \frac{\delta \rho_2}{\delta u} + G_1 D^{-1} \circ \frac{\delta \rho_1}{\delta u},$$

where the coefficients a_i have the form

$$a_1 = -4\frac{u_2}{u_1}, \qquad a_2 = 6\frac{u_2^2}{u_1^2} - 2\frac{u_3}{u_1} - \frac{4}{3}\frac{P}{u_1^2},$$

$$a_3 = -2\frac{u_4}{u_1} + 8\frac{u_3 u_2}{u_1^2} - 6\frac{u_2^3}{u_1^3} + 4P\frac{u_2}{u_1^3} - \frac{2}{3}\frac{P'}{u_1},$$

$$a_4 = \frac{u_5}{u_1} - 2\frac{u_3^2}{u_1^2} + 8\frac{u_3 u_2^2}{u_1^3} - 4\frac{u_4 u_2}{u_1^2} - 3\frac{u_2^4}{u_1^4} + \frac{4}{9}\frac{P^2}{u_1^4} + \frac{4}{3}P\frac{u_2^2}{u_1^4} + \frac{10}{9}P'' - \frac{8}{3}P'\frac{u_2}{u_1^2}.$$

It turns out [145] that equation (4.88) also has a weakly nonlocal operator of order 6.

Proposition 4.3.1. *The formula*

$$\mathcal{R}_2 = D^6 + b_1 D^5 + b_2 D^4 + b_3 D^3 + b_4 D^2 + b_5 D + b_6$$

$$- \frac{1}{2} u_x D^{-1} \circ \frac{\delta \rho_3}{\delta u} + G_1 D^{-1} \circ \frac{\delta \rho_2}{\delta u} + G_2 D^{-1} \circ \frac{\delta \rho_1}{\delta u}, \qquad (4.89)$$

where

$$b_1 = -6\frac{u_2}{u_1}, \qquad b_2 = -9\frac{u_3}{u_1} - 2\frac{P}{u_1^2} + 21\frac{u_2^2}{u_1^2},$$

$$b_3 = -11\frac{u_4}{u_1} + 60\frac{u_3 u_2}{u_1^2} + 14P\frac{u_2}{u_1^3} - 57\frac{u_2^3}{u_1^3} - 3\frac{P'}{u_1},$$

$$b_4 = -4\frac{u_5}{u_1} + 38\frac{u_4 u_2}{u_1^2} + 22\frac{u_3^2}{u_1^2} + 99\frac{u_2^4}{u_1^4} - 155\frac{u_3 u_2^2}{u_1^3} + \frac{34}{3}P\frac{u_3}{u_1^3} - 44P\frac{u_2^2}{u_1^4}$$

$$+ \frac{4}{3}\frac{P^2}{u_1^4} + 12P'\frac{u_2}{u_1^2} - P'',$$

$$b_5 = -2\frac{u_6}{u_1} + 29\frac{u_4 u_3}{u_1^2} + 80P\frac{u_2^3}{u_1^5} + \frac{23}{3}P'\frac{u_3}{u_1^2} - 104\frac{u_2 u_3^2}{u_1^3} - 70\frac{u_4 u_2^2}{u_1^3} + 241\frac{u_2^3 u_3}{u_1^4}$$

$$+ 14\frac{u_5 u_2}{u_1^2} + \frac{20}{3}P\frac{u_4}{u_1^3} - \frac{170}{3}P\frac{u_2 u_3}{u_1^4} + \frac{4}{3}\frac{P'P}{u_1^3} - 22P'\frac{u_2^2}{u_1^3} + 2P''\frac{u_2}{u_1}$$

$$- \frac{16}{3}P^2\frac{u_2}{u_1^5} - 108\frac{u_2^5}{u_1^5},$$

$$b_6 = \frac{u_7}{u_1} - 6\frac{u_2 u_6}{u_1^2} + \frac{8}{9}P^2\frac{u_2^2}{u_1^6} - 195\frac{u_3^2 u_2^2}{u_1^4} + 6P\frac{u_3^2}{u_1^4} + \frac{142}{3}P\frac{u_2^4}{u_1^6} + \frac{28}{9}P'P\frac{u_2}{u_1^4}$$

$$+101\frac{u_4 u_3 u_2}{u_1^3} + \frac{34}{3}P\frac{u_4 u_2}{u_1^4} - 72\frac{u_2^6}{u_1^6} - \frac{28}{9}P'''u_2 + \frac{38}{3}P''\frac{u_2^2}{u_1^2} - \frac{19}{3}P'\frac{u_4}{u_1^2}$$

$$-\frac{122}{3}P'\frac{u_2^3}{u_1^4} - 10\frac{u_4^2}{u_1^2} + 22\frac{u_3^3}{u_1^3} - \frac{178}{3}P\frac{u_3 u_2^2}{u_1^5} + \frac{14}{9}P''''u_1^2 + \frac{113}{3}P'\frac{u_3 u_2}{u_1^3}$$

$$-\frac{2}{3}P\frac{u_5}{u_1^3} - \frac{17}{3}P''\frac{u_3}{u_1} - \frac{4}{3}P^2\frac{u_3}{u_1^5} - 89\frac{u_4 u_2^3}{u_1^4} + 236\frac{u_3 u_2^4}{u_1^5} - 13\frac{u_5 u_3}{u_1^2}$$

$$+25\frac{u_5 u_2^2}{u_1^3} - \frac{7}{9}\frac{P'^2}{u_1^2} - \frac{8}{27}\frac{P^3}{u_1^6} - \frac{4}{9}\frac{P''P}{u_1^2},$$

defines a weakly nonlocal recursion operator for equation (4.88). *Operators* $\mathcal{R}_1$ *and* $\mathcal{R}_2$ *commute and are connected by the equation of the elliptic curve*

$$\mathcal{R}_2^2 = \mathcal{R}_1^3 - \phi\mathcal{R}_1 - \theta, \tag{4.90}$$

where

$$\phi = \frac{16}{27}\Big((P'')^2 - 2P'''P' + 2P''''P\Big),$$

$$\theta = \frac{128}{243}\Big(-\frac{1}{3}(P'')^3 - \frac{3}{2}(P')^2 P'''' + P'P''P''' + 2P''''P''P - P(P''')^2\Big).$$

Remark 4.3.20. *It is easy to verify that the expressions* ϕ *and* θ *are constants for any polynomial* $P(u)$ *such that* $\deg P \leq 4$. *Under fractional linear Möbius transformations*

$$u \to \frac{\alpha u + \beta}{\gamma u + \delta}$$

in equation (4.88), *the polynomial* $P(u)$ *is transformed according to the same rule as the polynomial in the differential* $\omega = \dfrac{du}{\sqrt{P(u)}}$. *The expressions* ϕ *and* θ *are invariants with respect to the action of the Möbius group.*

Remark 4.3.21. *The ratio* $\mathcal{R}_3 = \mathcal{R}_2\mathcal{R}_1^{-1}$ *satisfies equation* (4.21) (*see Proposition 4.2.2*) *and belongs to the skew field of the ring of differential operators* [43]. *However, this operator is not weakly nonlocal and it is not clear how to apply it even to the simplest symmetry* u_x.

4.3.4 Weakly nonlocal Hamiltonian operators for Krichever–Novikov equation

The relationship between the Hamiltonian and recursion operators is described by the following statement.

Lemma 4.3.22. *If $\mathcal{H}_1$ and $\mathcal{H}_2$ satisfy the relation (4.81), then $\mathcal{R} = \mathcal{H}_2\mathcal{H}_1^{-1}$ satisfies (4.21).*

Exercise 4.3.23. Verify that the recursion operator (2.11) for the KdV equation is proportional to the ratio of the Hamiltonian operators (3.4).

It can be verified that the recursion operators $\mathcal{R}_i$ for equation (4.88), presented in Section 4.3.3, can be written as

$$\mathcal{R}_1 = \mathcal{H}_1\mathcal{H}_0^{-1}, \qquad \mathcal{R}_2 = \mathcal{H}_2\mathcal{H}_0^{-1},$$

where operator $\mathcal{H}_0$ is defined by the formula (4.87), and the weakly nonlocal Hamiltonian operators $\mathcal{H}_1$ and $\mathcal{H}_2$ have the form

$$\mathcal{H}_1 = \frac{1}{2}(u_x^2 D^3 + D^3 \circ u_x^2) + (2u_{xxx}u_x - \frac{9}{2}u_{xx}^2 - \frac{2}{3}P)D + D \circ (2u_{xxx}u_x - \frac{9}{2}u_{xx}^2 - \frac{2}{3}P)$$
$$+ G_1 D^{-1} \circ G_1 + u_x D^{-1} \circ G_2 + G_2 D^{-1} \circ u_x,$$

$$\mathcal{H}_2 = \frac{1}{2}(u_x^2 D^5 + D^5 \circ u_x^2) + (3u_{xxx}u_x - \frac{19}{2}u_{xx}^2 - P)D^3 + D^3 \circ (3u_{xxx}u_x - \frac{19}{2}u_{xx}^2 - P)$$
$$+ hD + D \circ h + G_1 D^{-1} \circ G_2 + G_2 D^{-1} \circ G_1 + u_x D^{-1} \circ G_3 + G_3 D^{-1} \circ u_x.$$

In these formulas,

$$h = u_{xxxxx}u_x - 9u_{xxxx}u_{xx} + \frac{19}{2}u_{xxx}^2 - \frac{2}{3}\frac{u_{xxx}}{u_x}(5P - 39u_{xx}^2) +$$
$$\frac{u_{xx}^2}{u_x^2}(5P - 9u_{xx}^2) + \frac{2}{3}\frac{P^2}{u_x^2} + u_x^2 P'',$$

and $G_3 = \mathcal{R}_1(G_1) = \mathcal{R}_2(u_x)$ is the seventh order symmetry generator for equation (4.88):

$$G_3 = u_7 - 7\frac{u_2 u_6}{u_1} - \frac{7}{6}\frac{u_5}{u_1^2}(2P + 12u_3 u_1 - 27u_2^2) - \frac{21}{2}\frac{u_4^2}{u_1} + \frac{21}{2}\frac{u_4}{u_1^3}u_2(2P - 11u_2^2)$$
$$- \frac{7}{3}\frac{u_4}{u_1^2}(2P'u_1 - 51u_2 u_3) + \frac{49}{2}\frac{u_3^3}{u_1^2} + \frac{7}{12}\frac{u_3^2}{u_1^3}(22P - 417u_2^2) + \frac{2499}{8}\frac{u_2^4}{u_1^4}u_3$$
$$+ \frac{91}{3}P'\frac{u_2}{u_1^2}u_3 - \frac{595}{6}P\frac{u_2^2}{u_1^4}u_3 - \frac{35}{18}\frac{u_3}{u_1^4}(2P''u_1^4 - P^2) - \frac{1575}{16}\frac{u_2^6}{u_1^5} + \frac{1813}{24}\frac{u_2^4}{u_1^5}P$$
$$- \frac{203}{6}\frac{u_2^3}{u_1^3}P' + \frac{49}{36}\frac{u_2^2}{u_1^5}(6P'u_1^4 - 5P^2) - \frac{7}{9}\frac{u_2}{u_1^3}(2P'''u_1^4 - 5PP') + \frac{7}{54}\frac{P^3}{u_1^5}$$
$$- \frac{7}{9}P''\frac{P}{u_1} + \frac{7}{9}P''''u_1^3 - \frac{7}{18}\frac{P'^2}{u_1}.$$

Operators $\mathcal{H}_i$, $i = 1, 2$, were found in [145]. It was verified that they satisfy relation (4.81). The results of the recent paper [142] imply that the corresponding Poisson brackets satisfy the Jacobi identity (4.85) and that all three brackets defined by operators $\mathcal{H}_i$, $i = 0, 1, 2$, are pairwise compatible.

4.4   Integrable nonevolution equations

In this section, we generalize [148, 149] the basic concepts of the symmetry approach to the case of nonevolution equations of the form

$$u_{tt} = F(u, u_1, \ldots, u_n; u_t, u_{1t}, \ldots, u_{mt}), \qquad 2m < n, \tag{4.91}$$

where $u_{mt} = \partial^{m+1}u/\partial x^m \partial t$. Most of the contents are taken from the survey [149].

In this section, $\mathcal{F}$ denotes the field of functions of variables

$$u, \ u_x, \ u_{xx}, \ \ldots, u_i, \ldots, \qquad u_t, \ u_{1t} = u_{xt}, \ u_{2t} = u_{xxt}, \ \ldots, \ u_{it}, \ldots. \tag{4.92}$$

We call (4.91) an equation *of order* (n, m). The total x and t-derivatives have the form

$$D = \sum_{i=0}^{\infty} u_{i+1} \frac{\partial}{\partial u_i} + \sum_{j=0}^{\infty} u_{j+1t} \frac{\partial}{\partial u_{jt}}, \qquad D_t = \sum_{i=0}^{\infty} u_{it} \frac{\partial}{\partial u_i} + \sum_{j=0}^{\infty} D^j(F) \frac{\partial}{\partial u_{jt}}.$$

It is easy to see that $\mathrm{Ker}\, D = \mathbb{C}$.

The evolutionary vector fields commuting with D are written as

$$D_H = \sum_{i=0}^{\infty} D^i(H) \frac{\partial}{\partial u_i} + \sum_{j=0}^{\infty} D^j(D_t(H)) \frac{\partial}{\partial u_{jt}}.$$

For any function $H(u, u_1, \ldots, u_t, u_{1t}, \ldots) \in \mathcal{F}$, the differential operator H_* is defined by the formula

$$H_* = \sum_{i=0}^{\infty} \frac{\partial H}{\partial u_i} D^i + \sum_{j=0}^{\infty} \frac{\partial H}{\partial u_{jt}} D^j D_t.$$

The linearization operator (see Section 2.0.1) for equation (4.91) has the form

$$\mathcal{L} = D_t^2 - F_* = D_t^2 - U - V D_t,$$

where the differential operators

$$U = \mathfrak{u}_n D^n + \mathfrak{u}_{n-1} D^{n-1} + \cdots + \mathfrak{u}_0, \qquad \text{with} \quad \mathfrak{u}_i = \frac{\partial F}{\partial u_i}, \quad \mathfrak{v}_j = \frac{\partial F}{\partial u_{jt}} \qquad (4.93)$$
$$V = \mathfrak{v}_m D^m + \mathfrak{v}_{m-1} D^{m-1} + \cdots + \mathfrak{v}_0$$

are determined by the right-hand side of equation (4.91).

An equation

$$u_\tau = G(u, u_1, u_2, \ldots, u_r, \; u_t, u_{t1}, u_{t2}, \ldots, u_{ts}), \qquad (4.94)$$

compatible with (4.91), is called an infinitesimal symmetry of equation (4.91). Compatibility means the function G satisfies (cf. (2.4)) equation $\mathcal{L}(G) = 0$.

Equations of the form (4.91) were not considered in the surveys [3, 5], where only evolution equations were studied.

Remark 4.4.1. *Of course, equation* (4.91) *can be written as a system of two evolution equations*

$$u_t = v, \qquad v_t = F(u, u_1, u_2, \ldots, u_n, v, v_1, \ldots, v_m).$$

However, the leading coefficient of the Fréchet derivative of the right-hand side of such a system has a Jordan block structure, while in [5, Section 3.2.1] *this coefficient is assumed to be diagonal.*

Under the polynomiality assumption, a certain formalism was developed and applied to the classification of integrable equations (4.91) in [150, 151].

4.4.1 Formal symmetry and symplectic operator

Definition 4.4.2. A pseudo-differential series $\mathcal{R} = X + Y \, D_t$ with components

$$X = \sum_{-\infty}^{p} x_i D^i, \qquad Y = \sum_{-\infty}^{q} y_i D^i$$

is called *a formal symmetry of order* (p, q) for equation (4.91) if it satisfies a relation of the form

$$\mathcal{L}(X + Y \, D_t) = (\bar{X} + \bar{Y} \, D_t)\mathcal{L} \qquad (4.95)$$

for some series $\bar{X}, \bar{Y}$.

From (4.95) it follows that $\bar{Y} = Y$ and $\bar{X} = X + 2Y_t + [Y, V]$. If X and Y are differential operators (or ratios of differential operators), then operator $\mathcal{R}$ maps symmetries of equation (4.91) to symmetries, i.e. it is a recursion operator.

Lemma 4.4.3. *The relation* (4.95) *is equivalent to the identities*

$$X_{tt} - V X_t + [X, U] + (2Y_t + [Y, V]) \, U + Y U_t = 0, \qquad (4.96)$$

$$Y_{tt} + 2X_t + [Y, U] + [X, V] + ([Y, V] + 2Y_t) \, V + Y V_t - V Y_t = 0. \qquad (4.97)$$

Let $\mathcal{R}_1 = X_1 + Y_1 D_t$ and $\mathcal{R}_2 = X_2 + Y_2 D_t$ be two formal symmetries. Then, their product $\mathcal{R}_3 = \mathcal{R}_1 \mathcal{R}_2$, in which D_t^2 is replaced by $(U + V D_t)$, is a formal symmetry with components

$$X_3 = X_1 X_2 + Y_1 Y_2 U + Y_1 (X_2)_t, \qquad Y_3 = X_1 Y_2 + Y_1 X_2 + Y_1 Y_2 V + Y_1 (Y_2)_t.$$

Thus, the set of all formal symmetries forms an associative algebra which we denote by A_{fsym}. In the evolution case, this algebra is generated by a single generator of the form (4.20). For equations (4.91), the structure of the algebra A_{fsym} essentially depends on the numbers n and m.

For any equation (4.91) with $2m < n$, the terms of the highest degree in D in relations (4.96), (4.97) are comprised in the terms $[X, U]$ and $[Y, U]$. Therefore, the unknown coefficients at each step are found from relations of the form

$$D\left(\mathbf{u}_n^{-i/n} x_i\right) = A_i, \qquad D\left(\mathbf{u}_n^{-i/n} y_i\right) = B_i, \tag{4.98}$$

where the functions $A_i, B_i \in \mathcal{F}$ are already known. Since equations (4.98) are not solvable for arbitrary functions A_i and B_i,[16] there is an infinite sequence of obstacles to integrability (cf. Remark 4.2.8). Since $\operatorname{Ker} D = \mathbb{C}$, integration constants appear at each step.

Definition 4.4.4. We call an equation (4.91) formally integrable if it has a formal symmetry $\mathcal{R}$ of some order (p, q) with a complete set of arbitrary integration constants.

Remark 4.4.5. *Since operator $\mathcal{R}$ depends on the integration constants linearly, in fact, we have an infinite-dimensional vector space of formal symmetries.*

Proposition 4.2.2 shows that in the evolution case the obstacles to formal integrability do not depend on the choice of the integration constants, nor on the order of the formal symmetry.

Conjecture 4.4.6. *The same is true for equations (4.91) with $2m < n$.*

An idea of a possible proof of this conjecture is contained in [148, Theorem 2].

Definition 4.4.7. A pseudo-differential series $\mathcal{S} = P + Q D_t$ is called *a formal symplectic operator* for equation (4.91), if it satisfies the relation[17]

$$\mathcal{L}^+(P + Q D_t) + (\bar{P} + \bar{Q} D_t)\mathcal{L} = 0 \tag{4.99}$$

for some pseudo-differential series $\bar{P}, \bar{Q}$.

---

[16]These functions must be total x-derivatives.
[17]In the evolution case, we have $\mathcal{L} = D_t - F_*$ and (4.99) coincides with (4.32).

It is easy to verify that

$$\bar{P} = -P - 2Q_t - V^+Q - QV, \qquad \bar{Q} = -Q.$$

The operator relations for the components of the operator $\mathcal{S}$ are as follows:

$$P_{tt} + V^*P_t + 2Q_tU + QU_t = U^*PPU - (QV + V^*Q)\,U - V_t^*P, \qquad (4.100)$$

$$Q_{tt} + 2P_t + 2Q_tV + V^*Q_t = U^*QQU - (QV + V^*Q)\,V - \\ (PV + V^*P) - (V_t^*Q + QV_t). \qquad (4.101)$$

The analysis of orders in D of the terms in these relations leads (cf. Section 4.2.4) to the following statement:

Lemma 4.4.8. *For equations* (4.91), *where* $2m < n$ *and* n *is odd, there are no nontrivial formal symplectic operators.*

4.4.2 Examples

Example 4.4.9. The simplest integrable equation (4.91) of order $(3,1)$

$$u_{tt} = u_{xxx} + 3u_x u_{xt} + \left(u_t - 3u_x^2\right)u_{xx} \qquad (4.102)$$

originated in [152]. The formal symmetry of order $(n+1, n)$[18] for this equation has the following structure of higher order terms:

$$(l_{n+1} + m_n n u_x)\,D^{n+1} + \left[l_n + m_{n-1}(n-1)u_x + l_{n+1}(n+1)\left(\tfrac{1}{2}(n-4)u_x^2 + u_t\right)\right. \\ + m_n \left(\tfrac{1}{6}(n+1)\left(n^2 - 7n - 6\right)u_x^3 + n(n+1)u_t u_x + \tfrac{1}{2}\left(n^2 - n + 2\right)u_{xx}\right)\right]\,D^n + \cdots \\ + \left\{m_n D^n + \left[m_{n-1} + l_{n+1}(n+1)u_x + m_n(n+1)\left(\tfrac{n}{2}u_x^2 + u_t\right)\right]\,D^{n-1} + \cdots\right\}D_t,$$

where m_i, l_i are constants of integration. Denote by $\mathcal{Y}_i$ the formal symmetry corresponding to the choice $m_i = 1$, $m_j = 0$ for $j \neq i$, and $l_j = 0$ of the constants, and by $\mathcal{X}_i$ the operator with $l_i = 1$, $l_j = 0$ for $j \neq i$, and $m_j = 0$. One can verify that

$$\mathcal{Y}_{-1}^{-1} = \mathcal{Y}_{-2}, \quad \mathcal{Y}_{-1}^2 = \mathcal{X}_1, \quad \mathcal{X}_1^{-1} = \mathcal{X}_{-1}, \quad \mathcal{X}_1^i = \mathcal{X}_i$$

and, therefore, any formal symmetry $\mathcal{R}$ is uniquely representable in the form

$$\mathcal{R} = \sum_{-\infty}^{k} c_i \mathcal{Y}_{-1}^i, \qquad c_i \in \mathbb{C}.$$

Thus, in this case, the algebra A_{fsym} is generated by a single generator $\mathcal{Y}_{-1}$.

---

[18]We do not assume that the leading coefficients of the series X and Y are not zero and therefore, without loss of generality, we can suppose that the formal symmetry is of this order.

Example 4.4.10. As an example of integrable equation of order (4,0) we consider the potential Boussinesq equation

$$u_{tt} = u_{xxxx} - 3u_x u_{xx}. \tag{4.103}$$

Similarly to Example 4.4.9, there exist two types of formal symmetries:

$$\mathcal{Y}_i = -\tfrac{3}{8}(i+2)u_t\, D^{-1+i} - \tfrac{3}{16}(i-2)(i+1)u_{xt}\, D^{-2+i}$$
$$+ \tfrac{1}{32}\left[9(i-1)(i+2)u_x u_t - 2i\left(i^2 - 3i + 8\right)u_{xxt}\right]D^{-3+i} + \cdots$$
$$+ \left\{D^i - \tfrac{3}{4}iu_x\, D^{-2+i} - \tfrac{3}{8}(i-2)iu_{xx}\, D^{-3+i}\right.$$
$$\left. + \left[\tfrac{9}{32}(i^2 - 3i + 2)u_x^2 - \tfrac{1}{16}(i-3)(i-2)(2i+1)u_{xxx}\right]D^{-4+i} + \cdots\right\}D_t$$

and

$$\mathcal{X}_i = D^i - \tfrac{3}{4}iu_x\, D^{-2+i} - \tfrac{3}{8}(i-2)iu_{xx}\, D^{-3+i}$$
$$+ \tfrac{1}{32}\left[9(i-3)iu_x^2 - 2(2i-7)(i-1)iu_{xxx}\right]D^{-4+i} + \cdots$$
$$+ \left\{-\tfrac{3}{8}iu_t\, D^{-5+i} - \tfrac{3}{16}(i-5)iu_{xt}\, D^{-6+i}\right.$$
$$\left. + \left[\tfrac{9}{32}(i-5)iu_x u_t - \tfrac{1}{16}(i-5)(i-4)iu_{xxt}\right]D^{-7+i} + \cdots\right\}D_t.$$

However, the algebraic relations between them are completely different:

$$\mathcal{X}_i = \mathcal{X}_1^i, \qquad \mathcal{Y}_i = \mathcal{Y}_1\mathcal{X}_1^{i-1}.$$

In addition,

$$\mathcal{Y}_1^2 = \mathcal{X}_6, \quad \mathcal{Y}_1^{-1} = \mathcal{Y}_{-5}, \quad \text{and} \quad \mathcal{Y}_{-2}^2 = 1.$$

We see that the algebra of formal symmetries is generated by $\mathcal{X}_1$ and $\mathcal{Y}_1$. The generators commute with each other[19] and are connected by the algebraic curve

$$\mathcal{X}_1^2 = \mathcal{Y}_1^6.$$

4.4.3 Integrability conditions

Here, we consider two classes of equations (4.91): equations of orders (3,1) and (4,1).

Equations of order (3, 1).

The first few conditions of the formal integrability for equations of the form

$$u_{tt} = f(u_3, u_{1t}, u_2, u_t, u_1, u), \qquad \frac{\partial f}{\partial u_3} \neq 0 \tag{4.104}$$

---

[19]To prove this fact, the homogeneity of formulas (4.103), (4.96), and (4.97) is used.

have the form of conservation laws $D_t(\rho_i) = D(\sigma_i)$, $i = 0, 1, \ldots$ (see [148, 153]), whose densities have the form

$$\rho_0 = \frac{1}{\sqrt[3]{u_3}}, \qquad \rho_1 = \frac{3u_2}{u_3} + \frac{2\sigma_1}{\sqrt[3]{u_3}} + \frac{\sigma_0 v_1}{u_3^{2/3}} + \frac{\sigma_0^2}{\sqrt[3]{u_3}}, \qquad \rho_2 = \frac{v_1}{u_3^{2/3}} - \frac{2\sigma_0}{\sqrt[3]{u_3}},$$

$$\rho_3 = \frac{v_1^3}{u_3^{4/3}} - \frac{27v_0}{\sqrt[3]{u_3}} + \frac{9u_2 v_1}{u_3^{4/3}} - \frac{9v_1 Du_3}{u_3^{4/3}} - \frac{6\sigma_2}{\sqrt[3]{u_3}} + \frac{3\sigma_1 v_1}{u_3^{2/3}} + \frac{6\sigma_1 \sigma_0}{\sqrt[3]{u_3}} + \frac{3\sigma_0^2 v_1}{u_3^{2/3}} + \frac{4\sigma_0^3}{\sqrt[3]{u_3}},$$

$$\rho_4 = -\frac{81u_1}{u_3^{2/3}} + \frac{27u_2^2}{u_3^{5/3}} + \frac{v_1^4}{u_3^{5/3}} + \frac{9u_2 v_1^2}{u_3^{5/3}} - \frac{27v_0 v_1}{u_3^{2/3}} - \frac{27u_2 Du_3}{u_3^{5/3}} - \frac{9v_1^2 Du_3}{u_3^{5/3}} + \frac{9\,(Du_3)^2}{u_3^{5/3}}$$
$$+ \frac{2\sigma_3}{\sqrt[3]{u_3}} + \frac{3\sigma_2 v_1}{u_3^{2/3}} + \frac{6\sigma_2 \sigma_0}{\sqrt[3]{u_3}} - \frac{3\sigma_1^2}{\sqrt[3]{u_3}} - \frac{6\sigma_1 \sigma_0 v_1}{u_3^{2/3}} - \frac{12\sigma_1 \sigma_0^2}{\sqrt[3]{u_3}} - \frac{7\sigma_0^4}{\sqrt[3]{u_3}} - \frac{5\sigma_0^3 v_1}{u_3^{2/3}}$$
$$- \frac{\sigma_0 v_1^3}{u_3^{4/3}} + \frac{27\sigma_0 v_0}{\sqrt[3]{u_3}} - \frac{9\sigma_0 u_2 v_1}{u_3^{4/3}} + \frac{9\sigma_0 v_1 Du_3}{u_3^{4/3}} - 27D_t\left(\frac{\sigma_0}{\sqrt[3]{u_3}}\right) - \frac{54v_1 D\sigma_0}{\sqrt[3]{u_3}}.$$

The following class

$$u_{tt} = u_3 + f(u_{1t}, u_2, u_t, u_1, u) \tag{4.105}$$

of equations (4.104) was considered in [148]. It turned out that, up to point transformations, there exist only two integrable equations of this type:

$$u_{tt} = u_3 + (3u_1 + k)u_{1t} + (u_t - u_1^2 - 2ku_1 + 6\wp)u_2 - 2\wp'u_t +$$
$$6\wp'u_1^2 + (\wp'' + k\wp')u_1, \tag{4.106}$$

and

$$u_{tt} = u_3 + \left[3\frac{u_t}{u_1} + \frac{3}{2}X(u)\right]u_{1t} - \frac{u_2^2}{u_1} - \left[3\frac{u_t^2}{u_1^2} + \frac{3}{2}X(u)\frac{u_t}{u_1}\right]u_2 +$$
$$c_2\left[u_1 u_t + \frac{3}{2}X(u)u_1^2\right],$$

where $\wp = \wp(u)$ is a solution of the equation $(\wp')^2 = 8\wp^3 + k^2\wp^2 + k_1\wp + k_0$, $X(u) = c_2 u + c_1$, and k_1, k_2, c_1, c_2 are arbitrary constants. Both equations are C-integrable.

Equations of order $(4, 1)$.

For equations of the form

$$u_{tt} = f(u_4, u_3, u_{1t}, u_2, u_t, u_1, u), \qquad \frac{\partial f}{\partial u_4} \neq 0 \tag{4.107}$$

the simplest conditions of the formal integrability are defined [153] by the following densities:

$$\rho_0 = \frac{1}{\sqrt[4]{u_4}}, \qquad \rho_1 = \frac{u_3}{u_4}, \qquad \rho_2 = \frac{v_1}{\sqrt{u_4}} - 2\frac{\sigma_0}{\sqrt[4]{u_4}},$$

$$\rho_3 = -\frac{4v_0}{\sqrt[4]{u_4}} + \frac{u_3 v_1}{u_4^{5/4}} - \frac{3v_1 Du_4}{2u_4^{5/4}} - \frac{2\sigma_1}{\sqrt[4]{u_4}},$$

$$\rho_4 = -\frac{32u_2}{u_4^{3/4}} - \frac{4v_1^2}{u_4^{3/4}} + \frac{12u_3^2}{u_4^{7/4}} + \frac{5\,(Du_4)^2}{u_4^{7/4}} - \frac{12u_3 Du_4}{u_4^{7/4}} - \frac{16\sigma_2}{\sqrt[4]{u_4}} - \frac{16\sigma_0^2}{\sqrt[4]{u}_{four}},$$

$$\rho_5 = \frac{16u_1}{\sqrt{u_4}} + \frac{3v_1^2 Du_4}{u_4^{3/2}} - \frac{3u_3^2 Du_4}{u_4^{5/2}} - \frac{2u_3\,(Du_4)^2}{u_4^{5/2}} + \frac{4u_3 D^2 u_4}{u_4^{3/2}}$$

$$+ \frac{8u_2 Du_4}{u_4^{3/2}} - \frac{8u_2 u_3}{u_4^{3/2}} + \frac{8v_0 v_1}{\sqrt{u_4}} + \frac{2u_3^3}{u_4^{5/2}} - \frac{2u_3 v_1^2}{u_4^{3/2}} - \frac{2\sigma_3}{\sqrt[4]{u_4}} - \frac{4\sigma_0 \sigma_1}{\sqrt[4]{u_4}}$$

$$+ 8D_t\left(\frac{\sigma_0}{\sqrt[4]{u_4}}\right) + \frac{16v_1\,(D\sigma_0)}{\sqrt[4]{u_4}} - \frac{3\sigma_0 v_1\,(Du_4)}{u_4^{5/4}} + \frac{2\sigma_0 u_3 v_1}{u_4^{5/4}} - \frac{8\sigma_0 v_0}{\sqrt[4]{u_4}}.$$

Remark 4.4.11. *If in relations* (4.96), (4.97) *the series X and Y have orders $n+1$ and n, respectively, then the density ρ_0 appears as an obstacle in finding the coefficient of D^{n+4} in* (4.96), *the density of ρ_1 corresponds to the coefficient of D^{n+3} in* (4.96), ρ_2 *matches the coefficient of D^{n+1} in* (4.97), ρ_3 *corresponds to the coefficient of D^n in* (4.97), ρ_4 *is originated by the coefficient of D^{n+2} in* (4.96), ρ_5 *corresponds to the coefficient of D^{n+1} in* (4.96), *etc.*

Open problem 4.4.12. *Find recurrence formulas for the sequence of these densities, similar to formula* (4.41).

The existence of a formal symplectic operator imposes additional conditions on the densities ρ_i (cf. Theorem 4.2.30). The first few conditions have the same nature as in the evolution case: some of the densities ρ_i must be total x-derivatives. Namely, $\rho_1 = D(\omega_1)$, $\rho_3 = D(\omega_3)$, $\rho_5 = D(\omega_5)$, where $\omega_i \in \mathcal{F}$.

The paper [153] contains a classification of integrable Lagrangian equations with Lagrangians of the form

$$L = \frac{1}{2}L_2(u_2, u_1, u)\,u_t^2 + L_1(u_2, u_1, u)\,u_t + L_0(u_2, u_1, u).$$

It is easy to see that the corresponding Euler–Lagrange equations have the form (4.107). The following Lagrangians:

$$L_1 = \frac{u_t^2}{2} + \epsilon\, u_1 u_t + \frac{u_2^2}{2} + \delta_2\frac{u_1^2}{2} + \delta_1\frac{q^2}{2} + \delta_0 q,$$

$$L_2 = \frac{u_t^2}{2} - u_1^2 u_t + \frac{u_2^2}{2} + \frac{u_1^4}{2},$$

$$L_3 = \frac{u_t^2}{2} + \frac{u_2^2}{2} + \frac{u_1^3}{2},$$

$$L_4 = \frac{u_t^2}{2} + a(q)\, u_1 u_t + \frac{u_2^2}{2u_1^4} + a'(q) u_1 \log u_1 + \frac{a^2(q)}{2} u_1^2 + d(q),$$

$$L_5 = \frac{u_t^2}{2} + \left(\frac{\gamma}{u_1} + \epsilon\, u_1\right) u_t + \frac{u_2^2}{2u_1^4} + \frac{\epsilon^2}{2} u_1^2 + \frac{\gamma^2}{2u_1^2} + \frac{\delta}{u_1}, \qquad |\gamma| + |\delta| \neq 0,$$

$$L_6 = \frac{u_1}{2} u_t^2 + (\epsilon u_1 + \beta) u_1 u_t + \frac{u_2^2}{2u_1^3} + \frac{\epsilon^2}{2} u_1^3 + \epsilon\beta\, u_1^2 + \frac{\delta}{u_1},$$

$$L_7 = \frac{u_t^2}{2u_1} + \frac{b(q)}{u_1} u_t + \frac{u_2^2}{2a(q)^4 u_1^5} + \frac{d_2(q)}{u_1},$$

$$L_8 = \frac{u_t^2}{2(u_1^2 - 1)} + \frac{u_2^2}{2(u_1^2 - 1)} + d(q) u_1^2 - \frac{d(q)}{3}, \qquad d'''(q) - 8d(q)d'(q) = 0,$$

nonequivalent with respect to contact transformations and the addition of total derivatives, were found. In this list, the Greek letters denote arbitrary constants. The integrability of L_4 was proved only for $a'(q) = 0$. All of the above canonical densities are trivial in this case, which usually indicates that the corresponding equation is linearizable.

For all other equations, weakly nonlocal recursion operators were found (see Section 4.4.4).

4.4.4   Weakly nonlocal recursion operators

The use of the weakly nonlocal ansatz (cf. (4.82))

$$\mathcal{R} = \mathcal{D} + \sum_k s_k D^{-1} \cdot \mathcal{C}_k, \tag{4.108}$$

where $\mathcal{D}$ is some differential operator, coefficients s_k are symmetries, and $\mathcal{C}_k$ are variational derivatives of conserved densities, is an efficient way to search for recursion operators also in the nonevolution case.

The only difference from the evolution case is that the variational derivative is not a function, but a differential operator of the form $pD_t + q$, $p, q \in \mathcal{F}$. Namely, if $\rho_* = P + QD_t$, then

$$\frac{\delta\rho}{\delta u} = P^+(1) + Q^+(1)D_t.$$

Example 4.4.13. Equation (4.106) has a unique nontrivial conserved density

$$\rho = u_t - u_x^2 + 2\wp(u). \tag{4.109}$$

A recursion operator for this equation can be written as

$$\mathcal{R} = D + (u_t - 2u_x^2 - ku_x + 2\wp) + u_x D^{-1}(D_t + 2u_{xx} + 2\wp').$$

The multiplier u_x in the nonlocal term is a symmetry, and the operator $D_t + 2u_{xx} + 2\wp'$ is the variational derivative of the density (4.109). The symmetries of equation (4.106) are obtained by applying this operator to the seed symmetries u_x and u_t.

Another weakly nonlocal recursion operator for (4.106) has the form

$$\bar{\mathcal{R}} = D_t + (u_{xx} - u_x^3 - ku_x^2 + 6\wp u_x + k\wp + \wp') + u_t D^{-1}(D_t + 2u_{xx} + 2\wp').$$

One can check that

$$\bar{\mathcal{R}}^2 = \mathcal{R}^3 + c_2\mathcal{R}^2 + c_1\mathcal{R} + c_0$$

for some constants c_i. In this sense, the situation is similar to the Krichever–Novikov equation (4.88) (see Section 4.3.3).

Example 4.4.14. Equation (4.103) possesses a recursion operator

$$\mathcal{R} = -\frac{9}{8}u_t + DD_t + \frac{3}{8}D^{-1} \cdot (u_{xt} - u_x D_t) - \frac{3}{8}u_x D^{-1}D_t$$

of the form (4.108). The two nonlocal terms are determined by the symmetries $s_1 = 1$, $s_2 = u_x$ and the conserved densities $\rho_1 = u_x u_t$, $\rho_2 = u_t$.

Applying the operator $\mathcal{R}$ to the symmetries $s_0^a = u_x$ and $s_0^b = u_t$, we get two series of symmetries with generators of the form

$$s_i^a = c_i\, u_{2i+1} + \cdots,$$

$$s_i^b = d_i\, u_{2it} + \cdots,$$

where $i = 0, 1, \ldots$, and c_i, d_i are some constants. In particular,

$$s_1^b = \mathcal{R}(u_x) = u_{xxt} - \frac{3}{2}u_x u_t,$$

$$s_2^a = \mathcal{R}(u_t) = u_{xxxxx} - \frac{15}{4}u_x u_{xxx} - \frac{15}{16}u_t^2 - \frac{45}{16}u_{xx}^2 + \frac{15}{16}u_x^3.$$

The recursion operator acts as follows:

$$\mathcal{R}(s_i^a) = s_{i+1}^b, \qquad \mathcal{R}(s_i^b) = s_{i+2}^a.$$

Therefore, symmetries of the type s_{3i+1}^a and s_{3i+2}^b do not occur.

Remark 4.4.15. *Since equation (4.103) is Lagrangian (with Lagrangian L_3), its linearization operator is self-adjoint: $\mathcal{L}^+ = \mathcal{L}$. For such equations, recursion and symplectic operators coincide.*

4.4.5   Discussion

The theoretical justification of the approach associated with the existence of a formal symmetry for equations of the form (4.91) is much less developed than in the case of evolution equations.

Analogues of Theorems 4.2.9 and 4.2.23 for equations of the form (4.91) are proved only in special cases (see [148, 153]). There is also no generalization of Definition 4.2.25 and Theorem 4.2.27 and, as a consequence, there is no analogue of Remark 4.2.26 and Theorem 4.2.30.

Open problem 4.4.16. *Prove that, for equations* (4.91) *of arbitrary order, all conditions of the existence of formal symmetry do not depend on its order and can be written as conservation laws. With the additional requirement of the existence of a formal symplectic operator, a "half" of the corresponding densities must be trivial.*

The problem of weak nonlocalities (see Definition 4.2.43) seems in the case of equations (4.91) to be more significant than for evolution equations.

Example 4.4.17. The well-known Boussinesq equation [154] has the form

$$q_{tt} = D(q_{xxx} + q\,q_x). \tag{4.110}$$

It is usually written as a system

$$q_t = p_x, \qquad p_t = q_{xxx} + q\,q_x.$$

Note that the equation and the system are not equivalent: in the case of the system there is an additional nonlocal variable $p = D^{-1}(q_t)$ which is not present in the equation. This fact leads to serious consequences: equation (4.110) does not have a formal symmetry with coefficients depending only on variables (4.92). In order for it to appear, we need to extend (4.92) by the potential variable u such that

$$u_x = -\frac{1}{3}q, \qquad u_t = -\frac{1}{3}p.$$

It is easy to see that u satisfies equation (4.103).

Thus, if classifying formally integrable equations of the form (4.107), equation (4.110) would be lost since it has only quasi-local symmetries. However, equation (4.103) which is related to (4.110) by a quasi-local transformation (see Definition 1.6.20) would be found. To classify equations with quasi-local symmetries, one can use the approach from the paper [117].

Chapter 5

Integrable hyperbolic equations of Liouville type

The open Toda lattices

$$(u_i)_{xy} = \sum_j A_i^j \exp(u_j), \tag{5.1}$$

where A_i^j is the Cartan matrix of a simple Lie algebra [155], provide examples of Liouville type systems.

For the Lie algebra of A_1-type the system coincides with the famous Liouville equation

$$u_{xy} = \exp u. \tag{5.2}$$

The Liouville equation possesses the following remarkable properties:

1. It has a local formula for the general solution

$$u(x, y) = \log\left(\frac{2f'(x)g'(y)}{(f(x) + g(y))^2}\right);$$

2. It admits a group of classical symmetries

$$x \to \phi(x), \qquad y \to \psi(y), \qquad u \to u - \log \phi'(x) - \log \psi'(y)$$

depending on two arbitrary functions of one variable;

3. It possesses the generalized first integrals

$$w = u_{xx} - \frac{1}{2}u_x^2, \qquad \bar{w} = u_{yy} - \frac{1}{2}u_y^2;$$

4. It has a noncommutative hierarchy of higher infinitesimal symmetries of the form

$$u_\tau = (D_x + u_x)\, P(w, w_x, \ldots, w_n) + (D_y + u_y)\, Q(\bar{w}, \bar{w}_y, \ldots, \bar{w}_m),$$

where w and $\bar{w}$ are generalized integrals, P and Q are arbitrary functions, n and m are arbitrary integers;

5. It has a terminated sequence of the Laplace invariants.

These are typical features of the so-called *equations of Liouville type* (another name is *Darboux integrable equations*) [156]–[160].

5.1 Generalized x and y-integrals

Consider hyperbolic nonlinear equations of the form

$$u_{xy} = F(x, y, u, u_x, u_y). \tag{5.3}$$

The corresponding total x and y-derivatives are defined by the recursive formulas

$$D = \frac{\partial}{\partial x} + \sum_{i=0}^{\infty} u_{i+1} \frac{\partial}{\partial u_i} + \sum_{i=1}^{\infty} \bar{D}^{i-1}(F) \frac{\partial}{\partial \bar{u}_i}$$

and

$$\bar{D} = \frac{\partial}{\partial y} + \sum_{i=0}^{\infty} \bar{u}_{i+1} \frac{\partial}{\partial \bar{u}_i} + \sum_{i=1}^{\infty} D^{i-1}(F) \frac{\partial}{\partial u_i},$$

where

$$u_0 = \bar{u}_0 = u, \quad u_1 = u_x, \quad \bar{u}_1 = u_y, \quad u_2 = u_{xx}, \quad \bar{u}_2 = u_{yy}, \ldots . \tag{5.4}$$

Although at first glance it seems that D is defined in terms of $\bar{D}$ and vice versa, in fact, vector fields D and $\bar{D}$ are correctly defined by these formulas. It is easy to see that $[D, \bar{D}] = 0$ by virtue of equation (5.3).

Definition 5.1.1. A function $W(x, y, u, u_1, \bar{u}_1 \ldots)$ is called *y-integral* for equation (5.3) if $\bar{D}(W) = 0$. Any function $W(x)$ is called a *trivial* y-integral. Similarly, a function $\bar{W}(x, y, u, u_1, \bar{u}_1, \ldots)$ such that $D(\bar{W}) = 0$ is called *x-integral*.

Remark 5.1.2. *Not so formal, for any solution $u(x, y)$ of equation (5.3) the function*

$$W\left(x, y, u(x, y), \frac{\partial u(x, y)}{\partial x}, \frac{\partial u(x, y)}{\partial y}, \ldots\right)$$

does not depend on the variable y.

Example 5.1.3. For the Liouville equation (5.2), we have

$$\bar{D} = \frac{\partial}{\partial y} + \sum_{i=0}^{\infty} \bar{u}_{i+1} \frac{\partial}{\partial \bar{u}_i} + \exp(u) \left(\frac{\partial}{\partial u_1} + u_1 \frac{\partial}{\partial u_2} + (u_2 + u_1^2) \frac{\partial}{\partial u_3} + \cdots \right).$$

It is easy to verify that $\bar{D}\left(u_2 - \frac{1}{2}u_1^2\right) = 0$.

Obviously, for any y-integral w and any function S expression

$$W = S\left(x, w, D(w), \cdots, D^k(w)\right) \tag{5.5}$$

is also a y-integral.

Lemma 5.1.4. *Any y-integral W is a function of the variables $x, y, u, u_1, u_2, \ldots,$ $u_k, \ldots$:*

$$W = W(x, y, u, u_1, u_2, \ldots, u_p).$$

The number p is called the order of W.

Proof. If the function $W = W(x, y, u, u_1, \ldots, u_n, \bar{u}_1, \ldots, \bar{u}_m)$, where $m \geq 1$, is the y-integral of equation (5.3), then

$$\left(\frac{\partial}{\partial y} + \bar{u}_1 \frac{\partial}{\partial u} + \bar{u}_2 \frac{\partial}{\partial \bar{u}_1} + \cdots + \bar{u}_{m+1} \frac{\partial}{\partial \bar{u}_m} \right) (W) +$$

$$\left(F \frac{\partial}{\partial u_1} + D(F) \frac{\partial}{\partial u_2} + \cdots + D^{n-1}(F) \frac{\partial}{\partial u_n} \right) (W) = 0.$$

Since the functions $F, D(F), \ldots, D^{n-1}(F)$ depend only on the variables x, y, u, $\bar{u}_1$, $u_1, u_2, \ldots, u_n$, the coefficient for $\bar{u}_{m+1}$ on the left-hand side should be zero and, therefore, $\frac{\partial W}{\partial \bar{u}_m} = 0$. $\square$

Proposition 5.1.1. [108] (i) *Any y-integral W has the form* (5.5), *where w is the integral of the lowest possible order. The minimal integral is uniquely determined up to a replacement of the form $w \to \phi(x, w)$.*
(ii) *If the order n of a minimal integral is greater than one, then there is a minimal integral w such that* $\frac{\partial^2 w}{\partial u_n^2} = 0$.[1]

Proof. Denote by w a y-integral of the smallest possible order n. Let $W = W(x, y, u, u_1, \ldots, u_m)$, $m \geq n$ is another y-integral. It can be written as

$$W = W(x, y, u, u_1, \ldots, u_{n-1}, w, w_1, \ldots, w_{m-n}), \qquad w_i = D^i(w).$$

---

[1] We formulate statements for y-integrals. Of course similar statements are true for x-integrals.

For any fixed values of the variables $x, w, w_1, \ldots, w_{m-n}$, this function is an y-integral of order less than n. Therefore, it is a constant and W does not depend on $y, u, u_1, \ldots, u_{n-1}$.

For the proof of Item (ii) let us differentiate the relation

$$\bar{D}w = \left(\frac{\partial}{\partial y} + \bar{u}_1 \frac{\partial}{\partial u} + F \frac{\partial}{\partial u_1} + D(F) \frac{\partial}{\partial u_2} + \cdots + D^{n-1}(F) \frac{\partial}{\partial u_n} \right) w = 0$$

with respect to the variable u_n twice and get

$$(\bar{D} + F_{u_1}) \frac{\partial w}{\partial u_n} = 0, \qquad (\bar{D} + 2F_{u_1}) \frac{\partial^2 w}{\partial u_n^2} = 0.$$

It follows that

$$\frac{\partial^2 w}{\partial u_n^2} \left(\frac{\partial w}{\partial u_n} \right)^{-2}$$

is a y-integral of minimal order and, therefore, a function of x and w. Therefore

$$H(x, w) \frac{\partial w}{\partial u_n} = g(x, y, u_1, \ldots, u_{n-1})$$

for some functions H and g. Then

$$W = \int H \, dw$$

is the minimal y-integral linear in u_n. $\qquad\qquad\qquad\qquad\qquad\qquad\square$

Definition 5.1.5. An equation of the form (5.3) is called *Darboux integrable* if it has nontrivial x and y-integrals.

Example 5.1.6. The wave equation

$$u_{xy} = 0$$

is Darboux integrable. First order integrals for this equation have the form

$$w = u_1, \qquad \bar{w} = \bar{u}_1.$$

Remark 5.1.7. *Example 5.1.6 shows (cf. Remark 1.4.8) that if a nonlinear hyperbolic equation reduces to the wave equation by some transformation, then it is natural to expect that the nonlinear equation has x and y-integrals.*

Example 5.1.8. For the Euler–Poisson equation

$$u_{xy} = \frac{u_y - u_x}{x - y}$$

minimal integrals have second order:

$$w = \frac{u_2}{x - y}, \qquad \bar{w} = \frac{\bar{u}_2}{x - y}.$$

Nonlinear examples of Darboux integrable equations are contained in Appendix 9.2.

5.2 Laplace invariants for linear hyperbolic operator

Consider an operator of the form

$$L_0 = \frac{\partial^2}{\partial x \partial y} + a_0(x,y)\frac{\partial}{\partial x} + b_0(x,y)\frac{\partial}{\partial y} + c_0(x,y). \tag{5.6}$$

It is easy to verify that

$$L_0 = \left(\frac{\partial}{\partial x} + b_0\right)\left(\frac{\partial}{\partial y} + a_0\right) - h_1 = \left(\frac{\partial}{\partial y} + a_0\right)\left(\frac{\partial}{\partial x} + b_0\right) - k_0,$$

where

$$h_1 = \frac{\partial a_0}{\partial x} + b_0 a_0 - c_0, \qquad k_0 = \frac{\partial b_0}{\partial y} + a_0 b_0 - c_0. \tag{5.7}$$

Definition 5.2.1. The functions (5.7) are called *main Laplace invariants* of the operator (5.6).

Lemma 5.2.2. *Operators L_0 and $\bar{L}$ of the form (5.6) are connected by a gauge transformation*

$$\bar{L} = \alpha(x,y)\, L_0\, \alpha(x,y)^{-1}$$

with some function α iff their main Laplace invariants coincide.

It is shown below that functions (5.7) are successive members of a sequence of invariants h_i, $i \in Z$. In this sequence, k_0 will play the role of h_0.

Let us define the *Laplace transformation* on the set of operators of the form (5.6). The equation $L_0(V) = 0$ is equivalent to a system of first order equations

$$\left(\frac{\partial}{\partial y} + a_0\right) V = V_1, \qquad \left(\frac{\partial}{\partial x} + b_0\right) V_1 = h_1 V.$$

If $h_1 \neq 0$, we can find V from the second equation and substitute it into the first. We obtain that V_1 satisfies the hyperbolic equation $L_1(V_1) = 0$, where

$$L_1 = \frac{\partial^2}{\partial x \partial y} + a_1(x,y)\frac{\partial}{\partial x} + b_1(x,y)\frac{\partial}{\partial y} + c_1(x,y).$$

Definition 5.2.3. The operator L_1 is said to be obtained from L_0 using the Laplace x-transformation.

It is easy to verify that

$$L_1 = h_1 \left(\frac{\partial}{\partial y} + a_0 \right) \frac{1}{h_1} L_0 \left(\frac{\partial}{\partial y} + a_0 \right)^{-1}.$$

The coefficients and the Laplace invariants of the operator L_1 are determined by the formulas

$$a_1 = a_0 - (\log h_1)_y, \qquad b_1 = b_0, \qquad c_1 = a_1 b_0 + (b_0)_y - h_1,$$

$$h_2 = (a_1)_x - (b_0)_y + h_1, \qquad k_1 = h_1.$$

Remark 5.2.4. *The invariant h_2 can be expressed in terms of the main invariants of the operator* (5.6):

$$h_2 = 2h_1 - k_0 - (\log h_1)_{xy}.$$

If $h_2 \neq 0$,, then we can apply the Laplace x-transformation to the operator L_1 and so on. As a result, we get a chain of hyperbolic operators

$$L_i = \frac{\partial^2}{\partial x \partial y} + a_i(x, y) \frac{\partial}{\partial x} + b_i(x, y) \frac{\partial}{\partial y} + c_i(x, y),$$

where $i \in \mathbb{N}$ and

$$a_i = a_{i-1} - (\log h_i)_y, \qquad b_i = b_0, \qquad c_i = a_i b_0 + (b_0)_y - h_i,$$

$$k_i = h_i, \qquad h_{i+1} = 2h_i - h_{i-1} - (\log h_i)_{xy}. \tag{5.8}$$

The Laplace y-transformation is defined similarly. The initial hyperbolic equation $L_0(V) = 0$ can be rewritten as a system

$$\left(\frac{\partial}{\partial x} + b_0 \right) V = V_{-1}, \qquad \left(\frac{\partial}{\partial y} + a_0 \right) V_{-1} = k_0 V.$$

If $k_0 \neq 0$, then V_{-1} satisfies the equation

$$L_{-1}(V) \overset{\text{def}}{=} \left(\frac{\partial^2}{\partial x \partial y} + a_{-1} \frac{\partial}{\partial x} + b_{-1} \frac{\partial}{\partial y} + c_{-1} \right) V_{-1} = 0.$$

It is said that the operator

$$L_{-1} = k_0 \left(\frac{\partial}{\partial x} + b_0 \right) \frac{1}{k_0} L_0 \left(\frac{\partial}{\partial x} + b_0 \right)^{-1}$$

is derived from L_0 using the Laplace y-transformation. Successively applying this transformation, we get the chain of operators

$$L_{-i} = \frac{\partial^2}{\partial x \partial y} + a_{-i}(x, y) \frac{\partial}{\partial x} + b_{-i}(x, y) \frac{\partial}{\partial y} + c_{-i}(x, y), \qquad i \in \mathbb{N},$$

where

$$a_{-i} = a_0, \qquad b_{-i} = b_{1-i} - (\log k_{1-i})_x, \qquad c_{-i} = a_0 b_{-i} + (a_0)_x - k_{1-i},$$

$$k_{-i} = h_{-i} = 2\, h_{1-i} - h_{2-i} - (\log h_{1-i})_{xy}. \qquad (5.9)$$

Definition 5.2.5. The functions h_i, $i \in \mathbb{Z}$, defined by formulas (5.8), (5.9) are called *Laplace invariants of the operator* (5.6).

The sequence of Laplace invariants is uniquely defined by the recurrent formula (see formulas (5.8) and (5.9))

$$h_i = 2\, h_{i-1} - h_{i-2} - (\log h_{i-1})_{xy}, \qquad i \in \mathbb{Z} \qquad (5.10)$$

and by the initial data

$$h_1 = \frac{\partial a_0}{\partial x} + a_0 b_0 - c_0, \qquad h_0 = \frac{\partial b_0}{\partial y} + a_0 b_0 - c_0. \qquad (5.11)$$

Remarkably, (5.10) is nothing but the integrable infinite A-Toda lattice [155].

If one of the Laplace invariants turns out to be identically equal to zero, then the next invariant is not defined and the sequence of invariants is terminated. If it breaks in both directions, that is

$$h_r = h_{-s} = 0 \qquad \text{for some} \qquad r \geq 1,\ s \geq 0,$$

then (5.10) turns into the open Toda lattice [155]. In this case, the equation $L_0(V) = 0$, as Laplace showed, can be solved explicitly (see, for example, [110]). In particular, the Laplace method allows one to solve the equation from Example (5.1.8).

5.3 Nonlinear hyperbolic equations of Liouville type

The linearization operator (see Section 2.0.1) for (5.3) is given by the formula

$$L = D\bar{D} - \frac{\partial F}{\partial u_1} D - \frac{\partial F}{\partial \bar{u}_1} \bar{D} - \frac{\partial F}{\partial u}. \qquad (5.12)$$

This is a linear hyperbolic operator of the form (5.6), where, however, the partial derivatives are replaced by the total derivative operators D and $\bar{D}$, and the coefficients are functions of x, y and a finite number of variables (5.4).

According to (5.11), we define the main Laplace invariants of equation (5.3) as

$$H_1 \overset{def}{=} -D\left(\frac{\partial F}{\partial u_1}\right) + \frac{\partial F}{\partial u_1}\frac{\partial F}{\partial \bar{u}_1} + \frac{\partial F}{\partial u}, \qquad H_0 \overset{def}{=} -\bar{D}\left(\frac{\partial F}{\partial \bar{u}_1}\right) + \frac{\partial F}{\partial u_1}\frac{\partial F}{\partial \bar{u}_1} + \frac{\partial F}{\partial u}.$$

The invariants H_i for $i > 1$ and for $i < 0$ are determined from the recurrent formula

$$D\bar{D}(\log H_i) = -H_{i+1} - H_{i-1} + 2H_i, \qquad i \in \mathbb{Z}. \tag{5.13}$$

For the Liouville equation (5.2), we have $H_0 = H_1 = \exp u$. It is easy to check that $H_2 = H_{-1} = 0$.

Definition 5.3.1. We call equation (5.3) a *Liouville-type equation* if there exist integers $r \geq 1$ and $s \geq 0$ such that

$$H_r = H_{-s} \equiv 0.$$

Remark 5.3.2. *Definition 5.3.1 is more constructive than the Darboux definition 5.1.5 because the Laplace invariants are calculated in terms of the left-hand side of equation (5.3) using a simple recurrent formula (5.13), while to verify the Darboux integrability, one must look for x and y-integrals.*

Remark 5.3.3. *For linear hyperbolic equations, the Laplace invariants coincide with the Laplace invariants of the corresponding linear operator defined in Section 5.2.*

Exercise 5.3.4. Verify that the linear equations from Examples 5.1.6 and 5.1.8 are equations of Liouville type.

Example 5.3.5. For the equation

$$u_{xy} = \frac{1}{u}\sqrt{1 - u_x^2}\sqrt{1 - u_y^2} \tag{5.14}$$

we have $H_2 = H_{-1} = 0$.

Example 5.3.6. For the equation

$$u_{xy} = u\,u_y \tag{5.15}$$

the invariants H_3 and H_0 are equal to zero.

Example 5.3.7. In the case of the equation

$$u_{xy} = -\frac{2k}{x+y}\sqrt{u_y}\sqrt{u_y}, \qquad k \in \mathbb{N} \tag{5.16}$$

we have $H_{k+1} = H_{-k} = 0$.

The following statement establishes the equivalence of Definitions 5.3.1 and 5.1.5 for scalar nonlinear hyperbolic equations.

Theorem 5.3.8. [161, 162] *An equation* (5.3) *has nontrivial y and x-integrals*

$$W(x, y, u, u_1, u_2, \ldots, u_p), \qquad \bar{W}(x, y, u, \bar{u}_1, \bar{u}_2, \ldots, \bar{u}_{\bar{p}}) \qquad (5.17)$$

iff it is a Liouville-type equation. Moreover, $r \leq \bar{p}$, $s \leq p - 1$.

The following two functions appear in almost all explicit formulas related to Liouville-type equations.

Proposition 5.3.1. *For any Liouville-type equation* (5.3) *there exist functions $\psi(x, y, u, u_1, \ldots, u_p)$ and $\bar{\psi}(x, y, u, \bar{u}_1, \ldots, \bar{u}_{\bar{p}})$ such that*

$$\frac{\partial F}{\partial u_1} = \bar{D} \log \psi(x, y, u, u_1, \ldots, u_p), \qquad \frac{\partial F}{\partial \bar{u}_1} = D \log \bar{\psi}(x, y, u, \bar{u}_1, \ldots, \bar{u}_{\bar{p}}).$$

Exercise 5.3.9. Prove that for any integrals (5.17) one can take

$$\psi = \frac{1}{W_{u_p}}, \qquad \bar{\psi} = \frac{1}{\bar{W}_{\bar{u}_{\bar{p}}}}$$

for ψ and $\bar{\psi}$.

Remark 5.3.10. *Another statement of this kind is* [108, 110]*:*

$$\frac{\partial F}{\partial u_1} \frac{\partial F}{\partial \bar{u}_1} + \frac{\partial F}{\partial u} = \bar{D}\, \phi(x, y, u, u_1, \ldots, u_q) = D\, \bar{\phi}(x, y, u, \bar{u}_1, \ldots, \bar{u}_{\bar{q}})$$

for some functions ϕ and $\bar{\phi}$.

The set of the minimal possible numbers $p, \bar{p}, q, \bar{q}$ from Proposition 5.3.1 and Remark 5.3.10 is an important characteristic of any Liouville-type equation.

The following formula describes x-symmetries of the form

$$u_\tau = G(x, y, u, u_1, u_2, \ldots)$$

for Liouville type equations.

Theorem 5.3.11. *For any Liouville-type equation* (5.3) *any evolution equation*

$$u_\tau = \mathcal{M} \left[Q\left(x, w, D(w), \cdots, D^k(w)\right) \right], \qquad k \geq 0, \qquad (5.18)$$

where

$$\mathcal{M} = \bar{\psi} \frac{1}{H_1} D \circ \frac{1}{H_2} \cdots D \circ \frac{1}{H_{r-1}} D \circ \frac{\psi H_1 \cdots H_{r-1}}{\bar{\psi}}, \qquad (5.19)$$

Q is an arbitrary function, and w is a minimal y-integral, is infinitesimal x-symmetry.[2]

---

[2]If $r = 1$ then $\mathcal{M} = \psi$.

Lemma 5.3.12. *The coefficients of the operator* $\mathcal{M}$ *do not depend on the derivatives* $\bar{u}_1, \bar{u}_2, \ldots$.

Remark 5.3.13. *For a generic function* Q, *the evolution equation* (5.18) *is not integrable in the sense of Section 4.2. For different functions* Q, *the flows* (5.18), *generally speaking, do not commute with each other.*

Remark 5.3.14. *A formula similar to* (5.18) *is also available for* y-*symmetries of the form*

$$u_\tau = \bar{G}(x, y, u, \bar{u}_1, \bar{u}_2, \ldots).$$

For a generalization of Theorem 5.3.11 to the case of multi-component Darboux integrable systems, see [163].

According to Theorem 5.3.11, the operator $\mathcal{M}$ maps any y-integral to a symmetry.

Lemma 5.3.15. *Let* $W(x, y, u, u_1, u_2, \ldots, u_p)$ *be a* y-*integral of equation* (5.3). *Then the differential operator* W_* *maps any* x-*symmetry to some* y-*integral.*

Proof. Linearizing the relation $\bar{D}(W) = 0$, we obtain that $\bar{D}W_*(G) = 0$ for any solution (see Section 2.0.1) of the linearized equation $\mathcal{L}(G) = 0$. The statement of the lemma follows from the fact that the generator of any symmetry satisfies the linearized equation. $\square$

The following operator can be used to construct a general solution of the Liouville-type equation (5.3) [110]. Moreover, explicit calculating of its coefficients is the most efficient way to find y-integrals.

Theorem 5.3.16. *For any equation* (5.3) *of Liouville-type, all coefficients of the differential operator*

$$\mathcal{N} = \frac{\bar{\psi}}{\psi} H_0 H_{-1} \cdots H_{1-s} D \circ \frac{1}{H_{1-s}} \circ D \cdots \frac{1}{H_0} \circ D \circ \frac{1}{H_1} \cdots D \circ \frac{1}{H_{r-1}} D \circ \frac{\psi H_1 \cdots H_{r-1}}{\bar{\psi}}$$
$$(5.20)$$

are y-*integrals.*

Example 5.3.17. For the Liouville equation (5.2), we have

$$\mathcal{N} = \exp(u)D \circ \exp(-u)D \circ \exp(-u)D \circ \exp(u) = D^3 + 2wD + w_x,$$

where $w = u_{xx} - \dfrac{1}{2}u_x^2$. The operator $\mathcal{M}$ is given by

$$\mathcal{M} = \exp(-u)D \circ \exp(u) = D + u_x.$$

If $Q(x, w, \ldots) = w$, then the corresponding symmetry is the integrable evolution equation

$$u_\tau = u_{xxx} - \frac{1}{2}u_x^3. \qquad (5.21)$$

An attempt to fully classify Darboux integrable equations was undertaken in [110]. The proof of the classification statement took more than 150 pages and was not published. For equations, whose minimal y and x-integrals have orders, which do not exceed 2, the result coincides with the well-known Goursat list [157]. After a while, O. Kaptsov pointed out to the authors a gap in the classification [164] for the case of integrals of order 3, which has not yet been patched up.

The discovery of each new equation with integrals of order greater than 2 is a significant advancement in the classical problem of classifying Darboux integrable equations. At the moment, three such equations are known. In Appendix 2, Darboux integrable equations known to the author are collected.

5.4 Differential substitutions and equations of Liouville type

In this section, we consider differential substitutions (see Section 1.6.2) of the form

$$\hat{u} = P(x, u, u_1, \ldots, u_k),$$

connecting some evolution equations

$$u_t = \Phi(x, u, u_1, \ldots, u_n)$$

and

$$\hat{u}_t = \Psi(x, \hat{u}, \hat{u}_1, \ldots, \hat{u}_n).$$

Let (5.3) be a Liouville-type equation with a minimal y-integral

$$\hat{u} = w(x, u, u_1, \ldots, u_p),$$

and

$$u_\tau = f(x, u, u_1, \ldots, u_n)$$

be an x-symmetry (see Section 4.1.1) of equation (5.3). Since the operators of the total derivatives $\bar{D}$ and D_τ commute, we have

$$\bar{D}D_\tau(\hat{u}) = D_\tau \bar{D}(\hat{u}) = 0,$$

that is the expression $D_\tau(\hat{u})$ is a y-integral. According to Proposition 5.1.1,

$$\hat{u}_\tau = Q(x, \hat{u}, \hat{u}_1, \ldots, \hat{u}_n)$$

for some function Q. Thus, the minimal y-integral of equation (5.3) defines a differential substitution from any x-symmetry of equation (5.3) to some evolution equation. Similarly, the minimal x-integral defines a differential substitution from y-symmetries.

Example 5.4.1. The Liouville equation has the symmetry (5.21). The minimal y-integral

$$\hat{u} = u_{xx} - \frac{1}{2}u_x^2$$

defines the differential substitution from (5.21) to the KdV equation

$$\hat{u}_\tau = \hat{u}_{xxx} + 3\,\hat{u}\hat{u}_x.$$

The minimal x-integral generates the same substitution, since the Liouville equation is symmetric with respect to the permutation x and y.

Example 5.4.2. Consider the Liouville-type equation (5.15). Its minimal y and x-integrals

$$w = u_x - \frac{1}{2}u^2, \qquad \bar{w} = \frac{u_{yyy}}{u_y} - \frac{3}{2}\frac{u_{yy}^2}{u_y^2}$$

give rise to two differential substitutions, the first of which is the Miura transformation.

The general formulas of Theorems 5.3.11 and 5.3.16 lead to operators

$$\mathcal{M} = D^2 + uD + u_x, \qquad \bar{\mathcal{M}} = u_y, \qquad \mathcal{N} = D^3 + 2wD + w_x.$$

The simplest x and y-symmetries

$$u_\tau = \mathcal{M}(w) = u_{xxx} - \frac{3}{2}u^2 u_x$$

and

$$u_\tau = \bar{\mathcal{M}}(\bar{w}) = u_{yyy} - \frac{3}{2}\frac{u_{yy}^2}{u_y}$$

are well-known integrable evolution equations mKdV and Schwartz–KdV. It is easy to verify that

$$w_\tau = w_{xxx} + 3\,ww_x$$

and

$$\bar{w}_\tau = \bar{w}_{yyy} + 3\,\bar{w}\bar{w}_y.$$

Moreover, for any x-symmetry of equation (5.15) given by the formula

$$u_\tau = \left(D^2 + uD + u_x\right) H(x, w, w_x, \ldots, w_n) \tag{5.22}$$

the minimal y-integral w defines a substitution from (5.22) to some evolution equation $w_\tau = Q(x, w, w_x, \ldots)$.

Let us find the function Q explicitly:

$$w_\tau = \left(D - u \right) u_\tau = \left(D - u \right) \left(D^2 + uD + u_x \right) H(x, w, w_x, \ldots, w_n) =$$
$$\left(D^3 + 2wD + w_x \right) H(x, w, w_x, \ldots, w_n) = \mathcal{N}\, H(x, w, w_x, \ldots, w_n).$$

Reasoning as above, we arrive at the following statement:

Proposition 5.4.1. *Let* $w(x, u, u_x, \ldots, u_k)$ *be the minimal y-integral of an equation of Liouville type. Then equation*

$$u_\tau = \mathcal{M}\, H(x, w, w_x, \ldots), \tag{5.23}$$

where H is an arbitrary function, is related to the evolution equation

$$\hat{u}_\tau = w_* \mathcal{M} \left(H(x, \hat{u}, \hat{u}_x, \ldots) \right) \tag{5.24}$$

by the differential substitution

$$\hat{u} = w(x, u, u_x, \ldots, u_k). \tag{5.25}$$

Here w_* denotes the Fréchet derivative of the integral w. The possible freedom $w \to f(x, w)$ in the choice of the minimal y-integral corresponds to arbitrary point transformations in equation (5.24).

Corollary 5.4.3. *From this proposition it follows that the coefficients of the differential operator*

$$\mathcal{S} = w_* \mathcal{M} \tag{5.26}$$

are y-integrals.

Remark 5.4.4. *In all known examples, with a consistent choice of the minimal integral w and functions $\psi, \bar{\psi}$ from Proposition 5.3.1, the operator $w_* \mathcal{M}$ coincides with the operator $\mathcal{N}$ given by the formula* (5.20). *If this is the case, then from* (5.19), (5.20) *it follows that*[3]

$$w_* = \frac{\bar{\psi}}{\psi} H_0 H_{-1} \cdots H_{1-s} D \frac{1}{H_{1-s}} D \cdots \frac{1}{H_0} D \frac{1}{\psi}.$$

This formula was proved in [162, 110] *under the assumption that the order k of the minimal integral is equal to* ord $\mathcal{N}$ − ord $\mathcal{M}$.

5.4.1 Differential substitutions of the first order

We associate the hyperbolic equation

$$u_{xy} = -\frac{P_u}{P_{u_x}} u_y \tag{5.27}$$

---

[3]If $H_0 = 0$, then $w_* = \frac{\bar{\psi}}{\psi} D(\frac{1}{\psi})$.

with arbitrary differential substitution

$$v = P(x, u, u_x)$$

of first order. It is easy to check that for equation (5.27) the function P is a minimal y-integral and that $H_0 = 0$.

Observation 5.4.5. *For the most famous differential substitutions of first order, equation* (5.27) *is an equation of Liouville type.*

Theorem 5.4.6. [108, Lemma 4.1], [165] *Equation* (5.27) *is a Liouville-type equation iff (up to transformations $u \to f(x, u)$) the function $P(x, u, u_x)$ is determined from the relation*

$$u_x = \alpha(x, P) \, u^2 + \beta(x, P) \, u + \gamma(x, P), \qquad (5.28)$$

where α, β and γ are arbitrary functions. Equation (5.27) *with such a function P has the x-integral*

$$\bar{W} = \frac{u_{yyy}}{u_y} - \frac{3}{2}\frac{u_{yy}^2}{u_y^2}.$$

Example 5.4.7. For the Miura transformation from Example 5.4.2 we have $u_x = P + \dfrac{1}{2}u^2$.

Example 5.4.8. For the well-known differential substitution [49]

$$v = u_x + \exp(u) + \exp(-u)$$

the corresponding hyperbolic equation has the form

$$u_{xy} = \Big(\exp(-u) - \exp(u) \Big) u_y. \qquad (5.29)$$

The function P for equation (5.29), after the transformation $u \to \ln u$, satisfies the relation $u_x = -u^2 + Pu - 1$ of the form (5.28).

Example 5.4.9. For the Cole–Hopf substitution (1.37), we have $u_x = P\,u$.

Exercise 5.4.10. Explicitly describe pairs of evolution equations (1.60) and (1.61), connected by the substitution from Theorem 5.4.6.

5.5   Pre-Hamiltonian operators

In this section, $\mathcal{F}$ denotes the differential field of functions of variables

$$x, \quad u_0 = u, \quad u_1 = u_x, \quad u_2 = u_{xx}, \quad \dots.$$

The Lie bracket

$$\left[f,\, g\right] \overset{def}{=} g_*\left(f\right) - f_*\left(g\right) \tag{5.30}$$

endows $\mathcal{F}$ with the structure of a Lie algebra (see (2.3)). Bracket (5.30) corresponds to the commutator of the flows of the evolution equations $u_{t_1} = f$ and $u_{t_2} = g$. Since the commutator of symmetries is a symmetry, the set of all x-symmetries of the form

$$u_\tau = S(x, u, u_1, u_2, \dots) \tag{5.31}$$

for any hyperbolic equation forms a Lie subalgebra in $\mathcal{F}$ (as usual, we identify symmetries with their right-hand sides).

The operator $\mathcal{N} = D^3 + 2uD + u_1$, corresponding to the Liouville equation (see Example 5.3.17), has the following remarkable property: its image is a Lie subalgebra in $\mathcal{F}$. Namely, it is easy to check that for any $f, g \in \mathcal{F}$

$$\left[\mathcal{N}(f), \mathcal{N}(g)\right] = \mathcal{N}\left(D(f)\,g - D(g)\,f + g_*\mathcal{N}(f) - f_*\mathcal{N}(g)\right).$$

It can be verified that the formula

$$\left[f,\, g\right]_1 \overset{def}{=} g_*\mathcal{N}(f) - f_*\mathcal{N}(g) + D(f)\,g - D(g)\,f$$

defines a new Lie bracket on $\mathcal{F}$. The operator $\mathcal{N}$ defines a homomorphism of this Lie algebra to the algebra with the bracket (5.30).

Definition 5.5.1. A differential operator $\mathcal{N}$ is called *pre-Hamiltonian* if its image $\mathrm{Im}\,\mathcal{N}$ is a Lie subalgebra in $\mathcal{F}$.

Remark 5.5.2. *It can be shown that for any Hamiltonian operator $\mathcal{H} : \mathcal{F} \to \mathcal{F}$ (see [12]) its image is a subalgebra in $\mathcal{F}$. Therefore, the pre-Hamiltonian operators can be regarded as a nonskewsymmetric generalization of Hamiltonian operators.*

Remark 5.5.3. *If $\mathcal{N}$ is a pre-Hamiltonian operator, then for any function $f \in \mathcal{F}$ the operator $\mathcal{N} \circ f$ is also pre-Hamiltonian. Therefore, we can reduce the leading coefficient of any pre-Hamiltonian operator to 1. We will call such an operator normalized.*

Exercise 5.5.4. Verify that for any function $S(x, u, u_x)$ the first order operator

$$\mathcal{M} = D + D\left(\log \frac{\delta S}{\delta u}\right) \tag{5.32}$$

is pre-Hamiltonian.

Remark 5.5.5. *It is asserted in [142] that if the kernel of the operator $\mathcal{M}^+$ consists of variational derivatives, then the operator $\mathcal{M}$ is pre-Hamiltonian. The operator (5.32) is a special case when the order of the operator is equal to one.*

The following statements were recently proved in [169].

Theorem 5.5.6. *For any hyperbolic equation of Liouville type, the operator (5.19), is a pre-Hamiltonian operator defined on functions in variables $y, x, u, u_1, \ldots$ (see Lemma 5.3.12).*

Let w be the minimal y-integral for a Liouville-type equation.

Theorem 5.5.7. *The $\mathcal{S}$-operator (5.26) is a pre-Hamiltonian operator defined on functions in the variables $x, w, w_1, \ldots$ (see Corollary 5.4.3).*

As it was mentioned in Remark 5.4.4, for all known Liouville-type equations the operator $\mathcal{N}$, given by formula (5.20), coincides with the operator $\mathcal{S}$ defined by formula (5.26). According to Theorem 5.5.7, this operator has to be pre-Hamiltonian.

5.5.1 Examples of pre-Hamiltonian operators related to Liouville type equations

Remark 5.5.5 shows that there are a lot of pre-Hamiltonian operators and their coefficients can contain arbitrary functions. However, the pre-Hamiltonian operators $\mathcal{M}$ and $\mathcal{N}$, generated by hyperbolic Liouville-type equations, have a very special structure.

Open problem 5.5.8. *Describe all such pre-Hamiltonian operators.*

Consider the operators $\mathcal{N}$ of the form (5.20) for hyperbolic equations

$$u_{xy} = F(x, y, u, u_x, u_y), \qquad \frac{\partial^2 F}{\partial u_x \partial u_y} \neq 0, \qquad (5.33)$$

with minimal y and x-integrals w and $\bar{w}$ of order two.[4]

Let w be the minimal y-integral linear (see Item (ii) of Proposition 5.1.1) with respect to the second derivative:

$$w = A(y, x, u, u_x)\, u_{xx} + B(y, x, u, u_x),$$

and

$$\mathcal{N} = D^3 + K_2 D^2 + K_1 D + K_0, \qquad K_i = K_i(x, w, w_x, \ldots)$$

be the corresponding normalized operator $\mathcal{N}$. From formula (5.20) we can deduce that:

---

[4] All such equations were described in [157].

Lemma 5.5.9. *The coefficient K_2 has the form*

$$K_2 = \frac{w_1}{w - z_0(x)} + z_1(x)w + z_2(x) + \frac{z_3(x)}{w - z_0(x)}.$$

Using the shift $w \mapsto w + z_0(x)$, we will assume that $z_0 = 0$.

Lemma 5.5.10. *The coefficients K_1 and K_2 have the following structure:*

$$K_1 = \frac{2w_2}{w} - \frac{2w_1^2}{w^2} + \left(a_1 + \frac{a_2}{w} + \frac{a_3}{w^2}\right)w_1 + a_4 w^2 + a_5 w + a_6 + \frac{a_7}{w} + \frac{a_8}{w^2},$$

$$K_0 = \frac{w_3}{w} - \frac{3w_1 w_2}{w^2} + \frac{2w_1^3}{w^3} + \left(b_1 + \frac{b_2}{w} + \frac{b_3}{w^2}\right)w_2 + \left(\frac{b_4}{w^2} + \frac{b_5}{w^3}\right)w_1^2 +$$

$$\left(b_6 w + b_7 + \frac{b_8}{w} + \frac{b_9}{w^2} + \frac{b_{10}}{w^3}\right)w_1 + b_{11} w^3 + b_{12} w^2 + b_{13} w + b_{14} + \frac{b_{15}}{w} + \frac{b_{16}}{w^2} + \frac{b_{17}}{w^3},$$

where a_i, b_i are some functions of x.

Theorem 5.5.11. *An operator $\mathcal{N}$ of this form is pre-Hamiltonian iff*

$$K_2 = \frac{w_1}{w} + z_1 w + z_2, \quad K_1 = \frac{2w_2}{w} - \frac{2w_1^2}{w^2} + \left(2z_1 + \frac{z_2}{w}\right)w_1 + y_1 w^2 + \left(2z_1'(x) + z_1 z_2\right)w + y_2,$$

$$K_0 = \frac{w_3}{w} - \frac{3w_1 w_2}{w^2} + \frac{2w_1^3}{w^3} + \left(z_1 + \frac{z_2}{w}\right)w_2 - \frac{z_2 w_1^2}{w^2} +$$

$$\left(2y_1 w + 2z_1' + z_1 z_2 + \frac{y_2 - 2z_2' - z_2^2}{w}\right)w_1 + \frac{3}{2}y_1'(x)w^2 +$$

$$\left(2z_2 z_1' + z_2' z_1\right)w + z_2'' - z_2 z_2' - z_2^3 + y_2 z_2,$$

where the functions z_1, z_2, y_1, y_2 are connected by differential relations

$$2y_1 z_1' = z_1 y_1', \qquad z_1'' - 3z_1 z_2' - z_1' z_2 - z_1 z_2^2 + y_2 z_1 = 0. \qquad \square \qquad (5.34)$$

For further simplification of the structure of the pre-Hamiltonian operator, we use the admissible transformation $w = \alpha(x)\tilde{w}$. As a result, we get the normalized operator $\tilde{\mathcal{N}} = \alpha^{-1}\mathcal{N}\alpha$. It is easy to verify that

$$\tilde{z}_1 = \alpha z_1, \quad \tilde{y}_1 = \alpha^2 y_1, \quad \tilde{z}_2 = z_2 + \frac{2\alpha'}{\alpha}, \quad \tilde{y}_2 = \frac{\alpha y_2 + 5\alpha' z_2 + 5\alpha''}{\alpha}. \qquad (5.35)$$

After another admissible transformation $\tilde{x} = \varphi(x)$, we get the operator $\tilde{\mathcal{N}} = \sigma^3 \mathcal{N}\sigma^{-3}$,

$$z_1(x) = \frac{\tilde{z}_1(\varphi(x))}{\sigma(x)}, \quad y_1(x) = \frac{\tilde{y}_1(\varphi(x))}{\sigma(x)^2}, \quad z_2(x) = \frac{\tilde{z}_2(\varphi(x)) - 3\sigma'(x)}{\sigma(x)},$$

$$y_2(x) = \frac{\tilde{y}_2(\varphi(x)) - 10\sigma'(x)\tilde{z}_2(\varphi(x)) + 19\sigma'(x)^2 - 8\sigma(x)\sigma''(x)}{\sigma(x)^2}, \qquad (5.36)$$

where $\sigma(x) = \dfrac{1}{\varphi'(x)}$.

From formulas (5.34)–(5.36) it follows that z_1 and y_1 can be reduced to constant, and z_2 can be brought to zero.

If $z_1(x) \neq 0$, we can assume that $z_1 = 1, z_2 = 0$. Then y_1 becomes a constant according to (5.34). This constant is essential (i.e. cannot be changed by transformations (5.35), (5.36)). From (5.34) it follows that $y_2 = 0$.

If $z_1 = 0$, $y_1(x) \neq 0$, we can use the normalization $y_1 = 1, z_2 = 0$. In this case, $y_2(x)$ remains an arbitrary function.

If $z_1 = y_1 = 0$, then we can put $z_2 = y_2 = 0$.

Example 5.5.12. Using [110, section 3, example 2], one can verify that the pre-Hamiltonian operator $\mathcal{N}$ for equation (5.14) is given by the formulas of Theorem 5.5.11 with $z_1 = z_2 = y_2 = 0, y_1 = 1$. The operator $\mathcal{M}$ for this equation has the form (5.32), where

$$S = \frac{\sqrt{1 - u_x^2}}{u}.$$

Example 5.5.13. Consider the special case of the Lainé equation [166]

$$u_{xy} = 2\left(u^2 + u_y + u\sqrt{u^2 + u_y}\right) \times \left(\frac{\sqrt{u_x} + u_x}{u - x} - \frac{u_x}{\sqrt{u^2 + u_y}}\right). \tag{5.37}$$

This equation has integrals

$$w = \frac{D(\mu) - \mu^2}{1 + x\mu}, \qquad \text{where} \quad \mu = \frac{1 + \sqrt{u_x}}{u - x},$$

$$\bar{w} = \bar{D}(\log \bar{\mu}) - x\bar{\mu}, \qquad \text{where} \quad \bar{\mu} = \frac{u + \sqrt{u^2 + u_y}}{u - x}.$$

Since the equation is not symmetric with respect to the permutation $x \leftrightarrow y$, the operators $\mathcal{N}$ and $\bar{\mathcal{N}}$ are different. Both are described by Theorem 5.5.11, where $z_1 = 3, y_1 = 2, y_2 = 0$ for $\bar{\mathcal{N}}$ and $z_1 = 3x, y_1 = 2x^2, y_2 = 0$ for $\mathcal{N}$.

A pre-Hamiltonian operator of a different type arises in the following example.

Example 5.5.14. Consider the equation [167]

$$u_{xy} = \frac{1}{6u + y}B^2(B - 1)\bar{B}(\bar{B} - 1)^2 + \frac{1}{6u + x}\bar{B}^2(\bar{B} - 1)B(B - 1)^2, \tag{5.38}$$

where $B = B(u_x)$ and $\bar{B} = \bar{B}(u_y)$ are solutions of the cubic equations

$$\frac{1}{3}B^3 - \frac{1}{2}B^2 = u_x, \qquad \frac{1}{3}\bar{B}^3 - \frac{1}{2}\bar{B}^2 = u_y. \tag{5.39}$$

For this equation we have $H_3 = H_{-2} = 0$. The minimal y and x-integrals w and $\bar{w}$ are of the third order:

$$w = D \left\{ \ln \left[u_2 - \frac{B^4(B-1)^2}{6u+y} - \frac{B^2(B-1)^4}{6u+x} \right] - \ln B(B-1) \right\} -$$
$$- \left[\left(\frac{1}{6u+y} + \frac{1}{6u+x} \right) B - \frac{1}{6u+x} \right] B(B-1),$$

$$\bar{w} = \bar{D} \left\{ \ln \left[\bar{u}_2 - \frac{\bar{B}^4(\bar{B}-1)^2}{6u+x} - \frac{\bar{B}^2(\bar{B}-1)^4}{6u+y} \right] - \ln \bar{B}(\bar{B}-1) \right\} -$$
$$- \left[\left(\frac{1}{6u+y} + \frac{1}{6u+x} \right) B - \frac{1}{6u+x} \right] B(B-1).$$

The functions ψ and $\bar{\psi}$ (see Proposition 5.3.1) are given by the formulas

$$\psi = u_2 - \frac{B^4(B-1)^2}{6u+y} - \frac{B^2(B-1)^4}{6u+x}, \qquad \bar{\psi} = \bar{u}_2 - \frac{\bar{B}^4(\bar{B}-1)^2}{6u+x} - \frac{\bar{B}^2(\bar{B}-1)^4}{6u+y}.$$

The Laplace invariants for equation (5.38) have the form

$$H_1 = \frac{\bar{B}(\bar{B}-1)}{B(B-1)} \left[\frac{\bar{B}-1}{(6u+y)(B-1)^2} - \frac{\bar{B}}{(6u+x)B^2} \right] \psi,$$

$$H_0 = \frac{B(B-1)}{\bar{B}(\bar{B}-1)} \left[\frac{B-1}{(6u+x)(\bar{B}-1)^2} - \frac{B}{(6u+y)\bar{B}^2} \right] \bar{\psi},$$

$$H_2 = \frac{2(6u+y)(6u+x)}{\bar{B}(\bar{B}-1)\left[(6u+x)B^2(\bar{B}-1) - (6u+y)\bar{B}B(B-1)^2\right]^2} \psi\bar{\psi},$$

$$H_{-1} = \frac{2(6u+y)(6u+x)}{B(B-1)\left[(6u+y)\bar{B}^2(B-1) - (6u+x)B(\bar{B}-1)^2\right]^2} \psi\bar{\psi}.$$

The corresponding operator $\mathcal{N}$ can be written in an extremely simple factorized form:

$$\mathcal{N} = D(D+w)(D+w)(D+2w)(D+3w). \tag{5.40}$$

It is easy to verify that the operator $\mathcal{N}$ is pre-Hamiltonian.

In the paper [110], pre-Hamiltonian operators "similar" to (5.40) were constructed. Namely, the operator (5.40) is homogeneous if we prescribe the weight 1 to the derivation D and $i+1$ to the variables w_i. All pre-Hamiltonian operators $\mathcal{N}$ of orders 2–6 with polynomial coefficients of such homogeneity are given below. It turned out that all of them can be fully factorized.

- Second order operator:
$$\mathcal{N}_1^{(2)} = D(D+w);$$

- Third order operator:

$$\mathcal{N}_1^{(3)} = D\left(D+w\right)\left(D+w\right);$$

- Fourth order operators:

$$\mathcal{N}_1^{(4)} = D\left(D+w\right)\left(D+w\right)\left(D+w\right),$$
$$\mathcal{N}_2^{(4)} = D\left(D+w\right)\left(D+w\right)\left(D+2w\right);$$

- Fifth order operators:

$$\mathcal{N}_1^{(5)} = D\left(D+w\right)\left(D+w\right)\left(D+w\right)\left(D+w\right),$$
$$\mathcal{N}_2^{(5)} = D\left(D+w\right)\left(D+w\right)\left(D+2w\right)\left(D+3w\right);$$

- Sixth order operators:

$$\mathcal{N}_1^{(6)} = D\left(D+w\right)\left(D+w\right)\left(D+w\right)\left(D+w\right)\left(D+w\right),$$
$$\mathcal{N}_2^{(6)} = D\left(D+w\right)\left(D+w\right)\left(D+w\right)\left(D+w\right)\left(D+2w\right),$$
$$\mathcal{N}_3^{(6)} = D\left(D+w\right)\left(D+w\right)\left(D+2w\right)\left(D+3w\right)\left(D+3w\right),$$
$$\mathcal{N}_4^{(6)} = D\left(D+w\right)\left(D+w\right)\left(D+2w\right)\left(D+3w\right)\left(D+4w\right).$$

The operator $\mathcal{N}_2^{(5)}$ coincides with (5.40).

In the paper [168], these examples were continued to infinite series of pre-Hamiltonian operators of arbitrary order. In [170] some examples of pre-Hamiltonian operators with matrix coefficients were constructed.

By analogy with compatible Hamiltonian operators, pairs of compatible pre-Hamiltonian operators were considered in [142].

Open problem 5.5.15. *Develop a geometric theory of compatible pre-Hamiltonian operators. Study algebraic structures arising in the case when the coefficients of these operators are polynomial.*

5.6 Integrable multi-component Liouville-type hyperbolic systems

Consider multi-component systems of the form

$$\vec{\mathbf{u}}_{xy} = \vec{\mathbf{F}}(x, y, \vec{\mathbf{u}}, \vec{\mathbf{u}}_x, \vec{\mathbf{u}}_y) \qquad \vec{\mathbf{u}} = (u^1, \ldots, u^N). \tag{5.41}$$

The definition of the Darboux integrability is transferred to such systems, in fact, unchanged. Namely, the existence of N independent y and x-integrals is required. We call y-integrals $W_1, \ldots, W_N$ *independent* if the integrals $D^j(W_i)$, where $i = 1, \ldots, N$, $j = 0, 1, \ldots$, are functionally independent.

Most of the definitions, constructions, and statements about Liouville-type equations can be generalized to the case of systems (5.41). However, when trying to determine Laplace invariants, one serious difficulty arises. For systems (5.41) the linearization operator (5.12) has matrix coefficients of dimension $N \times N$.

A direct generalization of the construction of Section 5.3 leads to the following. The main Laplace invariants for the operator

$$L = D\bar{D} + aD + b\bar{D} + c, \qquad a, b, c \in \text{Mat}_N \tag{5.42}$$

are $N \times N$ matrices defined by the formulas

$$H_1 = D(a) + ba - c, \qquad H_0 = \bar{D}(b) + ab - c.$$

The matrices H_i for $i > 1$ are recurrently determined from the following system of equations:

$$\bar{D}H_i - H_i a_{i-1} + a_i H_i = 0, \tag{5.43}$$

$$H_{i+1} = 2H_i + D(a_i - a_{i-1}) + [b, a_i - a_{i-1}] - H_{i-1}, \tag{5.44}$$

where $a_0 = a$. Obviously, in the scalar case, these formulas coincide with the corresponding formulas from (5.8).

Suppose that H_i for $i \leq k$ and a_i for $i \leq k - 1$ are already known. Then we define a_k from (5.43) and after that we find H_{k+1} from (5.44). However, if $\det H_k = 0$, then the matrix a_k either does not exist or is not unique, but is defined up to a matrix α such that $\alpha H_k = 0$. In the latter case, the existence and properties of the next Laplace invariants can significantly depend on the choice of α.

The degeneration of $\det H_k = 0$ for some k is typical for open Toda lattices.

Example 5.6.1. Consider the A_2-Toda lattice

$$u_{xy} = -2 \exp u + \exp v, \qquad v_{xy} = \exp u - 2 \exp v.$$

The linearization operator has the form

$$D\bar{D} + \begin{pmatrix} 2 \exp u & -\exp v \\ -\exp u & 2 \exp v \end{pmatrix}.$$

In this case, the invariant

$$H_0 = \begin{pmatrix} -2 \exp u & \exp v \\ \exp u & -2 \exp v \end{pmatrix}$$

is a nondegenerate matrix. Using (5.43), we find

$$a_1 = \frac{1}{3} \begin{pmatrix} -4u_y + v_y & -2u_y + 2v_y \\ 2u_y - 2v_y & u_y - 4v_y \end{pmatrix},$$

and

$$H_1 = \begin{pmatrix} \exp u - 2\exp v & 2\exp u - \exp v \\ -\exp u + 2\exp v & -2\exp u + \exp v \end{pmatrix}.$$

We see that $\det H_1 = 0$.

The following general statement (see [171]) shows that the existence of integrals implies (cf. Theorem 5.3.8) the degeneration of Laplace invariants:

Theorem 5.6.2. *If a system* (5.41) *has nontrivial y and x-integrals*

$$W(x, y, \vec{u}, \vec{u}_x, \ldots, \vec{u}_p), \qquad \bar{W}(x, y, \vec{u}, \vec{u}_y, \ldots, \vec{u}_{\bar{p}}),$$

then $\det H_r = \det H_{-s} = 0$ *for some integers $r \le \bar{p}$ and $s \le p - 1$.*

According to this theorem, for Darboux integrable systems, some of the Laplace invariants must be degenerate and we are faced with the question

Question 5.6.3. *How to correctly define a sequence of Laplace invariants?*

In [110], it was observed that for this, it is useful together with the invariants H_i to consider their products

$$Z_k = H_k H_{k-1} \cdots H_1.$$

From (5.43) it follows that

$$Z_k (\bar{D} + a) = (\bar{D} + a_k) Z_k. \tag{5.45}$$

If the matrices H_i (and, therefore, Z_k) for $i \le k$ and the matrices a_i for $i \le k - 1$ have already been found, then we define a_k from (5.45) and, after that, find H_{k+1} from the relation

$$H_{k+1} = (D + b) a_k - (\bar{D} + a_k) b + H_k. \tag{5.46}$$

One can verify that this procedure is equivalent to the recurrence (5.43), (5.44). The matrix a_k is defined up to an arbitrary matrix α such that $\alpha Z_k = 0$.

Theorem 5.6.4. [110, 172] *Suppose that the invariants H_i with $i \le k$ are already known and for $i < k$ the following conditions are satisfied*

$$(\bar{D} + a) \left(\ker Z_i \right) \subset \ker Z_i \tag{5.47}$$

and

$$(D - b^T)\left(\ker Z_i^T\right) \subset \ker Z_i^T. \tag{5.48}$$

Then the matrix a_k exists iff the condition (5.47) is satisfied for $i = k$. Moreover, Z_{k+1} does not depend on the choice of the arbitrary matrix α appearing in the formula for a_k iff the condition (5.48) is satisfied for $i = k$.

Proof. From (5.45) it follows that the condition (5.47) with $i = k$ is necessary for the existence of a_k. Sufficiency follows from the Kronecker–Capelli theorem. The formula (5.44) implies the condition (5.48). □

Remark 5.6.5. *For operators (5.42) with $a = b = 0$, the conditions (5.47), (5.48) are satisfied iff (see [172]) the vector spaces $\ker Z_k$ and $\ker Z_k^T$ have bases consisting of vectors, which belong to $\ker \bar{D}$ and $\ker D$, respectively.*

Example 5.6.6. For the open A_3-Toda lattice

$$\begin{cases} (u_1)_{xy} = 2\exp u_1 - \exp u_2, \\ (u_2)_{xy} = -\exp u_1 + 2\exp u_2 - \exp u_3, \\ (u_3)_{xy} = -\exp u_2 + 2\exp u_3, \end{cases}$$

all matrices Z_k are uniquely determined and rank $Z_k = 4-k$. In particular, $Z_4 = 0$. The vector $\mathbf{e}_1 = (1,1,1)^T$ forms a basis in $\ker Z_2$. As a basis in $\ker Z_3$ one can choose the vectors $\mathbf{e}_1$ and $\mathbf{e}_2 = (1,0,-1)^T$. The bases in $\ker Z_2^T$ and $\ker Z_3^T$ are $\mathbf{f}_1 = (3,4,3)^T$ and $\mathbf{f}_1, \mathbf{f}_2 = (1,0,-1)^T$. Thus, for this Toda lattice, the vector spaces $\ker Z_k$ and $\ker Z_k^T$ admit constant bases and, therefore, the conditions (5.47), (5.48) are fulfilled.

Example 5.6.7. For the C_3-Toda lattice

$$\begin{cases} (u_1)_{xy} = 2\exp u_1 - \exp u_2, \\ (u_2)_{xy} = -\exp u_1 + 2\exp u_2 - \exp u_3, \\ (u_3)_{xy} = -2\exp u_2 + 2\exp u_3, \end{cases}$$

the matrices Z_k are uniquely determined and rank $Z_1 = 3$, rank $Z_2 = 2$, rank $Z_3 = 2$, rank $Z_4 = 1$, rank $Z_5 = 1$, and $Z_6 = 0$. All vector spaces $\ker Z_k$ and $\ker Z_k^T$ admit constant bases.

Example 5.6.8. In the case of the D_3-Toda lattice

$$\begin{cases} (u_1)_{xy} = 2\exp u_1 - \exp u_2 - \exp u_3, \\ (u_2)_{xy} = -\exp u_1 + 2\exp u_2, \\ (u_3)_{xy} = -\exp u_1 + 2\exp u_3, \end{cases}$$

we have rank $Z_k = 4 - k$. All vector spaces $\ker Z_k$ and $\ker Z_k^T$ have constant bases.

In [110] the following was formulated

Conjecture 5.6.9. *For any Toda lattice* (5.1) *the numbers i, for which the rank of Z_i goes down, coincide with the indices of the corresponding simple Lie algebra, and the number h such that $Z_h = 0$, is equal to its Coxeter number.*

For the classical simple Lie algebras, this conjecture was proved in [173]. Later A.M. Gurieva verified [174] it for all exceptional simple Lie algebras. In the following example, we present the formulas for Z_k in the A_n-case.

Example 5.6.10. [173] We write the A_n-Toda lattice with the Cartan matrix

$$A = \begin{pmatrix} 2 & -1 & 0 & 0 & \ldots & 0 & 0 & 0 \\ -1 & 2 & -1 & 0 & \ldots & 0 & 0 & 0 \\ 0 & -1 & 2 & -1 & \ldots & 0 & 0 & 0 \\ \cdot & \cdot & \cdot & \cdot & \ldots & \cdot & \cdot & \cdot \\ 0 & 0 & 0 & 0 & \ldots & -1 & 0 & 0 \\ 0 & 0 & 0 & 0 & \ldots & 2 & -1 & 0 \\ 0 & 0 & 0 & 0 & \ldots & -1 & 2 & -1 \\ 0 & 0 & 0 & 0 & \ldots & 0 & -1 & 2 \end{pmatrix}$$

as

$$D\bar{D}\mathbf{u} = AU\mathbf{c},$$

where

$$\mathbf{u} = (u_1, u_2, u_3, \ldots, u_{n-1}, u_n)^T, \qquad \mathbf{c} = (1, 1, 1, \ldots, 1, 1)^T,$$

$$U = \mathrm{diag}\Big(\exp(u_1),\ \exp(u_2), \ldots,\ \exp(u_n)\Big).$$

The linearization operator is given by

$$L = D\overline{D} - AU.$$

For this operator we have

$$Z_k = AJ^{1-k}S_k\left(J^T\right)^{1-k}, \qquad k = 1, 2, \ldots, n.$$

Here

$$S_k = \mathrm{diag}\left\{0, 0, \ldots, 0, \exp\Big(\sum_{i=1}^{k} u^i\Big), \exp\Big(\sum_{i=2}^{k+1} u^i\Big), \ldots, \exp\Big(\sum_{i=nk}^{n-1} u^i\Big), \exp\Big(\sum_{i=nk+1}^{n} u^i\Big)\right\}$$

and

$$
J = \begin{pmatrix}
1 & 1 & 1 & \cdots & 1 & 1 \\
0 & 1 & 1 & \cdots & 1 & 1 \\
\cdot & \cdot & \cdot & \cdots & \cdot & \\
0 & 0 & 0 & \cdots & 1 & 1 \\
0 & 0 & 0 & \cdots & 0 & 1
\end{pmatrix}.
$$

Clearly, $S_{n+1} = 0$. It is easy to see that rank $Z_k = n - k + 1$, $k = 1, 2, \ldots$. Thus, the number k such that rank $Z_{k+1} <$ rank Z_k coincides with the exponents $1, 2, \ldots, n$ for the Lie algebra A_n, and the number $k = n + 1$, for which $Z_k = 0$, equal to the Coxeter number.

Remark 5.6.11. *It would be interesting to understand the algebraic meaning of the matrices that arise in Example 5.6.10 and to generalize them to the case of arbitrary simple Lie algebra.*

For $i > 0$ the invariants H_{-i} and the products of $Z_{-i} = H_{-i}H_{-(i-1)} \ldots H_0$ are determined from the formulas

$$
D(Z_{1-i}) - Z_{1-i}b + b_{-i}Z_{1-i} = 0,
$$

$$
H_{-i} = 2H_{1-i} + \bar{D}(b_{-i} - b_{1-i}) + [a, b_{-i} - b_{1-i}] - H_{2-i}.
$$

The conditions (5.47) and (5.48) are replaced by

$$
(D + b)\left(\ker Z_{-i}\right) \subset \ker Z_{-i}, \qquad (\bar{D} - a^T)\left(\ker Z_{-i}^T\right) \subset \ker Z_{-i}^T.
$$

The termination of the sequence of the matrices Z_i can be chosen as the definition of Liouville-type systems (5.41).

Definition 5.6.12. Suppose that for a system of the form (5.41) all conditions (5.47) and (5.48) are fulfilled and there exist numbers $r \geq 1$ and $s \geq 0$ such that $Z_r = Z_{-s} \equiv 0$. Then (5.41) is called a *Liouville-type system*.

In the scalar case, Theorem 5.3.8 shows that equation (5.3) is Darboux integrable (see Definitions 5.1.5) iff it is a Liouville-type equation. For multicomponent systems, this is not the case. In [172], an example of a system that is Darboux integrable but not a Liouville-type system was constructed.

Open problem 5.6.13. *Prove that any Liouville-type system is Darboux integrable.*

Using Definition 5.6.12, we find all systems

$$
\begin{cases}
(u_1)_{xy} = 2\exp u_1 + k_1 \exp u_2, \\[2mm]
(u_2)_{xy} = k_2 \exp u_1 + 2\exp u_2
\end{cases}
\tag{5.49}
$$

with a nondegenerate ($k_1 k_2 \neq 4$) nondiagonal Cartan matrix that are of the Liouville type.

It is easy to verify that for any k_1, k_2 we have rank $Z_1 = 2$, rank $Z_2 = 1$. Next, $Z_3 = 0$ only if $k_1 = k_2 = -1$ (open A_2-Toda lattice).

If $Z_3 \neq 0$, then the condition (5.47) with $i = 3$ is satisfied only if $k_1 = -1$ or $k_2 = -1$. Without loss of generality, we set $k_1 = -1$. Then $Z_4 = 0$ only in the case $k_2 = -2$ (C_2-Toda lattice).

If $k_2 \neq -2$, then rank $Z_5 = 1$ and the condition (5.47) with $i = 5$ is satisfied only if $k_2 = -3$. In this case, $Z_6 = 0$ (G_2-Toda lattice).

Thus, we have proved that all Liouville-type systems (5.49) are exhausted by the Toda lattices corresponding to the simple Lie algebras of rank 2.

Exercise 5.6.14. Prove a similar statement in the case of systems of rank 3.

Chapter 6

Integrable nonabelian equations

6.1 ODE on free associative algebras

6.1.1 Equations with matrix variables

Consider systems of ordinary differential equations of the form

$$\frac{du_\alpha}{dt} = F_\alpha(\mathbf{u}), \qquad \mathbf{u} = (u_1, \ldots, u_N), \tag{6.1}$$

where $u_i(t)$ are $m \times m$ matrices, and F_α are (noncommutative) polynomials with constant scalar coefficients. As in Section 1.4, an infinitesimal symmetry is defined as a system

$$\frac{du_\alpha}{d\tau} = G_\alpha(\mathbf{u}), \tag{6.2}$$

compatible with (6.1).

Integrable systems with matrix variables are fundamental models in the theory of integrable systems, since from each of them one can obtain a number of different integrable systems using the reductions.[1]

Manakov's top

It is known that the system

$$u_t = u^2 v - v u^2, \qquad v_t = 0 \tag{6.3}$$

has polynomial symmetries of arbitrarily high degree in the case of matrices u and v of any dimension. In addition, it has a Lax representation (see Example

---

[1] By reduction we mean additional algebraic constraints on unknown matrices that do not contradict the system (6.1).

1.2.2), which generates polynomial first integrals. Namely, it turns out that for any $i, j \in \mathbb{N}$ system (6.3) has first integrals of the form

$$I = \operatorname{tr}(P), \tag{6.4}$$

where $P(u, v)$ is some noncommutative polynomial. These integrals have double homogeneity, that is, they have the form $\operatorname{tr} P_{i,j}$, where $P_{i,j}$ is a polynomial of degree i in the variable v and degree j in u. For example,

$$P_{2,2} = 2v^2u^2 + vuvu, \qquad P_{3,2} = v^3u^2 + v^2uvu, \qquad P_{2,3} = v^2u^3 + vuvu^2. \tag{6.5}$$

Remark 6.1.1. *From the property* $\operatorname{tr}(uv - vu) = 0$ *it follows that the polynomial P in the formula (6.4) is defined up to cyclic permutations of the factors in its monomials or, equivalently, up to adding a linear combination of commutators of matrix polynomials.*

Both well-known and new *multi-component* integrable systems can be obtained using reductions in system (6.3).

Example 6.1.2. If u is a matrix of dimension $m \times m$ such that $u^T = -u$, and v is a diagonal constant matrix, then the system (6.3) is equivalent to Euler's m-dimensional top. The integrability of this model by the inverse scattering method was established by S. V. Manakov [18].

Example 6.1.3. Consider the cyclic reduction

$$u = \begin{pmatrix} 0 & u_1 & 0 & 0 & \cdot & 0 \\ 0 & 0 & u_2 & 0 & \cdot & 0 \\ \cdot & \cdot & \cdot & \cdot & \cdot & \cdot \\ 0 & 0 & 0 & 0 & \cdot & u_{n-1} \\ u_n & 0 & 0 & 0 & \cdot & 0 \end{pmatrix}, \qquad v = \begin{pmatrix} 0 & 0 & 0 & \cdot & 0 & J_n \\ J_1 & 0 & 0 & \cdot & 0 & 0 \\ 0 & J_2 & 0 & \cdot & 0 & 0 \\ \cdot & \cdot & \cdot & \cdot & \cdot & \cdot \\ 0 & 0 & 0 & \cdot & J_{n-1} & 0 \end{pmatrix},$$

where u_k and J_k are some matrices. Since the number m is arbitrary, the dimension of these matrices is arbitrary as well. It is easy to see that, with such a reduction, system (6.3) turns out to be equivalent to the nonabelian Volterra chain

$$\frac{d}{dt}u_k = u_k u_{k+1} J_{k+1} - J_{k-1} u_{k-1} u_k, \qquad k \in \mathbb{Z}_n.$$

If we set $n = 3$, $J_1 = J_2 = J_3 = 1$ and $u_3 = -u_1 - u_2$, then the Volterra chain will turn into the matrix system [175, section 3.1]

$$u_t = u^2 + uv + vu, \qquad v_t = -v^2 - uv - vu. \tag{6.6}$$

There is the following integrable generalization [177] of system (6.3) to the case of any number N of the matrix variables:

$$\frac{du_\alpha}{dt} = \sum_{\beta \neq \alpha} \frac{u_\alpha u_\beta^2 - u_\beta^2 u_\alpha}{(\lambda_\alpha - \lambda_\beta)c_\beta} + \sum_{\beta \neq \alpha} \frac{u_\beta u_\alpha^2 - u_\alpha^2 u_\beta}{(\lambda_\alpha - \lambda_\beta)c_\alpha}.$$

Here,

$$\sum_1^N u_\alpha = C,$$

where C is a constant matrix.

Integrable matrix generalization of flow on an elliptic curve

Consider the following matrix ODE system:

$$\begin{cases} u_t = v^2 + cu + a\,\mathbf{1}, \\ v_t = u^2 - cv + b\,\mathbf{1} \end{cases}, \tag{6.7}$$

where u and v are $m \times m$ matrices, $\mathbf{1}$ is the identity matrix, and a, b and c are arbitrary scalar constants.

In the case $m = 1$, we have a system of two equations which can be written in the Hamiltonian form

$$u_t = -\frac{\partial H}{\partial v}, \qquad v_t = \frac{\partial H}{\partial u}$$

with Hamiltonian

$$H = \frac{1}{3}u^3 - \frac{1}{3}v^3 - cuv + bu - av.$$

For generic a, b, c, the relation $H = const$ defines an elliptic curve, and system (6.7) describes the motion of a point along this curve.

For an arbitrary m system (6.7) remains to be Hamiltonian with Hamiltonian

$$H = \mathrm{tr}\left(\frac{1}{3}u^3 - \frac{1}{3}v^3 - cuv + bu - av\right) \tag{6.8}$$

and a constant trace Poisson bracket (see Section 6.2.1).

Perhaps the nonabelian system (6.7) can serve as one of the cornerstones on which a theory of nonabelian elliptic functions can be build. We show that this system is integrable.

We first consider the homogeneous matrix system (6.7) [175]

$$u_t = v^2, \qquad v_t = u^2. \tag{6.9}$$

This system, like system (6.3), has interesting reductions.

Remark 6.1.4. *F. Calogero noticed* [176] *that the functions* $z_i = \lambda_i^{1/2}$*, where* λ_i *are the eigenvalues of the matrix* $u - v$*, satisfy the integrable system*

$$z_i'' = -z_i^5 + \sum_{j \neq i} \left[(z_i - z_j)^{-3} + (z_i + z_j)^{-3} \right], \qquad i = 1, \ldots, m.$$

System (6.9) has a Lax pair

$$L = \begin{pmatrix} 1 & 0 & 0 \\ 0 & \varepsilon & 0 \\ 0 & 0 & \varepsilon^2 \end{pmatrix} \lambda + \begin{pmatrix} 0 & 3\varepsilon u & 3v \\ v & 0 & (\varepsilon - 1)u \\ u & (2\varepsilon + 1)v & 0 \end{pmatrix}, \tag{6.10}$$

$$A = -\frac{1}{3} \begin{pmatrix} \varepsilon^2 & 0 & 0 \\ 0 & \varepsilon & 0 \\ 0 & 0 & 1 \end{pmatrix} \lambda + \frac{1}{3} \begin{pmatrix} 0 & 3\varepsilon^2 u & 3v \\ \varepsilon v & 0 & (\varepsilon + 2)u \\ u & (1 - \varepsilon)v & 0 \end{pmatrix}, \tag{6.11}$$

where

$$\varepsilon^2 + \varepsilon + 1 = 0. \tag{6.12}$$

This Lax representation can be extracted from the Lax pair

$$L = \lambda v + u, \qquad A = \frac{1}{\lambda} u^2$$

for system (6.3) and from the fact that (6.9) is related to (6.6) by a linear change of variables.

Proposition 6.1.1. *The Lax equation* $\bar{L}_t = [A, \bar{L}]$*, where*

$$\bar{L} = \lambda L + \lambda c P + a Q + b R, \tag{6.13}$$

the operators L *and* A *are defined by formula* (6.10) *and*

$$P = \begin{pmatrix} \varepsilon + 2 & 0 & 0 \\ 0 & -2\varepsilon - 1 & 0 \\ 0 & 0 & \varepsilon - 1 \end{pmatrix}, \qquad Q = \begin{pmatrix} 0 & 3(\varepsilon + 2) & 0 \\ 0 & 0 & -3 \\ \varepsilon - 1 & 0 & 0 \end{pmatrix},$$

$$R = \begin{pmatrix} 0 & 0 & 3(1 - \varepsilon) \\ 2\varepsilon + 1 & 0 & 0 \\ 0 & -3\varepsilon & 0 \end{pmatrix},$$

is equivalent to system (6.7).

It is easy to verify that all entries of the matrix $uv - vu$ are first integrals for system (6.7). In addition to these integrals, there is a sequence of integrals of the form (6.4).

It is clear that $\operatorname{tr} M_n$, where $M_n = (vu - uv)^n$, is an integral of degree $2n$. Nontrivial integrals are generated by the Lax representation. According to Lemma 1.2.1, we have

$$\left(\operatorname{tr}\left(\bar{L}^k\right)\right)_t = 0, \qquad k \in N. \tag{6.14}$$

Each of the expressions $\operatorname{tr}\left(\bar{L}^k\right)$ is a polynomial in λ whose coefficients are first integrals. These integrals depend polynomially on the parameters a, b, c.

Furthermore, after replacing ε^2 by $-\varepsilon - 1$, all obtained integrals are linear in ε. Since ε is any of the two solutions of the quadratic equation (6.12), the coefficients of ε^1 and ε^0 are integrals.

Formula (6.14) with $k = 1, 2$ leads to trivial integrals. In the case $k = 3$, the Hamiltonian (6.8) arises. Relations (6.14) for $k = 4, \ldots, 9$ generate integrals $\operatorname{tr}(T_i)$, where

$$\begin{aligned}
T_1 =\ & v^6 - 6v^3u^3 + 6v^2uvu^2 - 2vuvuvu + u^6 + 6v^4uc - 6vu^4c + 6v^4a - 6vu^3a \\
& -6v^3ub + 6u^4b + 9vuvuc^2 + 18v^2uac - 18vu^2bc - 18vuab + 9v^2a^2 + 9u^2b^2,
\end{aligned}$$

$$\begin{aligned}
T_2 =\ & v^5u^2 - 2v^4uvu + v^3uv^2u + 2vu^4vu - v^2u^5 - vu^3vu^2 + 3v^2uvu^2c - 3vuvuvuc \\
& +3v^3u^2a - 3v^2uvua - 3v^2u^3b + 3vu^2vub,
\end{aligned}$$

$$\begin{aligned}
T_3 =\ & v^5u^2vu - v^5uvu^2 - v^4u^2v^2u + v^4uv^2u^2 - v^3uv^2uvu + v^3uvuv^2u - v^2u^5vu \\
& +v^2u^4vu^2 - v^2u^2vu^4 + v^2uvu^5 - vu^3vu^2vu + vu^3vuvu^2 + 3v^2u^2vuvuc \\
& -3v^2uvuvu^2c + 3v^3u^2vua - 3v^3uvu^2a + 3v^2uvu^3b - 3v^2u^3vub,
\end{aligned}$$

$$\begin{aligned}
T_4 =\ & v^9 - 9v^6u^3 + 9v^5uvu^2 + 9v^4u^2v^2u - 9v^4uvuvu + 9v^3u^6 - 9v^3u^2v^3u \\
& +9v^3uv^2uvu - 9v^2u^4vu^2 + 9v^2u^3vu^3 - 3v^2uv^2uv^2u - 9v^2uvu^5 + 9vu^4vuvu \\
& +3vu^2vu^2vu^2 - 9v^3u^3vuc - 9v^3u^2vu^2c - 9v^3uvu^3c - 27v^3uvubc \\
& -9v^2u^3v^2uc + 9v^2u^2vuvuc + 36v^2uvuvu^2c + 9vu^7c + 81vu^3b^2c \\
& -18vuvuvuvuc - 9v^3u^2vua + 18v^3uvu^2a + 18v^2u^2v^2ua - 9v^2uvuvua \\
& +9vu^6a + 36v^3u^4b + 9v^2u^3vub - 18v^2u^2vu^2b - 18v^2uvu^3b + 9vu^2vuvub \\
& -9u^7b - 27v^2u^4ac - 27v^2u^3a^2 + 27v^3u^2b^2 + 54vu^5bc - 27vu^4vuc^2 \\
& +54vu^4ab - 27vu^3vuac - 27u^5b^2 + 81v^3ua^2c + 81v^2uvuac^2 - 81v^2ua^2b \\
& -81vu^2vubc^2 + 81vu^2ab^2 + 27vuvuvuc^3 - 162vuvuabc + 27u^3a^3 - 27u^3b^3 \\
& -81vua^3c - 81va^4 + 81ua^3b.
\end{aligned}$$

It can be verified that all integrals of the form (6.4) of degree not higher than 9 are linear combinations of integrals $H, M_2, \ldots, M_4$ and $\operatorname{tr}(T_i)$, $i = 1, \ldots, 4$.

Yu. Suris informed me that the integrals $H, T_1, \ldots, T_4$ together with the entries of the matrix $uv - vu$ generate a full set of functionally independent integrals necessary for the Liouville integrability of the Hamiltonian system (6.7) in the case 2×2 and 3×3 matrices.

6.1.2   Systems of differential equations on free associative algebra

The matrix systems considered in the previous section have symmetries for any dimension of matrices. This means that between these matrices there are no polynomial relations that have to be taken into account when proving integrability. In particular, when verifying whether (6.2) is a symmetry of the system (6.1), only the associativity of the matrix multiplication is important. All calculations work only with noncommutative polynomials of unknown matrices, but never with their entries. This suggests that the symbols $u_1, \ldots, u_N$ should be considered as generators of the free unital associative algebra $\mathcal{A}$ over $\mathbb{C}$.

The following questions arise immediately:

Question 6.1.5. *What do formulas similar to (6.1) mean?*[2]

Question 6.1.6. *How to define the functional* tr *in the formula* (6.4)?[3]

Definition 6.1.7. A $\mathbb{C}$-linear mapping $d : \mathcal{A} \to \mathcal{A}$ is called *derivation* if it satisfies the Leibniz identity $d(xy) = xd(y) + d(x)y$.

If we fix polynomials $F_i(u_1, \ldots, u_N)$ and set $d(u_i) = F_i$, then the derivation $d(z)$ is uniquely determined for any element $z \in \mathcal{A}$ by the Leibniz identity. It is clear that the polynomials F_i can be chosen arbitrarily.

Instead of the dynamical system (6.1), we consider the derivation $D_t : \mathcal{A} \to \mathcal{A}$ such that $D_t(u_i) = F_i$. The compatibility of systems (6.1) and (6.2) means that the corresponding derivations D_t and D_τ commute: $D_t D_\tau - D_\tau D_t = 0$.

To emphasize that the right-hand sides in the systems (6.1) are elements of $\mathcal{A}$, we will call such systems *nonabelian*.

First integrals of nonabelian systems

Definition 6.1.8. We call elements $f_1, f_2 \in \mathcal{A}$ *equivalent* and denote it as $f_1 \sim f_2$, if f_1 can be obtained from f_2 by cyclic permutations of the generators in the monomials of the polynomial f_2. Denote by $\mathrm{tr}\,(f)$ the equivalence class of the element f.

Remark 6.1.9. *Any commutator is equivalent to zero and vice versa, if $f_1 \sim f_2$, then there exist elements a_i, b_i such that $f_1 - f_2 = \sum_i [a_i, b_i]$. Therefore, we can consider* $\mathrm{tr}\,(f)$ *as an element of the quotient vector space* $\mathcal{T} = \mathcal{A}/[\mathcal{A}, \mathcal{A}]$. *It is easy to see that the derivation D_t is well defined on* $\mathcal{T}$.

---

[2]Certainly we are not going to assume that the generators of the algebra $\mathcal{A}$ depend on the parameter t.

[3]The assumption that $\mathcal{A}$ is equipped with a quadratic form having the desired properties is not a good idea.

Definition 6.1.10. An element $\operatorname{tr}(h) \in \mathcal{T}$ is called *first integral* of the nonabelian system (6.1) if $D_t(h) \sim 0$.

There is an obvious analogy between Definition 6.1.10 and the definition of local conserved density in the theory of evolutionary equations (see Remark 4.2.13): in both cases, the object being defined is an element of a quotient vector space on which the total derivatives are well defined.

From the standpoint of the symmetry approach, a nonabelian system (6.1) is *integrable* if it has an infinite set of symmetries linearly independent over $\mathbb{C}$.

Another criterion for integrability is the existence of an infinite set of linearly independent first integrals of the form (6.4).

Sometimes nonabelian systems (for example, the system (6.7)) possess the so-called full first integrals.

Definition 6.1.11. The element $J \in \mathcal{A}$ is called *a full first integral* for a non-abelian system (6.1) if $D_t(J) = 0$.

In the matrix case, this means that any entry of the matrix J is a first integral.

6.1.3 Quadratic homogeneous nonabelian systems

Consider nonabelian systems of the form

$$(u^i)_t = \sum_{j,k} C^i_{jk}\, u^j\, u^k, \tag{6.15}$$

where u^i, $i = 1, \ldots, N$ are the generators of the free associative algebra. Denote the set of coefficients $C^i_{jk} \in \mathbb{C}$, $i, j, k = 1, \ldots, N$ by $\mathbf{C}$.

Remark 6.1.12. *We can associate an N-dimensional algebra defined by the structural constants C^i_{jk} with any system (6.15).*

Open problem 6.1.13. *Construct series of simple algebras with the structural constants C^i_{jk} such that the corresponding equation (6.15) has a nontrivial symmetry of the form*

$$(u_i)_\tau = \sum_{j,k,m} B^i_{jkm}\, u^j\, u^k\, u^m.$$

This means that the algebra and the triple system $\{\cdot, \cdot, \cdot\}$ with structural constants B^i_{jkm} (see Section 7.1) are related by the identity

$$X \circ \{Y, Z, V\} + \{X, Y, Z\} \circ V = \{X \circ Y, Z, V\} + \{X, Y \circ Z, V\} + \{X, Y, Z \circ V\}.$$

The class of systems (6.15) is invariant with respect to the action of the group GL_N of linear transformations of the form

$$\hat{u}^i = \sum_j s^i_j\, u^j, \qquad s^i_j \in \mathbb{C}.$$

Definition 6.1.14. A function $F(\mathbf{C})$ is called a GL_N semi-invariant of degree k if

$$F(\hat{\mathbf{C}}) = (\det S)^k\, F(\mathbf{C}),$$

where S is a matrix with the entries s^i_j. Semi-invariants of degree 0 are called *invariants*.

Definition 6.1.15. The row vector $\mathbf{V}(\mathbf{C})$ is called a vector GL_N semi-invariant of degree k if

$$\mathbf{V}(\hat{\mathbf{C}}) = (\det S)^k\, \mathbf{V}(\mathbf{C})\, S.$$

Vector semi-invariants of degree 0 are called *vector invariants*.

Lemma 6.1.16. (i) *The product* $\mathrm{I} \times \mathbf{V}$ *of a semi-invariant* I *and a vector semi-invariant* $\mathbf{V}$ *of degrees* k_1 *and* k_2 *is a vector semi-invariant of degree* $k_1 + k_2$.
(ii) *Let* $\mathbf{V}_i$, $i = 1, \ldots, N$ *be vector semi-invariants of degrees* k_i. *Then, the determinant of the matrix formed by the vectors* $\mathbf{V}_i$ *is a semi-invariant of degree* $1 + \sum k_i$.

6.1.4 Two-component nonabelian systems

Consider nonabelian systems of the form

$$u_t = P(u, v), \qquad v_t = Q(u, v), \qquad P, Q \in \mathcal{A} \tag{6.16}$$

on the free associative algebra $\mathcal{A}$ with generators u and v.
We define a $\mathbb{C}$-linear involution $\star$ on $\mathcal{A}$ by the formulas

$$u^\star = u, \quad v^\star = v, \quad (a\,b)^\star = b^\star\, a^\star, \quad a, b \in \mathcal{A}. \tag{6.17}$$

In the matrix case, the transposition is an example of such an involution.

Definition 6.1.17. Two systems connected with each other by a linear transformation of the form

$$\hat{u} = \alpha u + \beta v, \qquad \hat{v} = \gamma u + \delta v, \qquad \alpha\delta - \beta\gamma \neq 0 \tag{6.18}$$

or by the involution (6.17) are called *equivalent*.

Definition 6.1.18. A system (6.16) equivalent to a system of the form

$$u_t = P(u,v), \qquad v_t = Q(v), \tag{6.19}$$

is called *triangular*. If $Q = 0$, then the system is called *strongly triangular*.

Obviously, if system (6.16) has polynomial symmetries or first integrals of the form (6.4), then the same holds for any other system equivalent to it.

Open problem 6.1.19. *Describe all nonequivalent nontriangular systems of the form (6.16) that have infinitely many linearly independent symmetries or first integrals of the form (6.4).*

The simplest class of systems (6.15) consists of two-component systems of the form

$$\begin{cases} u_t = \alpha_1 u\, u + \alpha_2 u\, v + \alpha_3 v\, u + \alpha_4 v\, v, \\ v_t = \beta_1 v\, v + \beta_2 v\, u + \beta_3 u\, v + \beta_4 u\, u. \end{cases} \tag{6.20}$$

Let us describe in invariant terms what Definition 6.1.17 means in this case. To do this, we find semi-invariants of the action of the group (6.18) on the coefficients of (6.20). According to Remark 6.1.12, such a description also leads to a classification of nonequivalent two-dimensional algebras.

Equivalence problem

The group GL_2 of transformations (6.18) acts on functions of eight coefficients α_i, β_i of systems (6.20). The corresponding infinitesimal action of the Lie algebra $\mathfrak{gl}_2$ is defined by the following vector fields:

$$X_{11} = \alpha_1 \frac{\partial}{\partial \alpha_1} - \alpha_4 \frac{\partial}{\partial \alpha_4} + \beta_2 \frac{\partial}{\partial \beta_2} + \beta_3 \frac{\partial}{\partial \beta_3} - 2\beta_4 \frac{\partial}{\partial \beta_4},$$

$$X_{22} = \beta_1 \frac{\partial}{\partial \beta_1} - \beta_4 \frac{\partial}{\partial \beta_4} + \alpha_2 \frac{\partial}{\partial \alpha_2} + \alpha_3 \frac{\partial}{\partial \alpha_3} - 2\alpha_4 \frac{\partial}{\partial \alpha_4},$$

$$X_{12} = -\beta_4 \frac{\partial}{\partial \alpha_1} + (\alpha_1 - \beta_3)\frac{\partial}{\partial \alpha_2} + (\alpha_1 - \beta_2)\frac{\partial}{\partial \alpha_3}$$
$$+ (\alpha_2 + \alpha_3 - \beta_1)\frac{\partial}{\partial \alpha_4} + (\beta_2 + \beta_3)\frac{\partial}{\partial \beta_1} + \beta_4 \frac{\partial}{\partial \beta_2} + \beta_4 \frac{\partial}{\partial \beta_3},$$

$$X_{21} = -\alpha_4 \frac{\partial}{\partial \beta_1} + (\beta_1 - \alpha_3)\frac{\partial}{\partial \beta_2} + (\beta_1 - \alpha_2)\frac{\partial}{\partial \beta_3} + (\beta_2 + \beta_3 - \alpha_1)\frac{\partial}{\partial \beta_4}$$
$$+ (\alpha_2 + \alpha_3)\frac{\partial}{\partial \alpha_1} + \alpha_4 \frac{\partial}{\partial \alpha_2} + \alpha_4 \frac{\partial}{\partial \alpha_3}.$$

The eight-dimensional vector space over $\mathbb{C}$ with the basis α_i, β_i decomposes into a direct sum of three subspaces that are invariant with respect to this action. The first one is four-dimensional and is spanned by linear polynomials

$$x_1 = \alpha_4, \qquad x_2 = \alpha_2 + \alpha_3 - \beta_1, \qquad x_3 = \beta_2 + \beta_3 - \alpha_1, \qquad x_4 = \beta_4.$$

The other two are two-dimensional and are spanned by polynomials

$$x_5 = \alpha_1 + \beta_3, \qquad x_6 = \beta_1 + \alpha_3,$$

and

$$x_7 = \alpha_1 + \beta_2 \qquad x_8 = \beta_1 + \alpha_2.$$

Denote these vector spaces by W_1, W_2 and W_3, respectively.

Lemma 6.1.20. *The vectors* $\mathbf{V}_1 = (x_5, x_6)$ *and* $\mathbf{V}_2 = (x_7, x_8)$ *are vector invariants.*

Scalar semi-invariants have a rather simple form in variables x_i. There is a single semi-invariant of degree 1 (see Definition 6.1.14):

$$I_0 = x_6 x_7 - x_5 x_8.$$

In the initial variables we have

$$I_0 = \alpha_1 \alpha_2 - \alpha_1 \alpha_3 - \alpha_3 \beta_2 - \beta_1 \beta_2 + \alpha_2 \beta_3 + \beta_1 \beta_3.$$

The following polynomials are semi-invariant of degree 2:

$$I_1 = x_2^2 x_3^2 + 4\, x_1 x_3^3 + 4\, x_2^3 x_4 + 18\, x_1 x_2 x_3 x_4 - 27\, x_1^2 x_4^2,$$

$$I_2 = (x_5 - x_7)^3\, x_1 - (x_5 - x_7)^2 (x_6 - x_8)\, x_2 - (x_5 - x_7)(x_6 - x_8)^2\, x_3 + (x_6 - x_8)^3\, x_4,$$

$$I_3 = (x_5^3 - x_7^3)\, x_1 + (x_7^2 x_8 - x_5^2 x_6)\, x_2 + (x_7 x_8^2 - x_5 x_6^2)\, x_3 + (x_6^3 - x_8^3)\, x_4,$$

$$I_4 = (x_5^2 - x_7^2)\, x_2^2 + (x_6^2 - x_8^2)\, x_3^2 + 3\,(x_5^2 - x_7^2)\, x_1 x_3 + 3\,(x_6^2 - x_8^2)\, x_2 x_4 + 9\,(x_7 x_8 - x_5 x_6)\, x_1 x_4.$$

Remark 6.1.21. *The semi-invariant* I_1 *is symmetric with respect to the involution* (6.17), *and the others are skew-symmetric.*

Proposition 6.1.2. *Any invariant* J *of the transformation group* (6.18) *is a function of four functionally independent invariants* $J_i = \dfrac{I_i}{I_0^2}$, $i = 1, 2, 3, 4$.

Proof. Invariants are solutions of the system of partial differential equations $X_{ij}(J) = 0$, $i, j = 1, 2$. It is easy to verify that the vector fields X_{ij} are linearly independent over the field of functions. Therefore, there are exactly four functionally independent invariants. A straightforward calculation of the rank of the Jacobi matrix shows that the invariants J_i, $i = 1, 2, 3, 4$, are functionally independent. □

Remark 6.1.22. *It can be verified that there are 9 linearly independent over* $\mathbb{C}$ *polynomial semi-invariants of degree 2. In addition, there are 7 linearly independent polynomial vector semi-invariants of degree 1. All polynomial semi-invariants of degree 2 can be obtained using the construction from Item* (ii) *of Lemma 6.1.16, applied to the semi-invariants of degree 1 and to the vector invariants from Lemma 6.1.20.*

Example 6.1.23. The vectors

$$\mathbf{V}_3 = \Big(- \beta_4(\alpha_3 - \alpha_2)^2 + (\beta_3 - \beta_2)(\alpha_1\alpha_2 - \alpha_1\alpha_3 - \alpha_2\beta_2 + \alpha_3\beta_3),$$
$$\alpha_4(\beta_3 - \beta_2)^2 - (\alpha_3 - \alpha_2)(\beta_1\beta_2 - \beta_1\beta_3 - \beta_2\alpha_2 + \beta_3\alpha_3) \Big)$$

and

$$\mathbf{V}_4 = \Big((\beta_2 - \beta_3)\alpha_4\beta_4 - (\alpha_2 - \alpha_3)(\alpha_2 - \beta_1)\beta_4 + \beta_3(\alpha_1\alpha_2 - \alpha_1\alpha_3 - \alpha_2\beta_2 + \alpha_3\beta_3),$$
$$- (\alpha_2 - \alpha_3)\alpha_4\beta_4 + (\beta_2 - \beta_3)(\beta_2 - \alpha_1)\alpha_4 - \alpha_3(\beta_1\beta_2 - \beta_1\beta_3 - \beta_2\alpha_2 + \beta_3\alpha_3) \Big)$$

are vector semi-invariants of degree 1.

Proposition 6.1.3. *A system* (6.20) *is triangular iff* $\mathbf{V}_3 = \mathbf{V}_4 = 0$.

Remark 6.1.24. *The nontriangularity of a system* (6.20) *is equivalent to the simplicity of the corresponding algebra* (*see Remark 6.1.12*).

The following statement is obvious.

Lemma 6.1.25. *A system* (6.20) *is strongly triangular iff the vectors* $(\alpha_1, \alpha_2, \alpha_3, \alpha_4)$ *and* $(\beta_4, \beta_3, \beta_2, \beta_1)$ *are linearly dependent.*

Using the invariants and semi-invariants found above, it is easy to construct a set of nonequivalent canonical forms such that every system (6.20) is equivalent to one of them.

It is natural to begin with the consideration of the vector invariants $\mathbf{V}_i$ from Lemma 6.1.20. If these vectors are linearly independent, then by a transformation (6.18) one can normalize $x_5 = x_8 = 1$, $x_6 = x_7 = 0$. After the normalization, there are no admissible transformations (6.18) and nonequivalent systems are characterized by different values of the scalar invariants.[4]

Exercise 6.1.26. Construct a complete set of canonical forms in cases when $\mathbf{V}_1$ and $\mathbf{V}_2$ are linearly dependent and when $\mathbf{V}_1 = \mathbf{V}_2 = 0$.

Examples of integrable nonabelian systems

Several nonabelian nontriangular systems with symmetries were found in [175]. The following straightforward strategy was used. At first, all systems having cubic symmetry were found. Then, systems were found that do not have cubic symmetries but possess a fourth degree symmetry, etc. In this case, by analogy with Remark 4.1.11, it seemed likely that new systems with symmetries of degree 5 would not arise.

Proposition 6.1.4. *Any nontriangular system* (6.20) *that has a nonzero symmetry of the form*

$$\begin{cases} u_\tau = \gamma_1 u\,u\,u + \gamma_2 u\,u\,v + \gamma_3 u\,v\,u + \gamma_4 v\,u\,u + \gamma_5 u\,v\,v + \gamma_6 v\,u\,v + \gamma_7 v\,v\,u + \gamma_8 v\,v\,v, \\ v_\tau = \delta_1 u\,u\,u + \delta_2 u\,u\,v + \delta_3 u\,v\,u + \delta_4 v\,u\,u + \delta_5 u\,v\,v + \delta_6 v\,u\,v + \delta_7 v\,v\,u + \delta_8 v\,v\,v \end{cases}$$

is equivalent to one of the following systems:

$$(a): \begin{cases} u_t = u\,u - u\,v, \\ v_t = v\,v - u\,v + v\,u, \end{cases} \qquad (b): \begin{cases} u_t = u\,v, \\ v_t = v\,u, \end{cases}$$

$$(c): \begin{cases} u_t = u\,u - u\,v, \\ v_t = v\,v - u\,v, \end{cases} \qquad (d): \begin{cases} u_t = -u\,v, \\ v_t = v\,v + u\,v - v\,u, \end{cases}$$

$$(e): \begin{cases} u_t = u\,v - v\,u, \\ v_t = u\,u + u\,v - v\,u, \end{cases} \qquad (f): \begin{cases} u_t = v\,v, \\ v_t = u\,u. \end{cases}$$

Remark 6.1.27. *It is remarkable that the requirement of the existence of only one cubic symmetry leads to a finite list of systems without free parameters in the coefficients (or, more precisely, all parameters can be eliminated by linear transformations* (6.18)).

---

[4]If we also include the involution (6.17) in the definition of the equivalence, then the last assertion requires clarification.

The following five systems with fourth degree symmetries were found in [175]:

$$\begin{cases} u_t = -u\,v, \\ v_t = v\,v + u\,v, \end{cases} \qquad \begin{cases} u_t = -v\,u, \\ v_t = v\,v + u\,v, \end{cases} \qquad \begin{cases} u_t = u\,u - 2v\,u, \\ v_t = v\,v - 2v\,u, \end{cases}$$

$$\begin{cases} u_t = u\,u - u\,v - 2v\,u, \\ v_t = v\,v - v\,u - 2u\,v, \end{cases} \qquad (g): \begin{cases} u_t = u\,u - 2u\,v, \\ v_t = v\,v + 4v\,u. \end{cases}$$

By means of the CRACK computer algebra system [178], T. Wolf verified that this is a complete list of nonequivalent nontriangular systems that do not have cubic symmetries, but have a symmetry of degree fourth.

Exercise 6.1.28. Using results of previous section, show that eleven systems presented above are nonequivalent.

Presumably, systems (a) and (g) have only one symmetry, while the other nine systems have an infinite set of symmetries. For some of them this is rigorously proved.

Any attempts to describe systems (6.20) with fifth degree symmetries by a direct computer computation seem hopeless. One of the reasons is that the coefficients of (6.20) turn out to be related by nontrivial algebraic relations. Even if they can be resolved, the coefficients turn out to be algebraic numbers.

Example 6.1.29. The nonabelian system

$$\begin{cases} u_t &= 11\sqrt{7}\,u\,u - 7\sqrt{7}\,v\,v, \\ v_t &= -4\sqrt{7}\,v\,u - 4\sqrt{7}\,u\,v + 30\,u\,u \end{cases}$$

has a fifth degree symmetry. Calculations show that this system apparently has no more symmetries.

Open problem 6.1.30. *Describe all nonequivalent nontriangular systems of the form* (6.20) *that have infinitely many polynomial symmetries or first integrals.*

The main defect of the nonabelian system from Example 6.1.29 is that in the scalar case it is not integrable in the traditional sense. In particular, it does not satisfy the Painlevé test and does not have polynomial integrals. The same can be said about systems (a) and (g), while the remaining systems from this section have a polynomial integral and satisfy the Painlevé test.

State of classification problem

We are going to get the complete classification of integrable nonabelian systems (6.20) under two natural assumptions:

1. The corresponding scalar system of ODEs

$$\begin{cases} u_t = a_1 u^2 + a_2 uv + a_3 v^2, \\ v_t = b_1 v^2 + b_2 uv + b_3 u^2 \end{cases} \tag{6.21}$$

where

$$a_1 = \alpha_1, \quad a_2 = \alpha_2 + \alpha_3, \quad a_3 = \alpha_4, \quad b_1 = \beta_1, \quad b_2 = \beta_2 + \beta_3, \quad b_3 = \beta_4, \tag{6.22}$$

is integrable in a common sense (see Section 6.1.5).

Definition 6.1.31. Suppose systems (6.20) and (6.21) are related by (6.22). Then,

    (i) We call a system (6.20) a *nonabelization* of the system (6.21);

    (ii) The system (6.21) is called the *abelian limit* of system (6.20).

Additionally to **(1)**, we assume that

2. The abelian limit of the symmetry hierarchy of system (6.20) coincides with the whole symmetry hierarchy of (6.21).[5]

Question 6.1.32. *What are integrable systems* (6.21)*?*

6.1.5   Integrable scalar quadratic homogeneous systems

Consider systems of the form (6.15), where $u^i(t)$ are scalar functions with values in $\mathbb{C}$, and $C_{jk}^i = C_{kj}^i$ are some constants. The main feature of integrable systems (6.15) is the existence of polynomial infinitesimal symmetries and/or first integrals (see Sections 1.4 and 1.5). Without loss of generality, symmetries and integrals may be considered homogeneous.

Another evidence of integrability is the absence of moving branch points and logarithmic singularities for solutions with complex time t. The so-called Painlevé approach is based on this assumption. We use the *Kovalevskaya–Lyapunov test* which is one of the incarnations of the Painlevé test.

---

[5]The approach based on the assumptions **1** and **2** can be used for finding integrable noncommutative generalizations of diverse polynomial integrable models.

Kovalevskaya–Lyapunov test

Any system (6.15) has special Kovalevskaya solutions of the form

$$u^i(t) = \frac{z^i}{t}.$$

The constants z_i are determined from the following system of algebraic equations

$$-z^i = \sum_{j,k} C^i_{jk} z^j z^k.$$

For each Kovalevskaya solution $\frac{z^i}{t}$, we consider its deformation in the form of a formal series

$$\bar{u}^i = \frac{z^i}{t} + \varepsilon \, p^i \, t^s + O(\varepsilon^2).$$

It is easy to see that the constants p^i satisfy the system of linear equations

$$s\, p^i = \sum_{j,k} C^i_{jk} \, (p^j z^k + p^k z^j), \qquad i = 1, \ldots, N. \tag{6.23}$$

The number s is called the *Kovalevskaya exponent*. We see that the exponent is an eigenvalue, and $(p^1, \ldots p^n)$ is the corresponding eigenvector of the matrix of the linear system (6.23) defined by the Kovalevskaya solution.

Definition 6.1.33. A system (6.15) satisfies the *Kovalevskaya–Lyapunov* test if, for any Kovalevskaya solution, all the exponents are integers.

Integrable scalar two-component systems

Definition 6.1.34. A system (6.21) is called integrable if it has a polynomial first integral and satisfies the Kovalevskaya–Lyapunov test.

Two systems connected with each other by a linear transformation (6.18) are called *equivalent*. In this section, we describe all (up to equivalence) integrable systems (6.21).

Let $I(u,v)$ be a homogeneous[6] first integral of a system (6.21). We write it in the factorized form:

$$I = \prod_{i=1}^{k}(u - \kappa_i v)^{n_i}, \qquad n_i \in \mathbb{N}, \qquad \kappa_i \neq \kappa_j \text{ for } i \neq j. \tag{6.24}$$

---

[6]Clearly, each homogeneous component of a polynomial integral for a homogeneous system is a first integral.

Remark 6.1.35. *A factor of the form* v^n, *which may exist in* I, *corresponds to* $\kappa = \infty$. *Using a linear transformation* (6.18) *of the form* $\hat{u} = u$, $\hat{v} = v + cu$, *one can always reduce* I *to an integral of the form* (6.24) *with finite roots* κ_i, $i = 1, \ldots, k$.

Proposition 6.1.5. *Suppose that at least one of the coefficients of the system* (6.21) *is not equal to zero. Then,* $k \leq 3$.

Proof. Substituting $u = \kappa_i v$ with $i = 1, \ldots, k$ into the expression $Q \overset{def}{=} \dfrac{dI}{dt}$, we get that if $Q = 0$, then each of the roots of κ_i satisfies a cubic equation

$$b_3 \kappa_i^3 + (b_2 - a_1) \kappa_i^2 + (b_1 - a_2) \kappa_i + a_3 = 0.$$

If $k > 3$, then all coefficients of the cubic polynomial are equal to zero. Equating to zero the coefficients of the highest powers of u and v in Q, we arrive at the relations

$$\sum_{i=1}^{k} n_i (a_1 - \kappa_i b_3) = 0, \qquad \sum_{i=1}^{k} n_i \left(b_1 - \frac{a_3}{\kappa_i} \right) = 0. \tag{6.25}$$

Since $a_3 = b_3 = 0$ and $\sum n_i \neq 0$, it follows from (6.25) that $a_1 = b_1 = 0$ and, therefore, all the coefficients of the system (6.21) are zero. $\square$

Remark 6.1.36. *Similarly, it can be proved that if a system of scalar equations*

$$\begin{cases} u_t &= P_1(u, v), \\ v_t &= P_2(u, v), \end{cases} \tag{6.26}$$

where P_i *are homogeneous polynomials of degree* d, *has a first integral* (6.24), *then* $k \leq d + 1$.

If a system (6.21) has a polynomial first integral, then it has an infinite set of polynomial symmetries linearly independent over $\mathbb{C}$. The following statement is well known and can be easily verified.

Lemma 6.1.37. *Let* I *be a first integral of a system* (6.21). *Then, for any* $N \in \mathbb{N}$, *the system*

$$\begin{cases} u_\tau &= I^N (a_1 u^2 + a_2 uv + a_3 v^2), \\ v_\tau &= I^N (b_1 v^2 + b_2 uv + b_3 u^2) \end{cases}$$

is a symmetry for the system (6.21).

The existence of a polynomial integral is not sufficient for the integrability of system (6.21). We now switch to the Kovalevskay–Lyapunov test. It turns out that with its help one can determine the possible degrees n_i in the formula (6.24).

In the following we look into the three cases of 3, 2 and 1 distinct roots κ_i in the integral (6.24) (see Proposition 6.1.5).

Consider the most nondegenerate case of three distinct roots κ_i. Using a linear transformation (6.18), we can always reduce the integral to the form

$$I = u^{k_1}(u - v)^{k_2}v^{k_3}, \tag{6.27}$$

where k_i are some positive integers defined up to permutations. Without loss of generality, we assume that k_1, k_2, k_3 do not have a nontrivial common factor.

Proposition 6.1.6. *A system* (6.21) *has a first integral* (6.27) *iff up to a scaling* $u \mapsto \mu u, v \mapsto \mu v$ *it has the following form:*

$$\begin{cases} u_t &= -k_3\, u^2 + (k_3 + k_2)\, uv, \\ v_t &= -k_1\, v^2 + (k_1 + k_2)\, uv. \end{cases} \tag{6.28}$$

Remark 6.1.38. *According to Lemma 6.1.37, system* (6.28) *has polynomial symmetries of degrees* $2 + (k_1 + k_2 + k_3)\, j$, *where* $j \in \mathbb{N}$.

Let us apply the Kovalevskaya–Lyapunov test to the system (6.28). Finding Kovalevskaya solutions and exponents for each of them, we arrive at the following statement.

Theorem 6.1.39. *Any Kovalevskaya exponent for the system* (6.28) *coincides with one of the following three:*

$$s_1 = \frac{k_2 + k_3}{k_1}, \qquad s_2 = \frac{k_3 + k_1}{k_2}, \qquad s_3 = \frac{k_1 + k_2}{k_3}. \tag{6.29}$$

If the system satisfies the Kovalevskaya–Lyapunov test, then the numbers s_i must be integers. Moreover, since $k_i \in \mathbb{N}$, they are positive.

Lemma 6.1.40. *Up to permutations and proportionality, there exist three collections of* k_1, k_2, k_3, *for which* s_i *are natural numbers:*

- *Case 1.* $k_1 = k_2 = k_3 = 1;$

- *Case 2.* $k_1 = k_3 = 1,\ k_2 = 2;$

- *Case 3.* $k_1 = 1,\ k_2 = 2,\ k_3 = 3.$

Remark 6.1.41. *Using Remark 6.1.36 and the Kovalevskaya–Lyapunov test, we can describe homogeneous integrable systems* (6.26) *with* $d > 2$. *For example, in the case of cubic systems, the system*

$$\begin{cases} u_t &= -u^3\, k_3 + \nu\, uv^2\, (k_2 + k_3 + k_4) + u^2 v\, (k_2 + k_3 - \nu k_3 - \nu k_4), \\ v_t &= -\nu\, v^3\, k_1 + uv^2\, (-k_1 + \nu k_1 + \nu k_2 - k_4) + u^2 v\, (k_1 + k_2 + k_4), \end{cases}$$

where $\nu \neq 0, 1, \infty$, has the generic integral

$$I = u^{k_1}(u-v)^{k_2}v^{k_3}(u-\nu v)^{k_4}.$$

The Kovalevskaya–Lyapunov test distinguishes the following 13 collections of $\{k_1, k_2, k_3, k_4\}$:

$$\{1, 3, 8, 12\}, \quad \{1, 2, 6, 9\}, \quad \{2, 3, 10, 15\}, \quad \{1, 1, 4, 6\}, \quad \{1, 4, 5, 10\},$$

$$\{1, 2, 3, 6\}, \quad \{1, 2, 4\}, \quad \{1, 2, 2, 5\}, \quad \{1, 1, 1, 3\}, \quad \{1, 3, 4, 4\},$$

$$\{1, 1, 2, 2\}, \quad \{2, 3, 3, 4\}, \quad \{1, 1, 1, 1\},$$

corresponding to integrable cases.

In the case of two distinct roots κ_i in the integral (6.24) we normalize

$$I = u^{k_1}v^{k_3}.$$

Systems (6.21) with such an integral have the form

$$\begin{cases} u_t &=& -b_2 q\, u^2 + a_2\, uv, \\ v_t &=& -a_2 q^{-1}\, v^2 + b_2\, uv, \end{cases}$$

where $q = \dfrac{k_3}{k_1}$. The Kovalevskaya–Lyapunov test leads to the condition $q = 1$. This means that for integrable cases

$$I = u\, v, \tag{6.30}$$

and we arrive at the systems of the form

$$\begin{cases} u_t &=& -b_2\, u^2 + a_2\, uv, \\ v_t &=& -a_2\, v^2 + b_2\, uv. \end{cases} \tag{6.31}$$

The constants a_2, b_2 in (6.31) can be normalized by a scaling of the form $u \mapsto \mu\, u$, $v \mapsto \nu\, v$ and by the involution $u \leftrightarrow v$ to $a_2 = b_2 = 1$ or to $a_2 = 1, b_2 = 0$.

If the integral (6.24) has a single root, then we can set

$$I = u. \tag{6.32}$$

The corresponding system

$$\begin{cases} u_t &=& 0, \\ v_t &=& b_1\, v^2 + b_2\, vu + b_3\, u^2 \end{cases} \tag{6.33}$$

satisfies the Kovalevskaya–Lyapunov test for arbitrary constants b_i.

Using transformations of the form $u \mapsto u$, $v \mapsto s_i v + s_2 u$, we can reduce the set of constants b_i to one of the following:

- $b_1 = b_2 = 1$, $b_3 = 0$,

- $b_1 = 1$, $b_2 = b_3 = 0$,

- $b_1 = b_3 = 0$, $b_2 = 1$,

- $b_1 = b_2 = 0$, $b_3 = 1$.

6.1.6 Nonabelization of integrable homogeneous scalar systems

It is clear that any nonabelization of a system (6.21) has the form

$$\begin{cases} u_t & = & a_1 u^2 + a_2 uv + a_3 v^2 + \alpha(uv - vu), \\ v_t & = & b_1 v^2 + b_2 vu + b_3 u^2 + \beta(vu - uv) \end{cases} \tag{6.34}$$

where α and β are arbitrary constants.

For each integrable scalar system (6.21) from Section 6.1.5, we find [179] all nonabelizations with an infinite set of symmetries whose abelian limits coincide with the scalar symmetries from Lemma 6.1.37.[7]

It is important to note that in such a formulation of the problem, we a priori know the degrees of symmetries of the nonabelizations that we are looking for.

Remark 6.1.42. *For systems (6.28) (see Remark 6.1.38) the degrees d_j, $j \in \mathbb{N}$ of symmetries in Cases 1–3 of Lemma 6.1.37 are given by the formulas:*

Case 1. $d_j = 2 + 3j$;

Case 2. $d_j = 2 + 4j$;

Case 3. $d_j = 2 + 6j$.

Technically, the calculations are performed as follows. An ansatz with undetermined coefficients for a nonabelian symmetry of minimal degree is considered. According to Remark 6.1.42, these degrees are 5, 6 and 8. The requirement that abelian limit of the symmetry is described by Lemma 6.1.37 reduces the number of unknown coefficients in the ansatz.

The compatibility condition of system (6.34) and the symmetry is equivalent to an overdetermined system of linear algebraic equations for the coefficients of symmetry. The matrix of this system depends on the unknown parameters α and

---

[7]This means that we search for a nonabelization of not a single scalar system but of its whole hierarchy of symmetries.

β from the formula (6.34). The requirement that the system has a nonzero solution leads to an algebraic system for these parameters which can be easily solved.

Finally, from all solutions (α, β) of the algebraic system, those that correspond to nonequivalent nonabelian systems are chosen.

The resulting list contains all the examples from the paper [175], except for the "wrong" systems (a) and (g) (see Section 6.1.4). In addition, we find several new examples of nonabelian systems (6.20) having symmetries.

Conjecture 6.1.43. *A nonabelian nontriangular system* (6.20) *possesses an infinite sequence of symmetries iff it is equivalent to one of those listed in the next subsection.*

Formulation of results

Consider nonabelizations

$$
\begin{cases}
u_t &= -k_3\, u^2 + (k_2 + k_3)\, uv + \alpha(uv - vu), \\
v_t &= -k_1\, v^2 + (k_1 + k_2)\, vu + \beta(vu - uv)
\end{cases}
\tag{6.35}
$$

of integrable systems (6.28). Here, α and β are arbitrary constants.

Remark 6.1.44. *The involution* (6.17) *transforms* α *and* β *from* (6.35) *according to the rule*

$$\alpha \mapsto -\alpha - k_2 - k_3, \qquad \beta \mapsto -\beta - k_2 - k_1.$$

Proposition 6.1.7. *In the case of* $k_1 = k_2 = k_3 = 1$, *there exist the following nonequivalent nonabelizations with fifth degree symmetries* (*see Remark 6.1.42*)[8]:

1. $\alpha = -1, \quad \beta = -1;$

2. $\alpha = 0, \quad \beta = -1;$

3. $\alpha = 0, \quad \beta = -2;$

4. $\alpha = 0, \quad \beta = 0;$

5. $\alpha = 0, \quad \beta = -3.$

Remark 6.1.45. *Three of these systems were found in* [175] *because, apart from the symmetry of degree 5, they possess symmetries of degree less than 5, having a zero abelian limit. The system* **1** *is equivalent to the system* (6.9). *The systems* **4** *and* **5** *are new.*

---

[8]It is assumed that the abelian limit of symmetry coincides with nonzero symmetry of the corresponding system (6.28).

Proposition 6.1.8. *In the case of $k_1 = k_3 = 1$, $k_2 = 2$, there exist the following nonequivalent nonabelizations with symmetries of sixth degree:*

1. $\alpha = -1$, $\beta = -1$;

2. $\alpha = 0$, $\beta = -2$;

3. $\alpha = 0$, $\beta = 0$;

4. $\alpha = 0$, $\beta = -4$.

Remark 6.1.46. *The systems **2**, **3** and **4** are new.*

Proposition 6.1.9. *In the case of $k_1 = 1$, $k_2 = 2$, $k_3 = 3$, there exist the following nonequivalent nonabelization with symmetries of degree 8:*

1. $\alpha = -2$, $\beta = 0$;

2. $\alpha = -4$, $\beta = 0$;

3. $\alpha = -6$, $\beta = 0$;

4. $\alpha = 0$, $\beta = -6$;

5. $\alpha = 0$, $\beta = 0$.

Remark 6.1.47. *All these systems are new.*

Let us find now integrable nonabelizations

$$\begin{cases} u_t &= -b_2\,u^2 + a_2\,uv + \alpha(uv - vu), \\ v_t &= -a_2\,v^2 + b_2\,vu + \beta(vu - uv) \end{cases} \tag{6.36}$$

of systems (6.31). According to Lemma 6.1.37 the degrees of symmetries for them are of the form $2 + 2j$, $j \in \mathbb{N}$. In particular, the simplest symmetry is of degree 4.

Proposition 6.1.10. *There exist the following nontriangular nonequivalent systems (6.36) with symmetries of degree 4:*

1. $a_2 = b_2 = 1$, $\alpha = 0$, $\beta = 0$;

2. $a_2 = b_2 = 1$, $\alpha = 0$, $\beta = -2$;

3. $a_2 = b_2 = 1$, $\alpha = 0$, $\beta = -1$;

4. $a_2 = 1$, $b_2 = 0$, $\alpha = 0$, $\beta = 1$.

Nonabelizations of systems with the integral (6.32) are given by the formula

$$
\begin{cases}
u_t &= \alpha(uv - vu), \\
v_t &= b_1 v^2 + b_2 vu + b_3 u^2 + \beta(vu - uv).
\end{cases}
\tag{6.37}
$$

The simplest symmetry is of degree 3.

Proposition 6.1.11. *There exist the following nonequivalent nontriangular systems* (6.37) *with symmetries of third degree:*

1. $b_1 = b_2 = 1,$ $b_3 = 0,$ $\alpha = 1,$ $\beta = 0;$

2. $b_1 = b_2 = 0,$ $b_3 = 1,$ $\alpha = 1,$ $\beta = 0.$

Conjecture 6.1.48. *In the case of $m \times m$ matrices, various integrable nonabelizations of a given system define the same flow on the field of all invariants of the GL_m-action $u \mapsto SuS^{-1},\ v \mapsto SvS^{-1}$.*

Laurent involutions

Sometimes it is useful to extend the free associative algebra $\mathcal{A}$ with generators u and v by means of new symbols u^{-1} and v^{-1} such that $uu^{-1} = u^{-1}u = vv^{-1} = v^{-1}v = 1$. We will call elements of the extended algebra *nonabelian Laurent polynomials*.

Interesting examples of integrable nonabelian systems with Laurent right-hand sides are known. For example, in the paper [180] the system

$$
u_t = uv - uv^{-1} - v^{-1}, \qquad v_t = -vu + vu^{-1} + u^{-1},
$$

proposed by M. Kontsevich, was investigated. It can be regarded as a nontrivial deformation of the system

$$
u_t = uv, \qquad v_t = -vu
\tag{6.38}
$$

by Laurent terms of a lower degree. The system (6.38) is related to the first system from Proposition 6.1.11 by the transformation $u \mapsto u + v,\ v \mapsto -u$.

Open problem 6.1.49. *Find integrable Laurent deformations of integrable nonabelian systems from Subsection "Formulation of results".*

In the context of our classification, the Laurent polynomials interest us in the following sense. It turns out that for a special class of polynomial nonabelian systems (6.20) the set of admissible transformations (6.18), (6.17) can be extended by invertible Laurent involutions.

Consider nonabelian systems of the form

$$\begin{cases} u_t &= -p\,u^2 + q\,uv, \\ v_t &= -a\,v^2 + b\,uv + c\,vu. \end{cases} \tag{6.39}$$

It is easy to verify that the composition of the transformation

$$u = \bar{u}, \qquad v = \bar{u}^{-1}\bar{v}\bar{u}$$

and involution (6.17) maps (6.39) to the system

$$\begin{cases} u_t &= -p\,u^2 + q\,uv, \\ v_t &= -a\,v^2 + (c+p)\,uv + (b-p)\,vu. \end{cases} \tag{6.40}$$

Thus, we have an involution $\tau : (6.39) \mapsto (6.40)$ on the set of systems of the form (6.39).

Remark 6.1.50. *For any nontriangular system of the form* (6.39) *we can express* v *in terms of* u_t, u *and* u^{-1} *from the first equation. Substituting the resulting expression into the second equation, we obtain a nonabelian second order equation for* u *with a Laurent right-hand side.*

The system (6.35) belongs to the class (6.39) if $\alpha = 0$. For this reason, in Propositions 6.1.7–6.1.9 we have chosen among equivalent systems those for which $\alpha = 0$.

In the case of Proposition 6.1.7, the involution τ acts on the parameter β as $\tau : \beta \mapsto -\beta - 3$ and therefore the systems $2 \leftrightarrow 3$ and $4 \leftrightarrow 5$ are dual with respect to τ.

For systems from Proposition 6.1.8 we have $\tau : \beta \mapsto -\beta - 4$. The systems $3 \leftrightarrow 4$ are dual, and the system $2 \leftrightarrow 2$ is self-dual.

In the case of Proposition 6.1.9, we find the duality $4 \leftrightarrow 5$. For other systems we have $\beta = 0$. In this case, we can apply the composition of the Laurent transformation $\pi : v = \bar{v}, u = \bar{v}^{-1}\bar{u}\bar{v}$ and the involution (6.17). As a result, we get the dualities $1 \leftrightarrow 2$, $3 \leftrightarrow 3$.

For systems from Proposition 6.1.10, we have $1 \leftrightarrow 2$, $3 \leftrightarrow 3$, $4 \leftrightarrow 4$.

The first system from Proposition 6.1.11 can be reduced to the triangular system

$$\begin{cases} u_t &= 0, \\ v_t &= v^2 + uv \end{cases} \tag{6.41}$$

by the Laurent invertible transformation $u \mapsto v^{-1}uv$.

Conjecture 6.1.51. *For systems related by one of the invertible Laurent transformations described above, polynomial symmetries are pairwise related by the same transformation.*

6.1.7   Integrable inhomogeneous nonabelian systems

Consider nonabelian systems of the form

$$\begin{cases} u_t = \alpha_1 u^2 + \alpha_2 u\,v + \alpha_3 v\,u + \alpha_4 v^2 + \gamma_1 u + \gamma_2 v + \gamma_3\,\mathbf{1}, \\ v_t = \beta_1 v^2 + \beta_2 v\,u + \beta_3 u\,v + \beta_4 u^2 + \gamma_4 u + \gamma_5 v + \gamma_6\,\mathbf{1} \end{cases} \qquad (6.42)$$

System (6.7) provides an example of a system from this class.

Definition 6.1.52. System (6.42) is called an *inhomogeneous generalization* of system (6.20), and the system (6.20) is called the *homogeneous limit* of system (6.42).[9]

We call two inhomogeneous generalizations of the same equation (6.20) *equivalent* if they are connected by a shift of the form

$$u \mapsto u + c_1\,\mathbf{1}, \qquad v \mapsto v + c_2\,\mathbf{1}, \qquad\qquad (6.43)$$

where c_i are scalar constants.

Definition 6.1.53. The system (6.42) is called *integrable inhomogeneous generalization* of integrable (in the sense of Section 6.1.6) system (6.20) if it has an infinite set of inhomogeneous symmetries whose homogeneous limits coincide with the symmetries of the system (6.20).

In this section, we find all integrable inhomogeneous generalizations (see Definition 6.1.53) for each system (6.20) from Propositions 6.1.7–6.1.11. It turns out that the abelian limits of these integrable inhomogeneous systems have first integrals, which are inhomogeneous generalizations of the integrals described in Section 6.1.5.

Inhomogeneous generalizations of systems from Proposition 6.1.7

In the case of systems 1–5 of Proposition 6.1.7, we reduce γ_2 and γ_4 to zero by a shift of the form (6.43).

Proposition 6.1.12. *For each of the systems 1–5, there is exactly one inhomogeneous generalization (6.42) having a fifth degree symmetry. It is defined by the condition $\gamma_5 = -\gamma_1$.*

The abelian limits of all these generalizations have the same cubic integral

$$I = v(v - u)u + \gamma_1 vu + \gamma_3 v - \gamma_6 u,$$

which generalizes the integral (6.27).

---

[9]The systems (6.42) and (6.20) coincide up to linear terms.

Inhomogeneous generalizations of systems from Proposition 6.1.8

In Cases 1–4 of Proposition 6.1.8, we also reduce γ_2 and γ_4 to zero by a shift (6.43).

Proposition 6.1.13. *For each of the systems 1–4 there is exactly one inhomogeneous generalization with a symmetry of the sixth degree. It is defined by two conditions $\gamma_5 = -\gamma_1$ and $\gamma_6 = \gamma_3$.*

All abelian limits of these generalizations have the first integral

$$I = 2u(u-v)^2v + 4\gamma_5\, u(u-v)v + \gamma_3\, (u-v)^2 + 2\gamma_5^2\, uv + 2\gamma_3\gamma_5\, (u-v).$$

Inhomogeneous generalizations of systems from Proposition 6.1.9

For systems 1–5 from Proposition 6.1.9, we again bring γ_2 and γ_4 to zero by a shift.

Proposition 6.1.14. *For each of Cases 1–5 of Proposition 6.1.9 there are exactly two inhomogeneous generalizations with symmetries of eighth degree. They are defined by the relations*

    a. $\quad \gamma_1 = -3\,\gamma_5, \quad \gamma_3 = \gamma_6 = 0;$

    b. $\quad \gamma_1 = \gamma_5 = \gamma_6 = 0.$

All abelian limits of the systems of the case **a** have the first integral

$$I = u(u-v)^2v^3 + 2\gamma_5\, u(u-v)v^3 + \gamma_5^2\, uv^3,$$

whereas the integral of the abelian limits in the case **b** is given by

$$I = 4u(v-u)^2v^3 + \gamma_3\, (v-u)(v+3u)v^2 + \gamma_3^2\, v^2.$$

Inhomogeneous generalizations of systems from Proposition 6.1.10

Consider systems 1–4 of Proposition 6.1.10. In Cases 1–3 we use the normalization $\gamma_2 = \gamma_4 = 0$.

Proposition 6.1.15. *In each of Cases 1–3 of Proposition 6.1.10 there is exactly one inhomogeneous generalization with symmetry of degree 4. It is defined by the relations*

$$\gamma_1 = -\gamma_5, \qquad \gamma_3 = \gamma_6 = 0.$$

The common integral for all abelian limits is

$$I = u\,v.$$

In Case 4, using shift (6.43), we reduce γ_1 and γ_5 to zero.[10]

Proposition 6.1.16. *An inhomogeneous generalization in Case 4, having symmetry of degree 4, exists only if*

$$\gamma_3 = \gamma_4 = \gamma_6 = 0.$$

Its abelian limit has an integral

$$I = u\,v + \gamma_2\,v.$$

Inhomogeneous generalizations of systems from Proposition 6.1.11

In Case 1 of Proposition 6.1.11, we reduce γ_4 and γ_5 to zero.

Proposition 6.1.17. *An inhomogeneous generalization with a third degree symmetry exists iff*
$$\gamma_1 = \gamma_2 = \gamma_3 = 0.$$
The integral for the abelian limit is $I = u$.

For the equation from Case 2, we reduce γ_4 to zero.

Proposition 6.1.18. *For an inhomogeneous generalization with a third degree symmetry, we obtain*
$$\gamma_1 = \gamma_2 = \gamma_3 = \gamma_5 = \gamma_6 = 0.$$
That is, nontrivial integrable inhomogeneous generalizations do not exist.

Remarks on integrability of matrix systems

All nonabelian systems found in Sections 6.1.6 and 6.1.7 become systems of ordinary differential equations with $2m^2$ variables, if we replace u and v with $m \times m$ matrices. Probably, all these systems are integrable for any m in one sense or another. We demonstrate by three examples that different types of integrability arise here.

System (6.7), equivalent to the inhomogeneous generalization of system 1 of Proposition 6.1.7, has the Lax pair (6.13) and is integrable by the inverse scattering method.

The first of systems from Proposition 6.1.11 can be integrated explicitly. By the invertible Laurent transformation, it is reduced to (6.41). Let $u = \mathbf{c}$, where $\mathbf{c}$ is a nondegenerate constant matrix. Then, the variable $w = v^{-1}$ satisfies the linear matrix equation

$$w_t = -w\,\mathbf{c} - \mathbf{1}.$$

---

[10]In this case, we cannot always reduce to zero the coefficients γ_2 and γ_4.

This equation has the general solution

$$w = -\mathbf{c}^{-1} + \mathbf{c}_1 \exp(-ct),$$

where $\mathbf{c}_1$ is an arbitrary constant matrix. The case of degenerate initial data for the original system is also easily integrated in quadratures.

The second system from Proposition 6.1.11 is linearized using the generalized factorization method (see Section 2.4.3) [73, 82]. From the system it follows that

$$v_{tt} = [v_t, v]. \tag{6.44}$$

In the matrix case, this equation can be linearized as follows. Let Y be a matrix solution of a linear equation

$$Y_t = Y(\mathbf{c}_1 t + \mathbf{c}_2),$$

where $\mathbf{c}_i$ are arbitrary constant matrices. Then, $v = -Y_t Y^{-1}$ is a general solution of equation (6.44).

Open problem 6.1.54. *Establish a type of integrability of all matrix systems from Sections 6.1.6 and 6.1.7.*

6.2 Nonabelian Hamiltonian formalism and Poisson brackets defined on traces of matrices

6.2.1 Trace Poisson brackets

In this section, we discuss the Hamiltonian formalism for systems of ODE with matrix variables. Our goal is to describe a class of Poisson brackets on functions of entries of $m \times m$ matrices x_α, $\alpha = 1, \ldots, N$ such that for Hamiltonians of the form (6.4) the equations of motion can be written in matrix form (6.1).

Consider Nm^2-dimensional Poisson brackets on functions of entries $x_{j,\alpha}^i$ of matrices x_α. Hereinafter, we use Latin indices, taking values from 1 to m, for matrix entries and Greek indices with values from 1 to N for numbering of the matrices.

Definition 6.2.1. A Poisson bracket defined on the matrix entries is called a *trace bracket* if

1. The bracket is GL_m-invariant;

2. For any two matrix polynomials $P_i(x_1, \ldots, x_N)$, $i = 1, 2$, with coefficients from $\mathbb{C}$ the bracket between their traces is equal to the trace of some matrix polynomial P_3.

Lemma 6.2.2. *For any Hamiltonian of the form $H = \operatorname{tr} P$, where P is a matrix polynomial, and for any trace Poisson bracket, equations of motion (1.22) can be written in the matrix form (6.1).*

Theorem 6.2.3. (i) *Any constant trace Poisson bracket has the form*

$$\{x_{i,\alpha}^j, x_{i',\beta}^{j'}\} = \delta_{i'}^j \delta_i^{j'} c_{\alpha\beta}; \tag{6.45}$$

(ii) *Any linear trace bracket can be written as*

$$\{x_{i,\alpha}^j, x_{i',\beta}^{j'}\} = b_{\alpha,\beta}^\gamma x_{i,\gamma}^{j'} \delta_{i'}^j - b_{\beta,\alpha}^\gamma x_{i',\gamma}^j \delta_i^{j'}; \tag{6.46}$$

(iii) *Any quadratic trace bracket is given by*

$$\{x_{i,\alpha}^j, x_{i',\beta}^{j'}\} = r_{\alpha\beta}^{\gamma\epsilon} x_{i,\gamma}^{j'} x_{i',\epsilon}^j + a_{\alpha\beta}^{\gamma\epsilon} x_{i,\gamma}^k x_{k,\epsilon}^{j'} \delta_{i'}^j - a_{\beta\alpha}^{\gamma\epsilon} x_{i',\gamma}^k x_{k,\epsilon}^j \delta_i^{j'}. \tag{6.47}$$

Furthermore,

(1) *A bracket of the form (6.45) is a trace Poisson bracket iff*

$$c_{\alpha\beta} = -c_{\beta\alpha};$$

(2) *Formula (6.46) defines a trace Poisson bracket iff*

$$b_{\alpha\beta}^\mu b_{\mu\gamma}^\sigma = b_{\alpha\mu}^\sigma b_{\beta\gamma}^\mu; \tag{6.48}$$

(3) *Formula (6.47) defines a trace Poisson bracket iff the relations*

$$r_{\alpha\beta}^{\sigma\epsilon} = -r_{\beta\alpha}^{\epsilon\sigma}, \tag{6.49}$$

$$r_{\alpha\beta}^{\lambda\sigma} r_{\sigma\tau}^{\mu\nu} + r_{\beta\tau}^{\mu\sigma} r_{\sigma\alpha}^{\nu\lambda} + r_{\tau\alpha}^{\nu\sigma} r_{\sigma\beta}^{\lambda\mu} = 0, \tag{6.50}$$

$$a_{\alpha\beta}^{\sigma\lambda} a_{\tau\sigma}^{\mu\nu} = a_{\tau\alpha}^{\mu\sigma} a_{\sigma\beta}^{\nu\lambda}, \tag{6.51}$$

$$a_{\alpha\beta}^{\sigma\lambda} a_{\sigma\tau}^{\mu\nu} = a_{\alpha\beta}^{\mu\sigma} r_{\tau\sigma}^{\lambda\nu} + a_{\alpha\sigma}^{\mu\nu} r_{\beta\tau}^{\sigma\lambda}, \tag{6.52}$$

and

$$a_{\alpha\beta}^{\lambda\sigma} a_{\tau\sigma}^{\mu\nu} = a_{\alpha\beta}^{\sigma\nu} r_{\sigma\tau}^{\lambda\mu} + a_{\sigma\beta}^{\mu\nu} r_{\tau\alpha}^{\sigma\lambda} \tag{6.53}$$

hold.

Remark 6.2.4. *Formula (6.48) means that $b_{\alpha\beta}^\sigma$ are structural constants of an associative algebra $\mathcal{A}$. A direct verification shows that bracket (6.46) is nothing but the Lie–Kirillov–Konstant bracket defined by the Lie algebra corresponding to the associative algebra $\operatorname{Mat}_m \otimes \mathcal{A}$.*

Remark 6.2.5. *Relations (6.49) and (6.50) mean that the tensor* **r** *satisfies the associative Yang–Baxter equation (also called as the Rota–Baxter equation)* [181, 97].

Open problem 6.2.6. *Find an explicit formula for homogeneous trace bracket of degree $k > 2$, similar to* (6.47).

Quadratic trace Poisson brackets

Under a linear transformation $x_i \mapsto g_i^j x_j$ of the collection of matrices, the constants in formula (6.47) transform according to the usual tensor law

$$r_{ij}^{kl} \mapsto g_i^\alpha g_j^\beta h_\gamma^k h_\epsilon^l r_{\alpha\beta}^{\gamma\epsilon}, \qquad a_{ij}^{kl} \mapsto g_i^\alpha g_j^\beta h_\gamma^k h_\epsilon^l a_{\alpha\beta}^{\gamma\epsilon}, \tag{6.54}$$

where $g_i^j h_j^k = \delta_i^k$.

Definition 6.2.7. Two brackets of the form (6.47) connected by a transformation (6.54) are called *equivalent*.

The identities (6.49)–(6.53) can be written in a tensor form. Let $\mathbf{V}$ be a linear space with a basis $\mathbf{e}_i$, $i = 1, \ldots, N$. Let us define the linear operators R and A on the space $\mathbf{V} \otimes \mathbf{V}$ by the formulas $R\,\mathbf{e}_i \otimes \mathbf{e}_j = r_{ij}^{pq} \mathbf{e}_p \otimes \mathbf{e}_q$, $A\,\mathbf{e}_i \otimes \mathbf{e}_j = a_{ij}^{pq} \mathbf{e}_p \otimes \mathbf{e}_q$. Then the identities (6.49)–(6.53) can be rewritten as

$$R^{12} = -R^{21}, \qquad R^{23} R^{12} + R^{31} R^{23} + R^{12} R^{31} = 0,$$

$$A^{12} A^{31} = A^{31} A^{12}, \qquad \sigma^{23} A^{13} A^{12} = A^{12} R^{23} - R^{23} A^{12},$$

$$A^{32} A^{12} = R^{13} A^{12} - A^{32} R^{13}.$$

Here all operators act on the space $\mathbf{V} \otimes \mathbf{V} \otimes \mathbf{V}$, by σ^{ij} we denote the permutation of i-th and j-th components in the tensor product, and A^{ij}, R^{ij} mean the operators A, R acting on the tensor product of the i-th and j-th components.

Transformation (6.54) corresponds to the conjugation $A \mapsto GAG^{-1}$, $R \mapsto GRG^{-1}$, where $G = g \otimes g$ and $g \in \mathrm{End}(\mathbf{V})$.

Definition 6.2.8. (cf. formula (3.1)) A vector $\Lambda = (\lambda_1, \ldots, \lambda_N)$ is called *admissible* for a bracket (6.47), if for any i, j the relations

$$(a_{ij}^{pq} - a_{ji}^{qp} + r_{ij}^{pq})\lambda_p \lambda_q = 0 \tag{6.55}$$

hold.

Lemma 6.2.9. *For any admissible vector Λ the argument shift $x_i \mapsto x_i + \lambda_i \mathbf{1}$ in* (6.47) *generates a linear Poisson bracket* (6.46), *where*

$$b_{ij}^p = (a_{ij}^{qp} + a_{ij}^{pq} + r_{ij}^{pq})\lambda_q,$$

compatible with (6.47).

Case a = 0 and anti-Frobenius algebras

An important subclass of the Poisson brackets (6.47) corresponds to the case of zero tensor **a**. The relations (6.49), (6.50) mean that the tensor **r** is a constant solution of the associative Yang–Baxter equation ([97], [182]). The following algebraic approach allows us to construct such solutions.

Definition 6.2.10. An associative algebra $\mathcal{A}$ with multiplication $\circ$ is called *anti-Frobenius algebra* if it possesses a nondegenerate antisymmetric bilinear form $(\,,\,)$ that satisfies the condition

$$(x,\, y \circ z) + (y,\, z \circ x) + (z,\, x \circ y) = 0 \tag{6.56}$$

for any $x, y, z \in \mathcal{A}$. In other words, the form $(\,,\,)$ defines a 1-cocycle on $\mathcal{A}$.

Remark 6.2.11. *For any linear functional l on $\mathcal{A}$ the antisymmetric form $(x, y) = l([x, y])$ satisfies* (6.56).[11]

Remark 6.2.12. *From relation* (6.56) *it follows that the unity of $\mathcal{A}$ belongs to the kernel of the form* $(\,,\,)$. *Therefore, anti-Frobenius algebras are not unital.*

Theorem 6.2.13. *There is a one-to-one correspondence between solutions of the system* (6.49), (6.50) *up to equivalence and exact representations of anti-Frobenius algebras up to isomorphism.*

Proof. Suppose a tensor **r** satisfies the relations (6.49), (6.50). We write the tensor in the form

$$r_{kl}^{ij} = \sum_{\alpha,\beta=1}^{p} g^{\alpha\beta} y_{k,\alpha}^i y_{l,\beta}^j, \qquad i,j,k,l = 1,\ldots,m, \tag{6.57}$$

where $g^{\alpha\beta} = -g^{\beta\alpha}$, the matrix $G = \{g^{\alpha\beta}\}$ is nondegenerate, and the number p is the smallest possible. Substituting this expression into (6.50), we obtain that there exists a tensor $\phi_{\alpha\beta}^\gamma$, such that $y_{k,\alpha}^i y_{j,\beta}^k = \phi_{\alpha\beta}^\gamma y_{j,\gamma}^i$. It turns out that the algebra $\mathcal{A}$ with basis $y_1, \ldots, y_p$ and with the product $y_\alpha \circ y_\beta = \phi_{\alpha\beta}^\gamma y_\gamma$ is associative, and the antisymmetric bilinear form $(y_\alpha, y_\beta) = g_{\alpha\beta}$, where $g_{\alpha\beta}$ are entries of the matrix G^{-1}, satisfies the identity (6.56). The matrices y_γ with entries $y_{j,\gamma}^i$ define a m-dimensional representation of the algebra $\mathcal{A}$.

Conversely, suppose we have an exact m-dimensional representation of an anti-Frobenius algebra $\mathcal{A}$. Let a basis in $\mathcal{A}$ consists of matrices $y_\gamma = \{y_{j,\gamma}^i\}$. Denote the matrix of a bilinear form on $\mathcal{A}$ by $\bar{G}$. Let $g^{\alpha\beta}$ be entries of the matrix inverse to $\bar{G}$. It can be verified that the tensor (6.57) satisfies the relations (6.49), (6.50). □

---

[11]Given algebra $\mathcal{A}$ a description of functionals l, for which the corresponding form is nondegenerate, is a subject of linear algebra while a search for all anti-Frobenius algebras from a certain class of associative algebras is very nontrivial. For all associative algebras of a fixed dimension this classification problem is wild.

Example 6.2.14. (cf. [183]) Let $\mathcal{A}$ be the associative algebra of $N \times N$ matrices with zero N-th row. It is easy to verify that for a generic element l of $\mathcal{A}^*$ the antisymmetric bilinear form $(x, y) = l([x, y])$ (see Remark 6.2.11) is nondegenerate. Such a form can be written as $(x, y) = \mathrm{tr}([x, y]\, k^T)$, where $k \in \mathcal{A}$. Let us choose $k_{ij} = 0$, $j \neq i$, $k_{ii} = \mu_i$, where $i, j = 1, \ldots, N - 1$, and $k_{iN} = 1$, $i = 1, \ldots, N - 1$. The corresponding bracket (6.47) is given by the following tensor $\mathbf{r}$:

$$r^{ii}_{Ni} = -r^{ii}_{iN} = 1, \qquad r^{ij}_{ij} = r^{ji}_{ij} = r^{ii}_{ji} = -r^{ii}_{ij} = \frac{1}{\mu_i - \mu_j}, \tag{6.58}$$

where $i \neq j$, $i, j = 1, \ldots, N - 1$. The remaining elements of the tensor $\mathbf{r}$ and all elements of the tensor $\mathbf{a}$ are supposed to be zero. It can be verified that the tensor (6.58) is equivalent to a tensor of the form

$$r^{\alpha\beta}_{\alpha\beta} = r^{\beta\alpha}_{\alpha\beta} = r^{\alpha\alpha}_{\beta\alpha} = -r^{\alpha\alpha}_{\alpha\beta} = \frac{1}{\lambda_\alpha - \lambda_\beta}, \qquad \alpha \neq \beta, \qquad \alpha, \beta = 1, \ldots, N. \tag{6.59}$$

Here, $\lambda_1, \ldots, \lambda_N$ are arbitrary pairwise distinct parameters. Formula (6.47) with zero tensor $\mathbf{a}$ defines the Poisson bracket on the entries of matrices $x_1, \ldots, x_N$ of arbitrary dimension m, satisfying Definition 6.2.1. For $m = 1$ we have the scalar Poisson bracket

$$\{x_\alpha, x_\beta\} = \frac{(x_\alpha - x_\beta)^2}{\lambda_\beta - \lambda_\alpha}, \qquad \alpha \neq \beta, \qquad \alpha, \beta = 1, \ldots, N.$$

If N is even, then the Poisson structure given by the tensor (6.59) is nondegenerate, that is, the rank of the Poisson tensor Π is equal to Nm^2. In the odd case $\mathrm{rank}\,\Pi = (N - 1)\, m^2$.

Remark 6.2.15. *The tensor* (6.59) *can be directly obtained from the anti-Frobenius algebra*

$$\mathcal{A}_{N,1} = \{A \in \mathrm{Mat}_N \mid \sum_i a_{ij} = 0 \quad \forall j = 1, \ldots, N\},$$

endowed with the bilinear form

$$(x, y) = \mathrm{tr}\Big([x, y] \cdot \mathrm{diag}\,(\lambda_1, \ldots, \lambda_N)\Big), \tag{6.60}$$

using the construction of Theorem 6.2.13.

The Poisson bracket from Remark 6.2.15 was generalized in [184]. Let M be a divisor of N. Consider the $N(N - M)$-dimensional algebra

$$\mathcal{A}_{N,M} = \{A \in \mathrm{Mat}_N \mid \sum_{i \equiv r \ (\mathrm{mod}\ M)} a_{ij} = 0 \quad \forall r = 1, \ldots, M, \ \forall j = 1, \ldots, N\}$$

with the bilinear form (6.60). It can be verified that if λ_i are pairwise distinct, then this form is nondegenerate [183]. The components of the tensor $\mathbf{r}$, corresponding to the algebra $\mathcal{A}_{N,M}$, are given by the formula

$$r^{\alpha\beta}_{\gamma\delta} = 0, \qquad \text{if} \quad \alpha \neq \delta \quad \text{or} \quad \beta \neq \gamma,$$

$$r^{\alpha\alpha}_{\varepsilon\alpha} = -r^{\alpha\alpha}_{\alpha\varepsilon} = \frac{1}{\lambda_\alpha - \lambda_\varepsilon}, \qquad \text{where} \quad \alpha \neq \varepsilon,$$

$$r^{\alpha\alpha}_{\gamma\delta} = 0, \qquad \text{if} \quad \gamma \neq \alpha \quad \text{or} \quad \delta \neq \alpha,$$

$$r^{\alpha\beta}_{\beta\alpha} = \frac{1}{\lambda_\alpha - \lambda_\beta} \left(\frac{\displaystyle\prod_{\beta'\equiv\beta,\,\beta'\neq\beta} (\lambda_\alpha - \lambda_{\beta'}) \prod_{\alpha'\equiv\alpha,\,\alpha'\neq\alpha} (\lambda_\beta - \lambda_{\alpha'})}{\displaystyle\prod_{\alpha'\equiv\alpha,\,\alpha'\neq\alpha} (\lambda_\alpha - \lambda_{\alpha'}) \prod_{\beta'\equiv\beta,\,\beta'\neq\beta} (\lambda_\beta - \lambda_{\beta'})} - 1 \right), \qquad \text{if} \quad \alpha \neq \beta,$$

$$r^{\alpha\beta}_{\gamma\delta} = \frac{1}{\lambda_\alpha - \lambda_\beta} \cdot \frac{\displaystyle\prod_{\gamma'\equiv\gamma,\,\gamma'\neq\gamma} (\lambda_\alpha - \lambda_{\gamma'}) \prod_{\delta'\equiv\delta,\,\delta'\neq\delta} (\lambda_\beta - \lambda_{\delta'})}{\displaystyle\prod_{\alpha'\equiv\alpha,\,\alpha'\neq\alpha} (\lambda_\alpha - \lambda_{\alpha'}) \prod_{\beta'\equiv\beta,\,\beta'\neq\beta} (\lambda_\beta - \lambda_{\beta'})} \qquad \text{in other cases.}$$

Here, the formula $x \equiv y$ means that $x = y \pmod{M}$.

Exercise 6.2.16. Write the scalar Poisson brackets given by these formulas explicitly. For example, in the case $N = 2M$ and $m = 1$ the corresponding Poisson bracket has the form

$$\{x_\alpha, x_\beta\} = \frac{(x_\alpha - x_{\alpha'})(x_\beta - x_{\beta'})(\lambda_{\alpha'} - \lambda_{\beta'})}{(\lambda_\alpha - \lambda_{\alpha'})(\lambda_\beta - \lambda_{\beta'})},$$

where for any $1 \leq \gamma \leq 2M$ the positive integer γ' is uniquely determined by the condition $|\gamma' - \gamma| = M$.

Open problem 6.2.17. *Describe all anti-Frobenius algebras $\mathcal{A}$ of the form*

$$\mathcal{A} = \mathcal{S} \oplus \mathcal{M},$$

where $\mathcal{S}$ is a semisimple associative algebra, and $\mathcal{M}$ is an $\mathcal{S}$-bimodule such that $\mathcal{M}^2 = \{0\}$.[12]

Classification of quadratic trace brackets with $N = 2$

If $N = 1$, nonzero solutions of system (6.49)–(6.53) do not exist. Consider the simplest nontrivial case $N = 2$.

---

[12]The above examples belong to this class.

Theorem 6.2.18. *In the case $N = 2$ any Poisson bracket* (6.47) *coincides up to transformations* (6.54) *and proportionality with one of the following:*

(1) $\qquad r_{22}^{21} = 1, \qquad r_{22}^{12} = -1;$

(2) $\qquad r_{22}^{21} = 1, \qquad r_{22}^{12} = -1, \qquad a_{21}^{11} = a_{22}^{12} = 1;$

(3) $\qquad r_{22}^{21} = 1, \qquad r_{22}^{12} = -1, \qquad a_{12}^{11} = a_{22}^{21} = 1;$

(4) $\qquad r_{21}^{22} = 1, \qquad r_{12}^{22} = -1;$

(5) $\qquad r_{21}^{22} = 1, \qquad r_{12}^{22} = -1, \qquad a_{11}^{21} = a_{12}^{22} = 1;$

(6) $\qquad r_{21}^{22} = 1, \qquad r_{12}^{22} = -1, \qquad a_{11}^{12} = a_{12}^{22} = 1;$

(7) $\qquad a_{22}^{11} = 1.$

In this list of nonequivalent cases, we present only nonzero components of the tensors **r** *and* **a**.

Proof. Solving the system (6.50) for the six components of the skew-symmetric tensor **r**, we obtain the following two solutions:

$$r_{22}^{21} = -r_{22}^{12} = x^2, \quad r_{11}^{21} = -r_{11}^{12} = y^2, \quad r_{12}^{21} = r_{21}^{21} = -r_{21}^{12} = -r_{12}^{12} = xy \qquad (6.61)$$

and

$$r_{21}^{22} = -r_{12}^{22} = x^2, \quad r_{21}^{11} = -r_{12}^{11} = y^2, \quad r_{21}^{12} = r_{21}^{21} = -r_{12}^{21} = -r_{12}^{12} = xy, \qquad (6.62)$$

where x and y are arbitrary parameters. Under a transformation (6.54), the parameters in (6.61) are changing according to the rule

$$x \mapsto \frac{1}{\Delta}(xg_{22} + yg_{12}), \qquad y \mapsto \frac{1}{\Delta}(xg_{21} + yg_{11}),$$

where $\Delta = g_{22}g_{11} - g_{12}g_{21}$. For the solution (6.62) we have

$$x \mapsto \frac{1}{\Delta}(-xg_{11} + yg_{21}), \qquad y \mapsto \frac{1}{\Delta}(xg_{12} - yg_{22}).$$

For a nonzero solution of the form (6.61), the remaining system of equations (6.51)–(6.53) for the components of the tensor **a**, besides the zero solution, has the following two solutions:

$$a_{21}^{11} = a_{22}^{12} = x^2, \quad a_{11}^{21} = a_{12}^{22} = -y^2, \quad a_{11}^{11} = a_{12}^{12} = -a_{21}^{21} = -a_{22}^{22} = xy \qquad (6.63)$$

and

$$a^{11}_{12} = a^{21}_{22} = x^2, \quad a^{12}_{11} = a^{22}_{21} = -y^2, \quad a^{11}_{11} = a^{21}_{21} = -a^{12}_{12} = -a^{22}_{22} = xy. \qquad (6.64)$$

In the case (6.62), the system (6.51)–(6.53) has the following two nonzero solutions:

$$a^{21}_{11} = a^{22}_{12} = x^2, \quad a^{11}_{21} = a^{12}_{22} = -y^2, \quad a^{11}_{11} = a^{12}_{12} = -a^{21}_{21} = -a^{22}_{22} = xy \qquad (6.65)$$

and

$$a^{12}_{11} = a^{22}_{21} = -x^2, \quad a^{11}_{12} = a^{21}_{22} = y^2, \quad a^{12}_{12} = a^{22}_{22} = -a^{11}_{11} = -a^{21}_{21} = xy. \qquad (6.66)$$

In the case of the zero tensor **r**, the system (6.51)–(6.53) has solutions

$$a^{11}_{22} = x^4, \qquad a^{11}_{12} = a^{11}_{21} = -a^{12}_{22} = -a^{21}_{22} = x^3 y,$$

$$a^{11}_{11} = a^{22}_{22} = -a^{12}_{12} = -a^{12}_{21} = -a^{21}_{12} = -a^{21}_{21} = x^2 y^2,$$

$$a^{22}_{12} = a^{22}_{21} = -a^{12}_{11} = -a^{21}_{11} = xy^3, \qquad a^{22}_{11} = y^4.$$

The parameters x and y are transformed as

$$x \mapsto \frac{1}{\Delta^2}(xg_{22} + yg_{12}), \qquad y \mapsto \frac{1}{\Delta^2}(xg_{21} + yg_{11}).$$

Using the transformations (6.54), we reduce the parameters in the solutions found above to the form $x = 1$, $y = 0$ and obtain the statement of Theorem 6.2.18. □

Remark 6.2.19. *It is easy to verify (see [97]) that there exist exactly two nonisomorphic anti-Frobenius subalgebras in* $\mathrm{Mat}_2(\mathbb{C})$. *They consist of matrices with one zero column and with one zero row. Cases 1 and 4 correspond to these subalgebras.*

Remark 6.2.20. *Poisson brackets in Cases 2 and 4 are nondegenerate. The corresponding symplectic structures were found in* [185] *(Example 5.7 and Lemma 7.1).*

The classification in the case $N = 3$ under the additional assumption $\mathbf{a} = 0$ is contained in [186].

Open problem 6.2.21. *Describe all trace brackets of the form (6.47) for* $N = 3$.

6.2.2 Nonabelian Poisson brackets on free associative algebras

The Poisson brackets, which we considered in the previous section, were defined on functions of the entries of the matrices x_α, $\alpha = 1, \ldots, N$ and admitted a restriction on the vector space of traces of matrix polynomials.

Question 6.2.22. *How to generalize these Poisson brackets to the case of free associative algebra?*

A Poisson structure on a commutative associative algebra A is given by a Lie bracket
$$\{\cdot, \cdot\} : A \times A \mapsto A,$$
satisfying the Leibniz rule
$$\{a, \, bc\} = \{a, \, b\} \, c + b \, \{a, \, c\}, \qquad a, b, c \in A.$$

As it is well known, a naive generalization of this definition to the case of a noncommutative associative algebra A is not informative due to the lack of examples of such brackets, other than the usual commutator [187].

We consider the version of the Hamiltonian formalism on the free associative algebra $\mathcal{A}$, proposed in [175]. The corresponding *nonabelian Poisson brackets* are defined *only* between traces of elements from $\mathcal{A}$. The traces are regarded (see Definition 6.1.8 and Remark 6.1.9) as elements of the quotient vector space $\mathcal{T} = \mathcal{A}/[\mathcal{A}, \mathcal{A}]$. The idea that in the noncommutative case the Poisson brackets should be defined on the vector space $\mathcal{T}$ was independently offered by different authors (see, for example, [188]).

The Poisson brackets will be determined by the formula (1.21), in which a Hamiltonian operator and trace gradients are to be defined.

Let $\mathcal{A}$ be the free associative algebra $\mathbb{C}[x_1, \ldots, x_N]$. For any element $a \in \mathcal{A}$ we denote by L_a (respectively R_a) the operator of left (respectively, right) multiplication by a:
$$L_a(X) = a \, X, \qquad R_a(X) = X \, a, \qquad X \in \mathcal{A}.$$

The associativity of the algebra $\mathcal{A}$ is equivalent to the identity $[L_a, R_b] = 0$ for any a and b. Moreover,
$$L_{ab} = L_a \, L_b, \qquad R_{ab} = R_b \, R_a, \qquad L_{a+b} = L_a + L_b, \qquad R_{a+b} = R_a + R_b.$$

Definition 6.2.23. Denote by $\mathcal{X}$ the associative algebra generated by all operators of left and right multiplications by the generators x_i. This algebra is called the algebra of *local operators*.

The formulas

$$L_a^\star = R_a, \qquad R_a^\star = L_a$$

define an involution on the algebra $\mathcal{X}$. This involution can be extended to $\mathcal{X} \otimes \mathfrak{gl}_N$ in the usual way.

Definition 6.2.24. A nonabelian Poisson bracket is a bracket of the form

$$\{f, g\} = \langle \mathrm{grad}_\mathbf{x}\, f,\ \Theta(\mathrm{grad}_\mathbf{x}\, g) \rangle, \qquad f, g \in \mathcal{T}, \tag{6.67}$$

where $\Theta \in \mathcal{X} \otimes \mathfrak{gl}_N$, that satisfies the conditions

$$\{f, g\} + \{g, f\} \sim 0, \tag{6.68}$$

$$\{f, \{g, h\}\} + \{g, \{h, f\}\} + \{h, \{f, g\}\} \sim 0. \tag{6.69}$$

Here, ~ 0 means the equality to zero in $\mathcal{T}$.

In Definition 6.2.24, it remains to introduce the concept of a gradient of an element of $\mathcal{T}$. Let $H(\mathbf{x}) \in \mathcal{A}$, where $\mathbf{x} = (x_1, \ldots, x_N)$. Then, $\mathrm{grad}_\mathbf{x}(H) \in \mathcal{A}^N$ is a vector

$$\mathrm{grad}_\mathbf{x}(H) = \Big(\mathrm{grad}_{x_1}(H), \ldots, \mathrm{grad}_{x_N}(H) \Big),$$

the components of which are uniquely defined by the formula

$$\frac{d}{d\epsilon} H(x_1, \ldots, x_k + \epsilon\,\delta_k, \ldots, x_N)|_{\epsilon=0} \sim \delta_k\, \mathrm{grad}_{x_k}\Big(H(\mathbf{x}) \Big),$$

where δ_i are additional nonabelian variables.

Example 6.2.25. Let $\mathcal{A} = \mathbb{C}[u, v]$. Let us find the gradient of the polynomial $H = u^2 v^2 - uvuv$. We have

$$\frac{d}{d\epsilon} H(u + \epsilon\,\delta_1, v)|_{\epsilon=0} = \delta_1 uv^2 + u\delta_1 v^2 - \delta_1 vuv - uv\delta_1 v \sim \delta_1\,(uv^2 + v^2 u - 2vuv).$$

Therefore, $\mathrm{grad}_u(H) = uv^2 + v^2 u - 2vuv = [v, [v, u]]$. Similarly, $\mathrm{grad}_v(H) = vu^2 + u^2 v - 2uvu = [u, [u, v]]$ and

$$\mathrm{grad}_\mathbf{x}(H) = \Big([v, [v, u]], \quad [u, [u, v]] \Big),$$

where $\mathbf{x} = (u, v)$.

Lemma 6.2.26. *If $f \in [\mathcal{A}, \mathcal{A}]$, then $\mathrm{grad}_\mathbf{x}(f) = 0$.*

Remark 6.2.27. *From this lemma it follows that the mapping $\mathrm{grad}_\mathbf{x} : \mathcal{T} \to \mathcal{A}^N$ is well defined and the formula (6.67) defines a bracket on the vector space $\mathcal{T}$.*

For any element $f \in \mathcal{A}$ (or for its equivalence class in $\mathcal{T}$) we denote by f_i the components of its gradient. The gradient itself is denoted by $\mathbf{f}$.

The key property of the gradient is described in the following statement:

Proposition 6.2.1. *For any element $f \in \mathcal{A}$ the identity*

$$\sum_{i=1}^{N} [f_i, x_i] = 0 \tag{6.70}$$

holds.

Exercise 6.2.28. Prove this statement.

Any nonabelian Hamiltonian system of equations on $\mathcal{A}$ has the form

$$\frac{d\mathbf{x}}{dt} = \Theta\left(\mathrm{grad}_{\mathbf{x}} H\right), \tag{6.71}$$

where $H(\mathbf{x}) \in \mathcal{A}/[\mathcal{A}, \mathcal{A}]$ is the Hamiltonian, and Θ is a Hamiltonian operator. According to Definition 6.2.24, the operator Θ is supposed to be an $N \times N$ matrix with entries being local operators. Obviously, the system (6.71) has the form (6.1), where $F_\alpha \in \mathcal{A}$.

Let us discuss the conditions (6.68) and (6.69).

Lemma 6.2.29. *If the operator Θ is skew-symmetric:*

$$\Theta_{ij}^{\star} = -\Theta_{ji}, \tag{6.72}$$

then the condition (6.68) is satisfied.

In the general case, the left-hand side of the relation (6.68) is a noncommutative polynomial in the variables f_i, g_i and x_i (the latter appear when we apply the Hamiltonian operator Θ). This polynomial is bilinear in the gradients $\mathbf{f}$ and $\mathbf{g}$. Obviously, it is equivalent to a unique polynomial of the form

$$\sum_{i=1}^{N} Q_i(\mathbf{x}, \mathbf{f}) \, g_i.$$

Condition (6.68) is satisfied iff the polynomials Q_i are equal to zero. If we assume that their arguments $x_1, \ldots, x_N, f_1, \ldots, f_N$ are algebraically independent, the condition (6.68) is exactly the fact that the operator Θ is skew-symmetric.

However, as a consequence of the identity (6.70), which relates $\mathbf{x}$ to $\mathbf{f}$, the polynomials Q_i must be equal to zero modulo (6.70). This leads to conditions for the coefficients of the operator Θ, weaker than (6.72). In particular, it turns out that the bracket (6.67) with a nonzero operator Θ can be identically zero!

Similar, but more complex problems arise when analyzing the condition (6.69).

Relations between nonabelian and trace Poisson brackets

Given a nonabelian Poisson bracket, we can define the corresponding trace bracket by the following natural way. The nonabelian bracket defines the Poisson bracket between traces of matrix polynomials. It can be extended to the entries of the matrices $x_1, \cdots, x_N$ as follows. We have

$$x_{i,\alpha}^{j} = \operatorname{tr}(e_j^i x_\alpha), \qquad x_{i',\beta}^{j'} = \operatorname{tr}(e_{j'}^{i'} x_\beta),$$

where e_j^i denotes the matrix units. Since e_j^i are constants and

$$\operatorname{grad}_{x_\beta}(x_\alpha) = \delta_{\alpha,\beta}$$

formula (6.67) implies

$$\{x_{i,\alpha}^{j}, x_{i',\beta}^{j'}\} = \operatorname{tr}(e_j^i \Theta_{\alpha,\beta}(e_{j'}^{i'})). \tag{6.73}$$

It is important to note that if nonabelian bracket satisfies (6.72), then the corresponding trace bracket (6.73) is skew-symmetric, but does not always satisfy the Jacobi identity!

Conversely, due to property 2 of Definition 6.2.1 a given trace Poisson bracket defines a nonabelian bracket.

Exercise 6.2.30. Prove that this nonabelian bracket satisfies conditions (6.68) and (6.69).

Open problem 6.2.31. *Prove that this nonabelian bracket has the form* (6.67) *with a local Hamiltonian operator* Θ.

Linear nonabelian Poisson brackets are given by Hamiltonian operators with the entries of the form

$$\Theta_{ij} = b_{ij}^k R_{x_k} + \bar{b}_{ij}^k L_{x_k}. \tag{6.74}$$

Open problem 6.2.32. *Describe all collections of constants* b_{ij}^k, $\bar{b}_{ij}^k$ *in* (6.74) *such that:*

- Problem 1: *The corresponding linear bracket* (6.67) *is null;*

- Problem 2: *The bracket* (6.67) *is a nonabelian Poisson bracket.*

The condition (6.72) is equivalent to the relations

$$\bar{b}_{ij}^k = -b_{ji}^k. \tag{6.75}$$

Proposition 6.2.2. *If the operator* Θ *is given by formulas* (6.74), (6.75), *where* b_{ij}^k *are structural constants of an associative algebra, then* (6.67) *is a nonabelian Poisson bracket.*

The corresponding trace bracket (6.73) has the form

$$\{x_{i,\alpha}^j, x_{i',\beta}^{j'}\} = b_{\alpha,\beta}^{\gamma} x_{i,\gamma}^{j'} \delta_{i'}^j - b_{\beta,\alpha}^{\gamma} x_{i',\gamma}^j \delta_i^{j'},$$

which coincides with (6.46). Conversely, any trace Poisson bracket (6.46) generates the nonabelian Poisson bracket with the Hamiltonian operator (6.74).

For quadratic nonabelian Poisson brackets, we have

$$\Theta_{i,j} = a_{ij}^{pq} L_{x_p} L_{x_q} + \bar{a}_{ij}^{pq} R_{x_p} R_{x_q} + r_{ij}^{pq} L_{x_p} R_{x_q}. \tag{6.76}$$

If the operator Θ satisfies condition (6.72), then

$$\bar{a}_{ij}^{pq} = -a_{ji}^{qp}, \qquad r_{ij}^{pq} = -r_{ji}^{qp}. \tag{6.77}$$

Proposition 6.2.3. *If the operator Θ is given by formulas (6.76), (6.77), where r_{ij}^{pq} and a_{ij}^{pq} satisfy the relations (6.49)–(6.53), then (6.67) is a nonabelian Poisson bracket.*

Remark 6.2.33. *The corresponding trace Poisson bracket (6.73) is given by (6.47). Conversely, any trace Poisson bracket (6.47) generates the nonabelian Poisson bracket with the Hamiltonian operator (6.76), (6.77).*

Some remarks on two-dimensional nonabelian Poisson brackets

Identity (6.70) provides the existence of nonzero operators Θ of the form (6.76) that define a null bracket (i.e. $\{f, g\} = 0$ for any elements $f, g \in \mathcal{T}$).

Lemma 6.2.34. *In the case $N = 2$, an operator of the form (6.76), (6.77) defines a null bracket iff*

$$\Theta_{1,1} = (k_1 L_{x_1} + k_2 L_{x_2} + k_3 R_{x_1} + k_4 R_{x_2})(L_{x_1} - R_{x_1}),$$

$$\Theta_{1,2} = (k_1 L_{x_1} + k_2 L_{x_2} + k_3 R_{x_1} + k_4 R_{x_2})(L_{x_2} - R_{x_2}),$$

$$\Theta_{2,1} = (r_1 L_{x_1} + r_2 L_{x_2} + r_3 R_{x_1} + r_4 R_{x_2})(L_{x_1} - R_{x_1}),$$

$$\Theta_{2,2} = (r_1 L_{x_1} + r_2 L_{x_2} + r_3 R_{x_1} + r_4 R_{x_2})(L_{x_2} - R_{x_2}),$$

where k_i, r_i are arbitrary constants.

There exist Hamiltonian operators (6.76), (6.77) for which the constants r_{km}^{ij} and a_{km}^{ij} do not satisfy the identities (6.49)–(6.53).

Example 6.2.35. In the case $N = 2$, consider operators (6.76), (6.77) with zero tensor **a**. According to Theorem 6.2.18, any nonzero solution of the system (6.49)–

(6.53) c $\mathbf{a} = 0$ is equivalent to the solution from Case 1 or from Case 4. In the case of nonabelian Poisson brackets there are the following generalizations of these cases:

$$r_{22}^{21} = 1, \quad r_{22}^{12} = -1, \quad r_{21}^{12} = r_{12}^{12} = -r_{12}^{21} = -r_{21}^{21} = k$$

and

$$r_{21}^{22} = 1, \quad r_{12}^{22} = -1, \quad r_{12}^{12} = r_{12}^{21} = -r_{21}^{21} = -r_{21}^{12} = k,$$

where k is an arbitrary parameter. The corresponding trace brackets (6.73) do not satisfy the Jacobi identity (1.20) for $k \neq 0$.

Lemma 6.2.36. *In the case $N = 2$, any nonabelian Hamiltonian operator (6.76), (6.77) with zero tensor* $\mathbf{a}$ *is equivalent* [13] *to one of those described in Example 6.2.35.*

Open problem 6.2.37. *Describe all nonequivalent nonabelian Hamiltonian operators (6.76) in the case $N = 2$.*

According to Proposition 6.2.2, any pair of compatible associative algebras (see Definition 3.1.5) generates a pair of compatible nonabelian linear Poisson brackets.

Example 6.2.38. Let $N = 2$. Consider the pair of compatible associative multiplications

$$x_1 \star x_1 = x_1, \quad x_1 \star x_2 = x_2 \star x_1 = x_2 \star x_2 = 0$$

and

$$x_1 \circ x_1 = x_2, \quad x_1 \circ x_2 = x_2 \circ x_1 = x_2 \circ x_2 = 0.$$

The corresponding linear nonabelian Poisson brackets $\{\cdot, \cdot\}_i$ are given by the Hamiltonian operators

$$\Theta_i = \begin{pmatrix} R_{x_i} - L_{x_i} & 0 \\ 0 & 0 \end{pmatrix}, \qquad i = 1, 2.$$

The pencil $\{\cdot, \cdot\}_1 + \lambda\{\cdot, \cdot\}_2$ of these Poisson brackets has a Casimir function $C = \mathrm{tr}\,(x_1 + \lambda x_2)^3$, which generates Hamiltonians $H_1 = -\mathrm{tr}\,(x_1^2 x_2)$ and $H_2 = \frac{1}{3}\mathrm{tr}\,(x_1^3)$ commuting with respect to both brackets (see Theorem 3.0.2). The formula

$$\frac{d\mathbf{x}}{dt} = \Theta_1\Big(\mathrm{grad}_\mathbf{x} H_1\Big) = \Theta_2\Big(\mathrm{grad}_\mathbf{x} H_2\Big)$$

gives a bi-Hamiltonian representation for the nonabelian Manakov system

$$\frac{dx_1}{dt} = x_1^2 x_2 - x_2 x_1^2, \qquad \frac{dx_2}{dt} = 0,$$

already mentioned in Examples 1.2.2 and 3.1.36 and in Section 6.1.1 (subsection "Manakov top").

---

[13]That is, it is connected by a linear transformation (6.54).

Nonabelian elliptic bracket in two variables

All Hamiltonian operators mentioned above are homogeneous. An interesting example of an inhomogeneous nonabelian Poisson bracket was recently found by A. Odesskii and V. Sokolov. This is a nonabelization of the Poisson bracket

$$\{x_1,\, x_2\} = x_1^3 + x_2^3 + k\, x_1 x_2 + 1, \tag{6.78}$$

where k is an arbitrary "elliptic" parameter. The bracket (6.78) is a $\mathbb{CP}$-version of the $q_{3,1}$ elliptic bracket (see [27]).

Example 6.2.39. Let $N = 2$. The nonabelian bracket (6.67) is defined by a Hamiltonian operator of the form

$$\Theta = \Theta^3 + k\,\Theta^2 + \Theta^0, \tag{6.79}$$

where Θ^i are homogeneous operators of order i, whose components are as follows:

$$\Theta^3_{1,1} = -R_{x_1}R_{x_2}L_{x_2} + R_{x_2}L_{x_2}L_{x_1},$$

$$\Theta^3_{1,2} = R_{x_1}R_{x_1}R_{x_1} + R_{x_2}L_{x_2}L_{x_2},$$

$$\Theta^3_{2,1} = -L_{x_1}L_{x_1}L_{x_1} - R_{x_2}R_{x_2}L_{x_2},$$

$$\Theta^3_{2,2} = -L_{x_1}L_{x_1}L_{x_2} - R_{x_1}L_{x_1}L_{x_2} - R_{x_1}R_{x_1}L_{x_2} +$$
$$R_{x_2}L_{x_1}L_{x_1} + R_{x_2}R_{x_1}L_{x_1} + R_{x_2}R_{x_1}R_{x_1},$$

$$\Theta^2_{1,1} = 0, \qquad \Theta^2_{1,2} = \frac{1}{2}\Big(R_{x_1}L_{x_2} + R_{x_1}R_{x_2}\Big),$$

$$\Theta^2_{2,1} = -\frac{1}{2}\Big(L_{x_2}L_{x_1} + R_{x_2}L_{x_1}\Big) \qquad \Theta^2_{2,2} = \frac{1}{2}\Big(-L_{x_2}L_{x_2} + R_{x_2}R_{x_2}\Big),$$

$$\Theta^0_{1,1} = \Theta^0_{2,2} = 0, \qquad \Theta^0_{1,2} = 1, \qquad \Theta^0_{2,1} = -1.$$

One can verify that this bracket satisfies conditions (6.68) and (6.69). If x_i are matrices of dimension 1×1, the bracket coincides with (6.78).

Remark 6.2.40. *In the paper* [189] *for any coprime n and m the* $\mathbb{CP}$ *elliptic brackets* $q_{n,m}$ *were generalized to the nonabelian case. Probably, homogeneous nonabelizations of* $q_{n,m}$-*brackets do not exist.*

6.2.3   Double Poisson brackets on free associative algebras

In the previous sections, we have observed that identities (6.49)–(6.53) describe both quadratic trace Poisson brackets and special quadratic nonabelian Poisson brackets (see Theorem 6.2.3 and Proposition 6.2.3). It turns out [191] that these identities also describe double Poisson brackets [190] on the free associative algebra $\mathbb{C}[x_1, \ldots, x_N]$.

Definition 6.2.41. [190]. A double Poisson bracket on an associative algebra A is a $\mathbb{C}$-linear mapping $\{\!\{,\}\!\} : A \otimes A \to A \otimes A$, satisfying the following conditions:

$$\{\!\{u, v\}\!\} = -\{\!\{v, u\}\!\}^\circ, \tag{6.80}$$

$$\{\!\{u, \{\!\{v, w\}\!\}\}\!\} + \sigma\{\!\{v, \{\!\{w, u\}\!\}\}\!\} + \sigma^2\{\!\{w, \{\!\{u, v\}\!\}\}\!\} = 0, \tag{6.81}$$

and

$$\{\!\{u, vw\}\!\} = (v \otimes 1)\{\!\{u, w\}\!\} + \{\!\{u, v\}\!\}(1 \otimes w). \tag{6.82}$$

Here,

$$(u \otimes v)^\circ \stackrel{def}{=} v \otimes u; \qquad \{\!\{v_1, v_2 \otimes v_3\}\!\} \stackrel{def}{=} \{\!\{v_1, v_2\}\!\} \otimes v_3$$

and

$$\sigma(v_1 \otimes v_2 \otimes v_3) \stackrel{def}{=} v_3 \otimes v_1 \otimes v_2.$$

Let us define the multiplication map $\mu : A \otimes A \to A$ by the formula $\mu(u \otimes v) = uv$. Consider the $\mathbb{C}$-bilinear operation $\{\cdot, \cdot\} \stackrel{def}{=} \mu(\{\!\{\cdot, \cdot\}\!\})$ on A.

Proposition 6.2.4. [190]   *Let $\{\!\{\cdot, \cdot\}\!\}$ be a double Poisson bracket on A. Then, the operation $\{\cdot, \cdot\}$ defines a nonabelian Poisson bracket*

$$\{\bar{a}, \bar{b}\} = \overline{\mu(\{\!\{a, b\}\!\})}, \tag{6.83}$$

on the space $A/[A, A]$. Here, $\bar{a}$ denotes the image of an element $a \in A$ under the natural projection $A \to A/[A, A]$.

Let $\mathcal{A} = \mathbb{C}[x_1, \ldots, x_N]$ be the free associative algebra. If double brackets $\{\!\{x_i, x_j\}\!\}$ between generators x_i are given, then the brackets between any two elements of $\mathcal{A}$ are uniquely determined by the identities (6.80)–(6.82). Constant, linear and quadratic double brackets are defined by

$$\{\!\{x_i, x_j\}\!\} = c_{ij} 1 \otimes 1,$$

$$\{\!\{x_i, x_j\}\!\} = b_{ij}^k x_k \otimes 1 - b_{ji}^k 1 \otimes x_k,$$

and

$$\{\!\{x_\alpha, x_\beta\}\!\} = r_{\alpha\beta}^{uv} x_u \otimes x_v + a_{\alpha\beta}^{vu} x_u x_v \otimes 1 - a_{\beta\alpha}^{uv} 1 \otimes x_v x_u,$$

respectively.

Proposition 6.2.5. *These formulas define double Poisson brackets iff the constants $c_{ij}, b_{ij}^k, r_{ij}^{pq}, a_{ij}^{pq}$ satisfy the identities of Theorem 6.2.3.*

6.3 Evolution equations on free associative algebras

In this section, we consider nonabelian evolution equations which are a natural generalization of evolution equations with matrix variables.

6.3.1 Matrix integrable equations

It is known that the matrix KdV equation

$$\mathbf{U}_t = \mathbf{U}_{xxx} + 3\,(\mathbf{U}\mathbf{U}_x + \mathbf{U}_x\mathbf{U}),\tag{6.84}$$

where $\mathbf{U}(x,t)$ is an unknown matrix of dimension $m \times m$, has infinitely many matrix symmetries for arbitrary m. The simplest of them is given by

$$\mathbf{U}_\tau = \mathbf{U}_{xxxxx} + 5\,(\mathbf{U}\mathbf{U}_{xxx} + \mathbf{U}_{xxx}\mathbf{U}) + 10\,(\mathbf{U}_x\mathbf{U}_{xx} + \mathbf{U}_{xx}\mathbf{U}_x)+$$
$$10\,(\mathbf{U}^2\mathbf{U}_x + \mathbf{U}\mathbf{U}_x\mathbf{U} + \mathbf{U}_x\mathbf{U}^2).$$

In the case $m = 1$, the matrix hierarchy of equation (6.84) coincides with the usual KdV (see Section 2.1.2).

Generalizing this example, consider [192] integrable matrix evolution equations of the form

$$\mathbf{U}_t = F(\mathbf{U}, \mathbf{U}_1, \ldots, \mathbf{U}_n), \qquad \mathbf{U}_i = \frac{\partial^i \mathbf{U}}{\partial x^i},\tag{6.85}$$

where F is a (noncommutative) polynomial or rational function. Such equations can be considered as matrix generalizations of scalar integrable equations (1.26). As elsewhere in Part 3 of this book, the existence of (matrix) higher symmetries of the form

$$\mathbf{U}_\tau = G(\mathbf{U}, \mathbf{U}_1, \ldots, \mathbf{U}_k)\tag{6.86}$$

is chosen as a criterion for integrability.

Open problem 6.3.1. *Apply the approach described in Section 6.1.6 to the systematic search for integrable matrix generalizations of well-known scalar integrable equations.*

Below we present several matrix equations which are integrable for any dimension of the matrices. The main goal here is to demonstrate that the matrix KdV equation is not a rarity. Most of the well-known scalar integrable models admit integrable matrix generalization [193, 194, 195].

For example, the mKdV equation (2.25) has two different integrable matrix generalizations

$$\mathbf{U}_t = \mathbf{U}_{xxx} + 3\mathbf{U}^2\mathbf{U}_x + 3\mathbf{U}_x\mathbf{U}^2,\tag{6.87}$$

and (see [194])

$$\mathbf{U}_t = \mathbf{U}_{xxx} + 3\mathbf{UU}_{xx} - 3\mathbf{U}_{xx}\mathbf{U} - 6\mathbf{UU}_x\mathbf{U}. \tag{6.88}$$

The matrix generalization of the NLS equation (1.13) is given by

$$\mathbf{U}_t = \mathbf{U}_{xx} - 2\,\mathbf{UVU}, \qquad \mathbf{V}_t = -\mathbf{V}_{xx} + 2\,\mathbf{VUV}. \tag{6.89}$$

The following integrable matrix system of the derivative NLS equation type (cf. Appendix 9.4)

$$\mathbf{U}_t = \mathbf{U}_{xx} + 2\,\mathbf{UVU}_x, \qquad \mathbf{V}_t = -\mathbf{V}_{xx} + 2\,\mathbf{V}_x\mathbf{UV} \tag{6.90}$$

was found in [196].

The Schwartz–KdV equation (1.71) admits the integrable matrix generalization

$$\mathbf{U}_t = \mathbf{U}_{xxx} - \frac{3}{2}\,\mathbf{U}_{xx}\mathbf{U}_x^{-1}\mathbf{U}_{xx}. \tag{6.91}$$

Equation (1.71) coincides with the special case $Q = 0$ of the Krichever–Novikov equation (4.7). Apparently, the Krichever–Novikov equation with a generic polynomial Q has no matrix generalizations.

The matrix Heisenberg equation has the form

$$\mathbf{U}_t = \mathbf{U}_{xx} - 2\,\mathbf{U}_x(\mathbf{U}+\mathbf{V})^{-1}\mathbf{U}_x \qquad \mathbf{V}_t = -\mathbf{V}_{xx} + 2\,\mathbf{V}_x(\mathbf{U}+\mathbf{V})^{-1}\mathbf{V}_x. \tag{6.92}$$

One of the most well-known matrix hyperbolic equations is the σ-model of the principle chiral field

$$\mathbf{U}_{xy} = \frac{1}{2}\,(\mathbf{U}_x\mathbf{U}^{-1}\mathbf{U}_y + \mathbf{U}_y\mathbf{U}^{-1}\mathbf{U}_x). \tag{6.93}$$

The following, possibly new, integrable system

$$\mathbf{U}_t = \lambda_1\mathbf{U}_x + (\lambda_2 - \lambda_3)\mathbf{W}^T\mathbf{V}^T,$$

$$\mathbf{V}_t = \lambda_2\mathbf{V}_x + (\lambda_3 - \lambda_1)\mathbf{U}^T\mathbf{W}^T, \tag{6.94}$$

$$\mathbf{W}_t = \lambda_3\mathbf{W}_x + (\lambda_1 - \lambda_2)\mathbf{V}^T\mathbf{U}^T$$

is a matrix generalization of the 3-waves model. Unlike the previous equations, it contains the operation of the matrix transposition ^{T}.

Let $\mathbf{e}_1, \ldots, \mathbf{e}_N$ be a basis of some associative algebra $\mathcal{B}$. Substituting instead of $\mathbf{U}$ the element

$$U(x,t) = \sum_{i=1}^{N} u^i(x,t)\,\mathbf{e}_i$$

into any of the above polynomial matrix integrable equations, we obtain an *integrable* system[14] with respect to the unknown functions $u^1, \ldots, u^N$. Indeed, in order to verify that a matrix equation (6.86) is a symmetry for equation (6.85), only the associativity of the product is required. Therefore, the matrix symmetries turn into the symmetries of the system for functions u^i generated by the algebra $\mathcal{B}$.

This construction leads to interesting examples of multi-component integrable systems in the case when $\mathcal{B}$ is the Clifford algebra or the group algebra of an associative ring. However, nonabelian equations, corresponding to free associative algebras (cf. Section 6.1.2) are the most fundamental.

6.3.2 Nonabelian evolution equations

Consider evolution equations with one variable on the free associative algebra $\mathcal{A}$. In this case, the generators of $\mathcal{A}$ are

$$U, \quad U_1 = U_x, \quad \ldots, \quad U_k, \quad \ldots . \tag{6.95}$$

Since $\mathcal{A}$ is assumed to be a free algebra, there are no algebraic relations between the generators.

As in the case of Definition 6.2.23, we denote by $\mathcal{X}$ the associative algebra generated by the operators of left and right multiplication by the generators (6.95) of algebra $\mathcal{A}$. We call $\mathcal{X}$ the algebra of *local operators*.

The formula

$$U_t = F(U, U_1, \ldots, U_n), \qquad F \in \mathcal{A}, \tag{6.96}$$

as usual (see Section 4.2.1), defines the derivation D_t of the algebra $\mathcal{A}$ commuting with the derivation

$$D = \sum_0^\infty U_{i+1} \frac{\partial}{\partial U_i}.$$

It is easy to see that D_t is given by the vector field

$$D_t = \sum_0^\infty D^i(F) \frac{\partial}{\partial U_i}.$$

We call (6.96) *a nonabelian evolution equation*. After replacing U with a matrix, equation (6.96) turns into a matrix evolution equation.

A generalization of the symmetry approach to the case of nonabelian evolution equations requires suitable definitions of the symmetry, the conservation law, the Fréchet derivative, and the formal symmetry.

---

[14]In principle, this system may be trivial in some sense.

As in Section 2.0.1, a symmetry is a nonabelian evolution equation

$$U_\tau = G(U, U_1, \ldots, U_m),$$

such that the corresponding vector field

$$D_\tau = \sum_0^\infty D^i(G)\frac{\partial}{\partial u_i}$$

commutes with D_t. The element $G \in \mathcal{A}$ is called *the symmetry generator*.

The condition $[D_t, D_G] = 0$ is equivalent to the relation $D_t(G) = D_G(F)$. It can be rewritten as

$$G_*(F) - F_*(G) = 0, \tag{6.97}$$

where the Fréchet derivative H_* of any element $H \in \mathcal{A}$ is defined as follows.

Let us extend the set of generators (6.95) by adding noncommutative symbols $V_0, V_1, \ldots$ and prolong the derivation D by the formula $D(V_i) = V_{i+1}$.

For a given element $H(U, U_1, U_2, \ldots, U_k) \in \mathcal{A}$, we find a polynomial

$$L_H = \frac{\partial}{\partial \varepsilon}H(U + \varepsilon V_0,\ U_1 + \varepsilon V_1,\ U_2 + \varepsilon V_2, \ldots)\Big|_{\varepsilon=0}$$

and represent it in the form $H_*(V_0)$, where H_* is a linear differential operator of order k with coefficients from $\mathcal{X}$. This operator is called the *Fréchet derivative*. For example, $(U_2 + UU_1)_* = D^2 + L_U D + R_{U_1}$.

Unlike the definition of symmetry, which is a direct generalization of the corresponding definition in the scalar case, the definition of conservation law has to be modified in the spirit of Definition 6.1.10.

Recall that in the scalar case, the conserved density is a function $\rho \in \mathcal{F}$ such that $D_t(\rho) = D(\sigma)$ for some function $\sigma \in \mathcal{F}$. The equivalent densities define the same functional, the value of which on any solitonic type solution of the equation does not depend on time (see Section 1.5). More accurately, the conserved density is an equivalence class in $\mathcal{F}$[15] which the total derivative D_t takes to the zero class.

In the nonabelian case, the conserved density $\rho \in \mathcal{A}$ is the equivalence class in $\mathcal{A}$ with respect to the following two operations:

- Addition to ρ of elements of the form $D(s)$, where $s \in \mathcal{A}$;

- Cyclic permutation of factors in monomials[16] of the polynomial ρ.

---

[15]Here $\rho_1 \sim \rho_2$ if $\rho_1 - \rho_2 = D(s)$, $s \in \mathcal{F}$.
[16]Or, which is the same, adding elements from $[\mathcal{A}, \mathcal{A}]$.

In other words, $\rho_1 \sim \rho_2$ if

$$\rho_1 - \rho_2 \in \mathcal{A}/(\operatorname{Im} D + [\mathcal{A},\,\mathcal{A}]).$$

It is clear that in the scalar case this definition coincides with Definition 4.2.12.

The equivalence class of an element ρ is called by the *trace* of ρ and is denoted by $\operatorname{tr}(\rho)$.

The definition is motivated by the fact that in the matrix case the functional

$$I(\mathbf{U}) = \int_{-\infty}^{\infty} \operatorname{trace}\big(\rho(\mathbf{U}, \mathbf{U}_x, \dots)\big)\, dx$$

is well defined on equivalence classes if all entries of the matrix $\mathbf{U}(x,t)$ with their derivatives rapidly decrease as $x \to \pm\infty$.

Definition 6.3.2. The equivalence class of the function ρ is called *conserved density* for equation (6.96) if $D_t(\rho) \sim 0$.

In this book, we are not concerned with the theory of Hamiltonian structures on the free associative algebra $\mathcal{A}$ with generators (6.95) (see, for example, [28]). We only note that the Poisson bracket (4.86) in the nonabelian case is defined on the vector space $\mathcal{A}/(\operatorname{Im} D + [\mathcal{A},\,\mathcal{A}])$ (cf. Section 6.2.2).

Formal symmetries

All definitions and results from Section 4.2.2 related to formal symmetries can be generalized to nonabelian equations of the form (6.96), where

$$F = U_n + f(U, U_1, \dots, U_{n-1}). \tag{6.98}$$

Definition 6.3.3. A formal pseudo-differential series of the form

$$\Lambda = D + l_0 + l_{-1}D^{-1} + \cdots, \qquad l_i \in \mathcal{X} \tag{6.99}$$

is called a *formal symmetry* (or a formal recursion operator) for a nonabelian equation (6.96), (6.98) if it satisfies the operator equation

$$D_t(\Lambda) - [F_*,\,\Lambda] = 0. \tag{6.100}$$

Remark 6.3.4. *We emphasize that the coefficients of the operator F_* and of the series Λ belong to the associative algebra $\mathcal{X}$ of local operators, but not the algebra $\mathcal{A}$ as it was in the scalar case. The same is true for recursion, Hamiltonian and other operators arising in the theory of nonabelian integrable equations.*

Example 6.3.5. In the case of the nonabelian KdV equation (6.84) one can take $\Lambda = \mathcal{R}^{1/2}$ as a formal symmetry. Here, $\mathcal{R}$ is the recursion operator

$$\mathcal{R} = D^2 + 2(L_U + R_U) + (L_{U_x} + R_{U_x})\, D^{-1} + (L_U - R_U)\, D^{-1}\, (L_U - R_U)\, D^{-1},$$

found in [192]. In the scalar case, this operator coincides with the usual recursion operator (4.38) for the KdV equation.

The analogs of Theorems 4.2.9 and 4.2.23 can be proved by the same arguments as these theorems themselves.

All definitions and constructions given above are easily generalized to the case of several nonabelian variables.

In the papers [192, 196] with the help of the simplest symmetry test (cf. Section 1.4.1 and Appendix 9.4) several problems of classifying scalar homogeneous nonabelian equations of third order and special systems of two equations of second order were solved.

Open problem 6.3.6. *Find homogeneous integrable nonabelizations (see Section 6.1.6) of known homogeneous scalar integrable systems.*

Open problem 6.3.7. *Find inhomogeneous integrable deformations (see Section 6.1.7) of known homogeneous nonabelian integrable systems.*

Open problem 6.3.8. *Find homogeneous integrable nonabelian equations (6.96), (6.98) of the fifth order (see Section 1.4.1). It seems necessary to assume that the right-hand side of the nonabelian equation also depends on the variables U^*, $U_1^*, \ldots$, where $*$ is a formal involution on the algebra $\mathcal{A}$.*[17]

Open problem 6.3.9. *Classify the quasilinear nonabelian systems of two second order equations (cf. Lemma (4.1.14)).*

---

[17]Even under this assumption it is not obvious that integrable generalizations of equations (1.42) and (1.43) exist.

Chapter 7

Integrable evolution systems and nonassociative algebras

One of the most remarkable observations by S. Svinolupov [197, 198, 199] is the fact that coefficients of some polynomial multi-component integrable systems can be expressed in terms of structural constants of well-known nonassociative structures, such as Jordan algebras, triple Jordan systems, etc. This allows us to clarify the nature of vector and matrix generalizations (see, for example, [200, 201, 202]) of scalar polynomial integrable systems and construct new examples of this kind [195].

In this section, we use the "naive" version of the symmetry approach (see Section 1.4.1): for a known scalar polynomial integrable equation and its simplest higher symmetry, we consider N-component generalizations of the pair (equation, symmetry) with the same homogeneity properties as the original scalar equation and symmetry.

7.1 Nonassociative algebraic structures related to integrability

Let $\mathbf{V}$ be a N-dimensional vector space over $\mathbb{C}$ with a basis $\mathbf{e}_1, \dots, \mathbf{e}_N$. For an algebra $\mathcal{A}$ with the product $\circ$, defined on $\mathbf{V}$, we denote by C^i_{jk} the structural constants:

$$\mathbf{e}_j \circ \mathbf{e}_k = \sum_{i=1}^{N} C^i_{jk} \mathbf{e}_i.$$

For a ternary system with the operation $\{\cdot, \cdot, \cdot\}$, we denote the structural constants by B^i_{jkm}:

$$\{\mathbf{e}_j, \mathbf{e}_k, \mathbf{e}_m\} = \sum_{i=1}^{N} B^i_{jkm} \mathbf{e}_i.$$

Let us denote by U the element

$$U = \sum_{i=1}^{N} u^i(x,t)\, \mathbf{e}_i, \tag{7.1}$$

where, in this section, u^i are (scalar) functions of the variables x and t. We also use the notation of Section 1.1.4.

7.1.1 Left-symmetric algebras

According to Definition 2.4.2, an algebra with identity $[X,Y,Z] = 0$ (see formula (1.6)) is called *left-symmetric*.

Let us give examples of left-symmetric algebras:

1. Any associative algebra is left-symmetric;

2. The operation
$$\mathbf{x} \circ \mathbf{y} = \langle \mathbf{x}, \mathbf{c} \rangle\, \mathbf{y} + \langle \mathbf{x}, \mathbf{y} \rangle\, \mathbf{c}, \tag{7.2}$$

   on a vector space $\mathbf{V}$ with a scalar product $\langle \cdot, \cdot \rangle$, where $\mathbf{c}$ is a fixed vector, defines a left-symmetric algebra;

3. Let A be an associative algebra and an operator $R : A \to A$ satisfies the modified classical Yang–Baxter equation
$$R([R(x), y] - [R(y), x]) = [x, y] + [R(x), R(y)].$$

   Then, the product
$$x \circ y = [R(x), y] - (xy + yx)$$

   is left symmetric.

7.1.2 Jordan algebras

Definition 7.1.1. An algebra $\mathcal{A}$ is called *Jordan* if the following identities hold for it:
$$X \circ Y = Y \circ X, \qquad X^2 \circ (Y \circ X) = (X^2 \circ Y) \circ X. \tag{7.3}$$

Remark 7.1.2. *For any associative algebra A the operation*

$$x \circ y = xy + yx$$

satisfies identities (7.3).

The following algebras:

1. The vector space of $m \times m$ matrices with the operation

$$\mathbf{X} \circ \mathbf{Y} = \mathbf{X}\mathbf{Y} + \mathbf{Y}\mathbf{X}; \qquad (7.4)$$

2. The vector space of all symmetric matrices with the same operation (7.4);

3. A N-dimensional vector space with the operation

$$\mathbf{x} \circ \mathbf{y} = \langle \mathbf{x}, \mathbf{c} \rangle \, \mathbf{y} + \langle \mathbf{y}, \mathbf{c} \rangle \, \mathbf{x} - \langle \mathbf{x}, \mathbf{y} \rangle \, \mathbf{c}, \qquad (7.5)$$

where $\langle \cdot, \cdot \rangle$ is a scalar product, and $\mathbf{c}$ is a fixed vector;

4. The exceptional Jordan algebra $H_3(O)$ of dimension 27

provide examples of simple[1] Jordan algebras [203].

7.1.3   Triple Jordan systems

Definition 7.1.3. A triple system $\{X, Y, Z\}$ is called *Jordan* if the following identities hold for it:

$$\{X, Y, Z\} = \{Z, Y, X\},$$

$$\{X, Y, \{V, W, Z\}\} - \{V, W, \{X, Y, Z\}\} = \{\{X, Y, V\}, W, Z\} - \{V, \{Y, X, W\}, Z\}.$$

Remark 7.1.4. *For any associative algebra the operation*

$$\{x, y, z\} = xyz + zyx$$

defines a triple Jordan system.

The following triple systems are triple Jordan ones [204]:

(a) The vector space of $m \times m$ matrices with respect to the operation

$$\{\mathbf{X}, \mathbf{Y}, \mathbf{Z}\} = \frac{1}{2}\left(\mathbf{X}\mathbf{Y}\mathbf{Z} + \mathbf{Z}\mathbf{Y}\mathbf{X}\right); \qquad (7.6)$$

(b) The vector space of all skew-symmetric matrices with the same operation (7.6);

---

[1]That is, this algebra has no nontrivial ideals.

(c) The vector space $\mathbf{V}$ equipped with a scalar product $\langle \cdot, \cdot \rangle$ under the operation

$$\{\mathbf{x}, \mathbf{y}, \mathbf{z}\} = \langle \mathbf{x}, \mathbf{y} \rangle \, \mathbf{z} + \langle \mathbf{y}, \mathbf{z} \rangle \, \mathbf{x} - \langle \mathbf{x}, \mathbf{z} \rangle \, \mathbf{y}; \tag{7.7}$$

(d) The vector space $\mathbf{V}$ under the operation

$$\{\mathbf{x}, \mathbf{y}, \mathbf{z}\} = \langle \mathbf{x}, \mathbf{y} \rangle \, \mathbf{z} + \langle \mathbf{y}, \mathbf{z} \rangle \, \mathbf{x}. \tag{7.8}$$

Remark 7.1.5. *There is the following generalization of the operation (7.8). The vector space of all matrices of dimension $n \times m$ is a triple Jordan system with respect to the operation*

$$\{\mathbf{X}, \mathbf{Y}, \mathbf{Z}\} = \mathbf{X}\,\mathbf{Y}^T\,\mathbf{Z} + \mathbf{Z}\,\mathbf{Y}^T\,\mathbf{X}, \tag{7.9}$$

where "T" denotes the matrix transposition.

There is close interconnections between Jordan algebras and triple Jordan systems.

Proposition 7.1.1. *For any triple Jordan system $\{X, Y, Z\}$ the product*

$$X \circ Y = \{X, C, Y\},$$

where C is an arbitrary fixed element, defines a Jordan algebra.

Proposition 7.1.2. *For any Jordan algebra, the formula*

$$\{X, Y, Z\} = (X \circ Y) \circ Z + (Z \circ Y) \circ X - Y \circ (X \circ Z) \tag{7.10}$$

defines a triple Jordan system.

Proposition 7.1.3. *For any triple Jordan system $\{X, Y, Z\}$ and a fixed element C, the formula*

$$\sigma(X, Y, Z) = \{X, \{C, Y, C\}, Z\} \tag{7.11}$$

defines a triple Jordan system σ.

7.2 Jordan KdV systems

Consider an equation of the form (6.96), where F is, generally speaking, a noncommutative and nonassociative polynomial of the element $U \in \mathcal{A}$, defined by formula (7.1), and its derivatives U_i. Suppose that the polynomial F is given, but the multiplication in the algebra $\mathcal{A}$ is unknown.

For example, consider the KdV type equation

$$U_t = U_{xxx} + 3\, U \circ U_x, \tag{7.12}$$

where $\circ$ is a multiplication in some algebra $\mathcal{A}$.

Question 7.2.1. *For which algebras $\mathcal{A}$ equation* (7.12) *is integrable?*

It is easy to see that equation (7.12) is equivalent to the following N-component system:

$$u_t^i = u_{xxx}^i + \sum_{k,j} C_{jk}^i\, u^k\, u_x^j, \qquad i,j,k = 1,\dots,N, \tag{7.13}$$

where C_{jk}^i are structural constants of the algebra $\mathcal{A}$. Vice versa, any system of the form (7.13) can be written in the form (7.12) for a suitable algebra $\mathcal{A}$.

The systems (7.13) look like natural multi-component generalizations of the KdV equation (1.12). An essential restriction here is the fact that all coefficients of third derivatives in (7.13) are equal to each other. Interesting integrable systems of the form

$$U_t = S\,(U_{xxx}) + 3\, U \circ U_x \tag{7.14}$$

with a constant linear operator $S \neq 1$ were constructed in [37]. An algebraic description of such systems, in addition to the algebra $\mathcal{A}$, involves a linear operator $S : \mathcal{A} \to \mathcal{A}$.

Open problem 7.2.2. *Find necessary integrability conditions for equation* (7.14) *in terms of the algebra $\mathcal{A}$ and the operator S.*

The linear transformations of the vector $\vec{u} = (u_1, \dots, u_N)$ preserve the class of systems of the form (7.13). Any reasonable description of integrable cases must be invariant with respect to such transformations.

It turns out that if the algebra $\mathcal{A}$ has a nontrivial ideals, then the system (7.13) is "triangular".

Remark 7.2.3. *Let I be a double-sided ideal in $\mathcal{A}$. Choose a basis in $\mathcal{A}$ such that $\mathbf{e}_{M+1}, \dots, \mathbf{e}_N$ is a basis in I. In this basis, we have $C_{jk}^i = 0$ for $i \leq M$ and for $j > M$ or for $k > M$. This means that the equations for $u_1, \dots, u_M$ form a closed subsystem of the same form* (7.13) *but of smaller dimension and that the whole system is triangular.*

The system (7.13) is homogeneous (see Section 1.4.1): the weights in the formula (1.38) are $\lambda_i = 2$, $\mu = 3$. Without loss of generality, we can assume that the polynomial symmetries of system (7.13) are also homogeneous. Any such homogeneous symmetry has the form

$$U_\tau = U_m + A_1(U, U_{n-2}) + A_2(U_1, U_{n-3}) + \cdots . \tag{7.15}$$

In this formula, the quadratic terms are given by unknown bilinear operations A_i, the cubic terms are defined by triple operations B_i, and so on.

Remark 7.2.4. *In principle, we could additionally assume that all these operations are expressed in terms of multiplication $\circ$ in $\mathcal{A}$: $A_1(X, Y) = c_1\, X \circ Y + c_2\, Y \circ X$, where c_i are undetermined constants, etc. But this is not necessary since the explicit formulas for $A_i, B_i, \ldots$ are derived from the compatibility condition*

$$D_t(D_\tau(U)) = D_\tau(D_t(U)). \tag{7.16}$$

Theorem 7.2.5. [198] *Suppose that the algebra $\mathcal{A}$ is finite-dimensional and commutative. Then equation (7.12) has a symmetry of the form (7.15), where $m \geq 5$, iff $\mathcal{A}$ is a Jordan algebra.*

Initially, the proof of Theorem 7.2.5 was obtained for systems (7.13) with the help of a straightforward component-wise tensor calculation. Calculations are drastically simplified if we use the algebraic approach and define the equation and its symmetry in terms of algebraic operations.

In the following theorem we do not assume that the algebra is finite-dimensional or commutative. For simplicity, we assume that equation (7.12) has a fifth order symmetry (7.15). For the scalar KdV equation, such a symmetry is given by the formula (1.36).

Theorem 7.2.6. *Suppose that the algebra $\mathcal{A}$ has the following property: if*

$$\left(X \circ Y - Y \circ X\right) \circ Z = 0 \tag{7.17}$$

for any Z, then $X \circ Y - Y \circ X = 0$. Then equation (7.12) possesses a fifth order homogeneous symmetry iff $\mathcal{A}$ is a Jordan algebra.

Proof. The fifth order symmetry can be written in the form[2]

$$U_\tau = U_5 + 5\, A_1(U, U_3) + 5\, A_2(U_1, U_2) + 5\, B_1(U, U, U_1),$$

where A_1 and A_2 are unknown bilinear operations, and B_1 is a triple operation. The compatibility condition (7.16) has to be satisfied identically with respect to the

---

[2]We put the coefficients 3 and 5 in the right-hand sides of equation (7.12) and the symmetry to avoid rational numbers in formulas for A_i and B_1.

independent variables $U_5, U_4, \ldots, U_0 = U$. Let us perform the scaling $U_i \mapsto z_i U_i$, $z_i \in \mathbb{C}$ in (7.16) and equate the coefficients of different monomials in the parameters z_i. Comparing the coefficients of $z_5 z_1$, we get $A_1(X, Y) = X \circ Y$. The terms containing $z_4 z_2$, give $A_2(X, Y) = X \circ Y + Y \circ X$. The terms with $z_3 z_1 z_0$ lead to the formulas

$$B_1(X, X, Y) = \frac{1}{2}\Big(2\, X \circ (X \circ Y) + (X \circ X) \circ Y\Big)$$

and to the identity (7.17).[3]

Thus, we have expressed the algebraic operations which define the symmetry through the multiplication in the algebra $\mathcal{A}$. In addition, it follows from the assumption of Theorem 7.2.6 and identity (7.17) that the algebra $\mathcal{A}$ is commutative. Now the Jordan identity (7.3) arises from a comparison of the coefficients of $z_2 z_0^3$. The remaining identities are satisfied by virtue of (7.3).   ☐

Exercise 7.2.7. Prove that equation (7.12) has a homogeneous symmetry of the form (7.15), where $m \geq 5$, iff the identities

$$(X \circ X) \circ (X \circ Y) = X \circ \Big((X \circ X) \circ Y\Big),$$

$$2\,(X \circ Y) \circ (X \circ Y) - 2X \circ \Big(Y \circ (X \circ Y)\Big) + (X \circ X) \circ (Y \circ Y) - \Big((X \circ X) \circ Y\Big) \circ Y = 0$$

and (7.17) are satisfied.

The famous Witten–Kontsevich conjecture [205] means that a generating function for the intersection indices of the characteristic classes of some special linear bundles over the moduli spaces of complex curves with marked points is a solution of the Korteweg–de Vries equation.

Open problem 7.2.8. *Formulate and prove a similar statement for the Jordan KdV equations.*

Integrable nontriangular Jordan KdV equations

According to Remark 7.2.3, the most interesting nontriangular integrable systems (7.13) correspond to simple Jordan algebras. The classification of such algebras is contained in [203]. All simple Jordan algebras are given in Section 7.1.2. Systems corresponding to simple Jordan algebras have the form:

(1) The matrix KdV equation

$$\mathbf{U}_t = \mathbf{U}_{xxx} + \mathbf{U}\mathbf{U}_x + \mathbf{U}_x\mathbf{U},$$

which coincide with (6.84) up to a scaling of $\mathbf{U}$;

---

[3]If identity (7.17) holds, then $[\mathcal{A}, \mathcal{A}]$ is a double-sided ideal. Therefore, any simple algebra with identity (7.17) is commutative.

(2) The matrix KdV equation under the reduction $\mathbf{U}^T = \mathbf{U}$;

(3) The vector KdV equation [195]

$$\mathbf{u}_t = \mathbf{u}_{xxx} + \langle \mathbf{c},\, \mathbf{u}\rangle\, \mathbf{u}_x + \langle \mathbf{c},\, \mathbf{u}_x\rangle\, \mathbf{u} - \langle \mathbf{u},\, \mathbf{u}_x\rangle\, \mathbf{c}, \qquad (7.18)$$

where $\mathbf{u}(x,t)$ is a N-dimensional vector, $\langle \cdot, \cdot \rangle$ is a scalar product, and $\mathbf{c}$ is a fixed constant vector.[4]

Open problem 7.2.9. *Find soliton and finite-gap solutions for the KdV equation associated with the exceptional Jordan algebra $H_3(O)$.*

7.3 Left-symmetric algebras and Burgers type systems

Consider equations of the form

$$U_t = U_2 + 2\, U \circ U_1 + B(U,U,U), \qquad (7.19)$$

where B is some triple operation. In the finite-dimensional case, such an equation is equivalent to a system of the form

$$u_t^i = u_{xx}^i + 2C_{jk}^i u^k u_x^j + B_{jkm}^i u^k u^j u^m, \qquad (7.20)$$

where $i,j,k = 1,\ldots,N$.

The system (7.20) is homogeneous with the weights $\lambda_i = 1$, $\mu = 2$. The scalar Burgers equation has a homogeneous third order symmetry (1.35). The general ansatz for such a symmetry in the case of equation (7.19) is given by

$$U_\tau = U_3 + 3A_1(U,U_2) + 3A_2(U_1,U_1) + 3B_1(U,U,U_1) + C_1(U,U,U,U). \qquad (7.21)$$

Theorem 7.3.1. [197] *Equation (7.19) has a symmetry (7.21) iff*

$$B(X,X,X) = X \circ (X \circ X) - (X \circ X) \circ X,$$

and $\circ$ is a left-symmetric product, i.e. the identity

$$X \circ (Y \circ Z) - (X \circ Y) \circ Z = Y \circ (X \circ Z) - (Y \circ X) \circ Z \qquad (7.22)$$

---

[4]The vector $\mathbf{c}$ can be reduced to the canonical form $(1,0,\ldots,0)^T$ by an orthogonal transformation.

is fulfilled. The operations in formula (7.21) are given by

$$A_1(X,Y) = X \circ Y, \qquad A_2(X,X) = X \circ X,$$

$$B_1(X,X,Y) = X \circ (X \circ Y) + Y \circ (X \circ X) - (Y \circ X) \circ X,$$

$$C_1(X,X,X,X) = X \circ (X \circ (X \circ X)) - X \circ ((X \circ X) \circ X) +$$
$$(X \circ X) \circ (X \circ X) - ((X \circ X) \circ X) \circ X.$$

Unlike the case of Jordan algebras, a classification of simple finite-dimensional left-symmetric algebras does not exist and we have no explicit description of all nontriangular systems of Burgers type.

Every associative algebra is left-symmetric. For associative algebras, the cubic terms in (7.19) are vanishing. The system corresponding to the associative matrix algebra is the matrix Burgers equation

$$\mathbf{U}_t = \mathbf{U}_{xx} + 2\,\mathbf{U}\,\mathbf{U}_x.$$

The vector Burgers equation [195]

$$\mathbf{u}_t = \mathbf{u}_{xx} + 2\langle \mathbf{u}, \mathbf{u}_x \rangle \, \mathbf{c} + 2\langle \mathbf{c}, \mathbf{u} \rangle \, \mathbf{u}_x + \langle \mathbf{u}, \mathbf{u} \rangle \langle \mathbf{c}, \mathbf{u} \rangle \, \mathbf{c} - \langle \mathbf{u}, \mathbf{u} \rangle \langle \mathbf{c}, \mathbf{c} \rangle \, \mathbf{u}$$

is generated by the left-symmetric algebra with the product (7.2).

7.4 Integrable equations associated with triple Jordan systems

In this section, we consider integrable polynomial systems with cubic nonlinearity, which are multi-component generalizations of the mKdV equation (2.25), NLS equation (2.31), and the derivative nonlinear Schrödinger equation

$$u_t = u_{xx} + 2\,(u^2 v)_x, \qquad v_t = -v_{xx} - 2\,(v^2 u)_x.$$

7.4.1 Systems of mKdV type

Consider systems of the form

$$u_t^i = u_{xxx}^i + \sum_{j,k,m} B_{jkm}^i u^k u^j u_x^m, \qquad i,j,k = 1,\ldots,N. \tag{7.23}$$

For $N = 1$, the formula (7.23) defines the mKdV equation (2.25).

Let B be a triple system with a basis $\mathbf{e}_1, \ldots, \mathbf{e}_N$, such that

$$\{\mathbf{e}_j, \mathbf{e}_k, \mathbf{e}_m\} = \sum_i B_{jkm}^i \mathbf{e}_i.$$

Then the system (7.23) can be written in algebraic form

$$U_t = U_{xxx} + 3\,B(U, U_x, U), \tag{7.24}$$

where the element $U(x, t)$ is defined by formula (7.1). Triple systems $B(X, Y, Z)$, such that $B(X, Y, Z) = B(Z, Y, X)$, are in the one-to-one correspondence with the systems (7.23).

Systems (7.23) are homogeneous with the weights $\lambda_i = 1$, $\mu = 3$. In addition, equation (7.24) is invariant with respect to the discrete involution $U \mapsto -U$. Without loss of generality, we can assume that symmetries also have these two properties.

Theorem 7.4.1. [206]   *For any triple Jordan system* $\{\cdot, \cdot, \cdot\}$*, equation* (7.24)*, where* $B(X, Y, X) = \{X, X, Y\}$*, has a symmetry of the form*

$$U_\tau = U_5 + 5\,B_1(U, U, U_3) + 5\,B_2(U, U_1, U_2) + 5\,B_3(U_1, U_1, U_1) + 5\,C(U, U, U, U, U_1). \tag{7.25}$$

A stronger statement was proved by I. Shestakov and V. Sokolov in [207].

Theorem 7.4.2. *Equation* (7.24) *has a symmetry of the form* (7.25) *iff*

$$B(X, Y, Z) = \{X, Z, Y\} + \{Z, X, Y\}, \tag{7.26}$$

where $\{\cdot, \cdot, \cdot\}$ *is a triple Jordan system.*

Proof. The compatibility condition

$$0 = (U_t)_\tau - (U_\tau)_t = P(U, U_1, \dots, U_5) \tag{7.27}$$

of equation (7.24) and symmetry (7.25) generates the defining polynomial P, which has to be identically equal to zero. After the scaling $U_i \mapsto z_i U_i$ in P, the coefficients of different monomials with respect to the parameters $z_0, \dots, z_5$ give identities that algebraic operations have to satisfy. Since equations (7.24) and (7.25) are homogeneous, the polynomial P is also homogeneous: if we assign the weight $i+1$ to the parameter z_i, then P is of the weight 9.

Equating to zero the coefficient of $z_0 z_1 z_5$ in the polynomial P, we find that

$$B_1(X, X, Y) = B(X, Y, X). \tag{7.28}$$

The coefficient of $z_0 z_2 z_4$ leads to the formula

$$B_2(X, Y, Z) = 2B(X, Y, Z) + 2B(X, Z, Y). \tag{7.29}$$

All other terms with z_5 and z_4 disappear in virtue of (7.28) and (7.29).

Equating to zero the coefficient of $z_1 z_2 z_3$, we get

$$B_3(X, X, X) = B(X, X, X), \tag{7.30}$$

while the coefficient of $z_3 z_1 z_0^3$ leads to

$$C(X, X, X, X, Y) = B(X, B(X, Y, X), X) + \frac{1}{2} B(X, Y, B(X, X, X)). \tag{7.31}$$

Thus, all coefficients of the symmetry (7.25) are expressed in terms of the triple system B. All fifth order identities $I_i = 0$, $i = 1, 2, 3, 4$, for B are originated from the coefficients of $z_0^3 z_1 z_3$, $z_0^3 z_2^2$, $z_0^2 z_1^2 z_2$, $z_0 z_1^4$. The expressions I_i have the form

$$I_1(X, Y, Z) = 2B(X, Z, B(X, X, Y)) - 3B(X, Z, B(X, Y, X)) + B(Y, Z, B(X, X, X)),$$

$$I_2(X, Y) = 2B(X, Y, B(X, X, Y)) - 3B(X, Y, B(X, Y, X)) + B(Y, Y, B(X, X, X)),$$

$$\begin{aligned}
I_3(X, Y, Z) = &\ 2B(X, Y, B(X, Y, Z)) - 6B(X, Y, B(X, Z, Y)) + 2B(X, Y, B(Y, X, Z)) - \\
&\ 4B(X, Z, B(X, Y, Y)) + 2B(X, Z, B(Y, X, Y)) - 2B(X, B(Y, Z, Y), X) + \\
&\ 2B(Y, Y, B(X, X, Z)) - 3B(Y, Y, B(X, Z, X)) + 4B(Y, Z, B(X, X, Y)) - \\
&\ 2B(Y, Z, B(X, Y, X)) + 4B(Y, B(X, Y, X), Z) + 2B(Y, B(X, Z, X), Y) + \\
&\ 2B(Z, Y, B(X, X, Y)) - 3B(Z, Y, B(X, Y, X)),
\end{aligned}$$

and

$$\begin{aligned}
I_4(X, Y) = &\ B(X, Y, B(Y, Y, Y)) + 2B(Y, Y, B(X, Y, Y)) - \\
&\ B(Y, Y, B(Y, X, Y)) - 2B(Y, B(X, Y, Y), Y).
\end{aligned}$$

It is clear that $I_2(X, Y) = I_1(X, Y, Y)$. Using the method of undetermined coefficients, we will show that the identity $I_4 = 0$ is a consequence of the identities $I_2 = 0$ and $I_3 = 0$.

First, we perform the linearization of all these identities. It is clear that they are equivalent to the relations $J_i(X, Y, Z, U, V) = 0$, $i = 2, 3, 4$, where J_2 denotes the coefficient of $k_1 k_2 k_3$ in $I_2(k_1 X + k_2 U + k_3 V, Y, Z)$, J_3 is the coefficient of $k_1 k_2 k_3 k_4$ in $I_3(k_1 X + k_2 U, k_3 Y + k_4 V, Z)$ and J_4 is the coefficient of $k_1 k_2 k_3 k_4$ in $I_4(X, k_1 Y + k_2 Z + k_3 U + k_4 V)$.

From the condition

$$Z \overset{\text{def}}{=} J_4(X, Y, Z, U, V) - \sum_{\sigma \in S_5} b_\sigma J_2\Big(\sigma(X), \sigma(Y), \sigma(Z), \sigma(U), \sigma(V)\Big) -$$

$$\sum_{\sigma \in S_5} c_\sigma J_3\Big(\sigma(X), \sigma(Y), \sigma(Z), \sigma(U), \sigma(V)\Big) = 0,$$

where σ is a permutation of the set $\{X, Y, Z, U, V\}$, we find the unknown coefficients b_σ and c_σ. Namely, we fix the ordering

$$U < V < X < Y < Z < B(\cdot, \cdot, \cdot)$$

and, taking into account the identity $B(X, Y, Z) = B(Z, Y, X)$, replace all expressions of the form $B(P, Q, R)$ with $B(R, Q, P)$ if $P > R$. After that, equating to zero the coefficients of similar terms in the relation $Z = 0$, we obtain an overdetermined system of linear equations for the coefficients b_σ and c_σ. Solving it using, for example, the computer algebra system "Mathematica", we find that

$$J_4(X, Y, Z, U, V) = \frac{1}{6}\Big(J_2(U, X, V, Y, Z) + J_2(U, X, Y, V, Z) + J_2(U, X, Z, V, Y) +$$
$$J_2(V, X, U, Y, Z) - J_3(U, V, X, Y, Z) - J_3(U, V, X, Z, Y) - J_3(U, V, Y, X, Z) -$$
$$J_3(U, V, Z, X, Y) - J_3(U, Y, V, X, Z) - J_3(U, Y, X, V, Z) - J_3(V, U, X, Y, Z) -$$
$$J_3(V, U, X, Z, Y) - J_3(V, U, Y, X, Z) - J_3(V, U, Z, X, Y) - J_3(V, Y, U, X, Z) -$$
$$J_3(X, U, V, Y, Z) - J_3(X, U, V, Z, Y) - J_3(X, U, Y, Z, V) - J_3(X, U, Z, Y, V) -$$
$$J_3(X, V, U, Y, Z) - J_3(X, V, U, Z, Y) - J_3(Y, U, X, Z, V)\Big).$$

Thus, the identity $J_4 = 0$ follows from $J_2 = J_3 = 0$ and therefore $I_4 = 0$ follows from $I_2 = I_3 = 0$.

Consider the triple system

$$\{X, Y, Z\} = \frac{1}{2}\Big(B(Y, Z, X) + B(Y, X, Z) - B(X, Y, Z)\Big). \tag{7.32}$$

Using the relation $B(X, Y, Z) = B(Z, Y, X)$, it is easy to verify that

$$B(X, Y, Z) = \{X, Z, Y\} + \{Z, X, Y\}. \tag{7.33}$$

It turns out that the pair of identities $J_2 = J_3 = 0$ is equivalent to the Jordan (see Definition 7.1.3) identity $\mathcal{J} = 0$, where

$$\mathcal{J}(X, Y, Z, U, V) = \{X, Y, \{U, V, Z\}\} - \{\{X, Y, U\}, V, Z\} -$$
$$\{U, V, \{X, Y, Z\}\} + \{U, \{Y, X, V\}, Z\},$$

rewritten in terms of the triple system B by means of (7.32). Using the method of undetermined coefficients, we can verify that the identity $\mathcal{J} = 0$ follows from $J_2 = J_3 = 0$ and, vice versa, each of the identities $J_2 = 0$ and $J_3 = 0$ follows from $\mathcal{J} = 0$. For example,

$$J_2(X, Y, Z, U, V) = -\mathcal{J}(U, X, V, Z, Y) + \mathcal{J}(U, X, Y, Z, V) - \mathcal{J}(U, Y, X, Z, V) +$$
$$\mathcal{J}(V, X, Y, Z, U) - \mathcal{J}(V, Y, U, Z, X) - \mathcal{J}(V, Y, X, Z, U) -$$
$$\mathcal{J}(X, U, V, Z, Y) + \mathcal{J}(X, U, Y, Z, V) - \mathcal{J}(X, V, U, Z, Y) +$$
$$\mathcal{J}(Y, U, V, Z, X) + \mathcal{J}(Y, V, U, Z, X) + \mathcal{J}(Y, V, X, Z, U).$$

Much longer formulas express J_3 in terms of $\mathcal{J}$ and $\mathcal{J}$ in terms of J_2, J_3.

In addition to the fifth order identities considered above, there are two identities of order 7. The coefficient of $z_0^6 z_2$ in the polynomial P gives

$$B(X, B(X, Y, X),\, B(X, X, X)) - B(X, B(X, Y, B(X, X, X)), X) = 0,$$

while the coefficient of $z_0^5 z_1^2$ leads to the identity

$$
\begin{aligned}
&2B(X, Y, B(X, X, B(X, Y, X))) - 2B(X, Y, B(X, Y, B(X, X, X))) - \\
&3B(X, Y, B(X, B(X, Y, X), X)) + 4B(X, B(X, Y, X), B(X, Y, X)) + \\
&2B(X, B(X, Y, Y), B(X, X, X)) - 2B(X, B(X, Y, B(X, X, Y)), X) + \\
&3B(X, B(X, Y, B(X, Y, X)), X) - 4B(X, B(X, B(X, Y, X), Y), X) - \\
&B(X, B(Y, Y, B(X, X, X)), X) + B(B(X, X, X), Y, B(X, Y, X)) = 0.
\end{aligned}
$$

Using the method of undetermined coefficients, one can verify that each of these seventh order identities follows from $\mathcal{J} = 0$. So, $\{X, Y, Z\}$ is a triple Jordan system iff a symmetry (7.25) exists. The statement of the theorem follows from the formula (7.33).   □

Since all simple triple Jordan systems are described in [204], Theorem 7.4.2 gives a complete description of nontriangular integrable systems (7.23). Perhaps, the systems corresponding to the exceptional simple triple Jordan systems are of a particular interest.

Open problem 7.4.3. *Construct Lax pairs, recursion (see, for example, [208]) and Hamiltonian operators for these systems.*

The integrable vector systems corresponding to the operations (7.7) and (7.8) have the form

$$\mathbf{u}_t = \mathbf{u}_{xxx} + \langle \mathbf{u}, \mathbf{u} \rangle \, \mathbf{u}_x \tag{7.34}$$

and

$$\mathbf{u}_t = \mathbf{u}_{xxx} + \langle \mathbf{u}, \mathbf{u} \rangle \, \mathbf{u}_x + \langle \mathbf{u}, \mathbf{u}_x \rangle \, \mathbf{u}, \tag{7.35}$$

respectively.

The triple Jordan system (7.6) generates the matrix mKdV equation

$$\mathbf{U}_t = \mathbf{U}_{xxx} + \mathbf{U}^2 \mathbf{U}_x + \mathbf{U}_x \mathbf{U}^2.$$

7.4.2   Systems of NLS type

Multi-component generalizations (see [200, 199]) of the nonlinear Schrödinger equation (1.13) are systems of $2N$ equations of the form

$$u_t^i = u_{xx}^i + 2 \sum_{j,k,m} b_{jkm}^i u^j v^k u^m, \qquad v_t^i = -v_{xx}^i - 2 \sum_{j,k,m} b_{jkm}^i v^j u^k v^m, \tag{7.36}$$

where $i = 1, \ldots, N$, and b^i_{jkm} are some constants. In terms of the corresponding triple system B we have

$$U_t = U_{xx} + B(U, V, U), \qquad V_t = -V_{xx} - B(V, U, V). \qquad (7.37)$$

The triple systems $B(X, Y, Z)$, such that $B(X, Y, Z) = B(Z, Y, X)$, are in one-to-one correspondence with the systems (7.36). The systems (7.37) are homogeneous: the weights of variables U, V and of operator D are equal to one.

Remark 7.4.4. *The system* (7.37) *admits the following scaling group* $U \mapsto \lambda U$, $V \mapsto \lambda^{-1} V$. *Therefore, without loss of generality, we can assume that symmetries are homogeneous and admit the same scaling.*

Any third order symmetry described in Remark 7.4.4 has the form

$$\begin{cases} U_\tau = U_{xxx} + 3\, B_1(U, V, U_x) + 3\, B_3(U, V_1, U), \\ V_\tau = U_{xxx} + 3\, B_2(V, U, V_x) + 3\, B_4(V, U_1, V). \end{cases}$$

Theorem 7.4.5. [199] *A system* (7.37) *has a symmetry of this form iff*

$$B_1(X, Y, Z) = B_2(X, Y, Z) = B(X, Y, Z), \qquad B_3(X, Y, X) = B_4(X, Y, X) = 0,$$

and $B(\cdot, \cdot, \cdot)$ *is a triple Jordan system.*

The simple triple Jordan systems lead to several interesting vector and matrix integrable systems (7.36).

Example 7.4.6. The triple system (7.6) corresponds, up to a scaling, to the matrix NLS equation (6.89).

Example 7.4.7. The well-known vector NLS equation [50]

$$\mathbf{u}_t = \mathbf{u}_{xx} + 2\langle \mathbf{u}, \mathbf{v} \rangle\, \mathbf{u}, \qquad \mathbf{v}_t = -\mathbf{v}_{xx} - 2\langle \mathbf{v}, \mathbf{u} \rangle\, \mathbf{v} \qquad (7.38)$$

corresponds to the triple Jordan system (7.8).

Example 7.4.8. Another integrable vector generalization

$$\mathbf{u}_t = \mathbf{u}_{xx} + 4\langle \mathbf{u}, \mathbf{v} \rangle\, \mathbf{u} - 2\, |\mathbf{u}|^2\, \mathbf{v}, \qquad \mathbf{v}_t = -\mathbf{v}_{xx} - 4\langle \mathbf{v}, \mathbf{u} \rangle\, \mathbf{v} + 2\, |\mathbf{v}|^2\, \mathbf{u}, \qquad (7.39)$$

of the NLS equation was found in [209]. It is generated by the triple Jordan system (7.7).

7.4.3 Systems of derivative NLS type

The derivative NLS equation is given by

$$u_t = u_{xx} + 2(u^2 v)_x, \qquad v_t = -v_{xx} + 2(v^2 u)_x.$$

Consider its multi-component generalizations of the form

$$U_t = U_{xx} + B(U, V, U)_x, \qquad V_t = -V_{xx} + B(V, U, V)_x, \tag{7.40}$$

where $B(X, Y, Z) = B(Z, Y, X)$. Systems (7.40) are homogeneous: the weights of U, V are equal to one, and the weight of D equals 2.

Remark 7.4.9. *The system (7.40) admits the scaling group $U \mapsto \lambda U$, $V \mapsto \lambda^{-1} V$ and the discrete involution $t \mapsto -t$, $x \mapsto -x$, $U \mapsto -V$, $V \mapsto -U$.*

Theorem 7.4.10. *A system (7.40) has a third order symmetry possessing the properties from Remark 7.4.9, iff $B(\cdot, \cdot, \cdot)$ is a triple Jordan system.*

The matrix (see [196]) and vector examples of integrable systems of the derivative NLS type can be constructed using formulas (7.6)–(7.8).

7.5 Integrable systems corresponding to new algebraic structures

7.5.1 Equations of potential mKdV type

Consider multi-component generalizations of the known integrable equation (see (4.52) and Example 5.3.17)

$$u_t = u_{xxx} + 3\, u_x^3. \tag{7.41}$$

This equation has the fifth order symmetry

$$u_\tau = u_{xxxxx} + 15 u_x^2 u_{xxx} + 15 u_x u_{xx}^2 + \frac{27}{2} u_x^5.$$

A multi-component generalization of equation (7.41) is given by

$$U_t = U_{xxx} + 3\, B(U_x, U_x, U_x), \tag{7.42}$$

where the operation B is symmetric with respect to all arguments:

$$B(X, Y, Z) = B\Big(\sigma(X), \sigma(Y), \sigma(Z)\Big), \qquad \sigma \in S_3.$$

Theorem 7.5.1. *Equation* (7.42) *has a symmetry of the form*

$$U_\tau = U_5 + 15\,B_1(U_1, U_1, U_3) + 15\,B_2(U_2, U_2, U_1) + \frac{27}{2}\,C(U_1, U_1, U_1, U_1, U_1)$$

iff the triple system $B(\cdot, \cdot, \cdot)$ *satisfies the identity*

$$
\begin{aligned}
&B(X, Y, B(Y, Y, Z)) + B(Y, Y, B(X, Y, Z)) + \\
&B(Y, Z, B(X, Y, Y)) - 3B(X, Z, B(Y, Y, Y)) = 0.
\end{aligned}
\tag{7.43}
$$

In this case, the symmetry is defined by the formulas

$$B_1(X, X, Y) = B(X, X, Y), \qquad B_2(X, X, Y) = B(X, X, Y),$$

$$C(X, X, X, X, X) = B(X, X, B(X, X, X)).$$

Open problem 7.5.2. *Construct nontrivial examples of triple systems with the identity* (7.43).

7.5.2 Systems of Olver–Sokolov type

Multi-component analogues of the integrable matrix system (6.90) have the form

$$U_t = U_{xx} + 2\,B(U, V, U_x), \qquad V_t = -V_{xx} + 2\,B(V_x, U, V), \tag{7.44}$$

where $B(\cdot, \cdot, \cdot)$ is a triple system. The systems (7.44) are homogeneous with the same weights as in Section 7.4.3.

Theorem 7.5.3. *A system* (7.44) *has a third order homogeneous symmetry iff* $B(\cdot, \cdot, \cdot)$ *satisfies the identities*

$$
\begin{aligned}
B\Big(X, Y, B(X, V, U)\Big) &= B\Big(X, B(Y, X, V), U\Big), \\
B\Big(X, B(Y, Z, V), Z\Big) &= B\Big(B(X, Y, Z), V, Z\Big), \\
B\Big(X, Y, B(Z, Y, V)\Big) &= B\Big(B(X, Y, Z), Y, V\Big).
\end{aligned}
\tag{7.45}
$$

Remark 7.5.4. *The latter identity means the multiplication*

$$X \circ Y = B(X, Z, Y)$$

is associative for any Z. *Moreover, by varying* Z, *we get a vector space of compatible associative* (*see Definition 3.1.5*) *products* [91].

In the paper [210] triple systems with identities

$$B\Big(X, Y, B(Z, U, V)\Big) = B\Big(X, B(Y, Z, U), V\Big) = B\Big(B(X, Y, Z), U, V\Big) \qquad (7.46)$$

were studied. Clearly, (7.45) follows from (7.46).

Since the triple matrix product

$$B(X, Y, Z) = X\,Y\,Z, \qquad X, Y, Z \in \mathrm{Mat}_n$$

satisfies (7.46), system (6.90) belongs to the class described by Theorem 7.5.3. Another example of a triple system satisfying (7.46) is given by the formula

$$B(X, Y, Z) = XY^T Z, \qquad X, Y, Z \in \mathrm{Mat}_{n,m}\,.$$

As far as I know, triple systems with identities (7.45) were not considered by algebraists.

Open problem 7.5.5. *Find all simple triple systems with identities* (7.45).

7.5.3 Systems of mKdV type with two algebraic operations

Consider systems of the form

$$u_t^i = u_{xxx}^i + A_{jk}^i u^k u_{xx}^j + B_{jkm}^i u^j u^k u_x^m, \qquad i, j, k = 1, \ldots, N. \qquad (7.47)$$

If all the constants A_{jk}^i are equal to zero, then we arrive at systems of the form (7.23). In general, A_{jk}^i and B_{jkm}^i can be considered as structural constants of an algebra and a triple system and therefore a system of the form (7.47) is associated with a pair of algebraic structures defined on the same N-dimensional vector space.

An example of an integrable system of the form (7.47) gives the matrix equation (6.88). If we rewrite this equation as

$$\mathbf{U}_t = \mathbf{U}_{xxx} + 3\,[\mathbf{U}, \mathbf{U}_{xx}] + 3\,[\mathbf{U}, [\mathbf{U}, \mathbf{U}_x]] - 3\,\mathbf{U}^2 \mathbf{U}_x - 3\,\mathbf{U}_x \mathbf{U}^2, \qquad (7.48)$$

then the first two nonlinear terms on the right-hand side turn out to be written in terms of the skew-symmetric operation $[X, Y]$ and the other two correspond to the same triple Jordan system as in (6.87).

Let us write the system (7.47) in the algebraic form

$$U_t = U_3 + 3\,U \circ U_2 + 3\,\{U, U_1, U\}. \qquad (7.49)$$

Suppose additionally that $X \circ Y = -Y \circ X$.

Remark 7.5.6. *It is possible to prove without any assumptions that if (7.49) possesses a fifth order symmetry, then the binary operation has to satisfy the identity*

$$(X \circ Y + Y \circ X) \circ Z = 0.$$

It can be shown [207] that for any pair of algebraic structures, consisting of a triple Jordan system $F(\cdot, \cdot, \cdot)$ and skew-symmetric product $X \circ Y$, which are related by identities

$$Y \circ F(U, Z, X) - F(Y \circ U, Z, X) - F(U, Y \circ Z, X) - F(U, Z, Y \circ X) = 0 \quad (7.50)$$

and

$$Z \circ \Big(2F(X, Y, U) - 2F(X, U, Y) - X \circ (U \circ Y)\Big) +$$
$$\mathcal{Y}(U \circ Y, X, Z) + X \circ \mathcal{Y}(Y, Z, U) - \quad (7.51)$$
$$\mathcal{Y}(X \circ Y, Z, U) - \mathcal{Y}(Y, X \circ Z, U) - \mathcal{Y}(Y, Z, X \circ U) = 0,$$

where

$$\mathcal{Y}(X, Y, Z) \stackrel{def}{=} Z \circ (X \circ Y) + X \circ (Y \circ Z) + Y \circ (Z \circ X),$$

one can construct a system (7.47), which has a homogeneous fifth order symmetry of the form

$$u_\tau = U_5 + 5A_2(U, U_4) + 5A_3(U_1, U_3) + 5A_4(U_2, U_2) +$$
$$5B_2(U, U, U_3) + 5B_3(U, U_1, U_2) + 5B_4(U_1, U_1, U_1) +$$
$$5C_1(U, U, U, U_2) + 5C_2(U, U_1, U_1, U) + 5D_1(U, U, U, U, U_1).$$

The triple system in equation (7.49) is expressed in terms of the triple Jordan system F by the formula

$$\{X, Y, Z\} = F(X, Z, Y) + F(Z, X, Y) + \frac{1}{2} Z \circ (X \circ Y) + \frac{1}{2} X \circ (Z \circ Y).$$

In the case of zero skew-symmetric operation $X \circ Y$, we arrive at the formula (7.26) from Theorem 7.4.2.

A particular case of algebraic structure with identities (7.50) and (7.51) is given by the so-called *Lie–Jordan algebra*, which were defined in [211] as an algebra with a bilinear operation $[X, Y]$ and a ternary operation $\{X, Y, Z\}$ satisfying the identities

$[X, Y] = -[Y, X],$

$\{X, Y, Z\} = \{Z, Y, X\},$

$[[X, Y], Z] = \{X, Y, Z\} - \{Y, X, Z\},$

$[U, \{X, Y, Z\}] = \{[U, X], Y, Z\} + \{X, [U, Y], Z\} + \{X, Y, [U, Z]\},$

$\{\{X, Y, Z\}, U, V\} = \{\{X, U, V\}, Y, Z\} - \{X, \{Y, V, U\}, Z\} + \{X, Y, \{Z, U, V\}\}.$

It is easy to verify that for any Lie–Jordan algebra the pair of operations $X \circ Y = [X, Y]$, $F[X, Y, Z] = \frac{1}{2}\{Z, Y, X\}$ satisfies identities (7.50) and (7.51).

Any associative algebra A is a Lie–Jordan algebra with respect to operations $[a, b] = ab - ba$, $\{a, b, c\} = abc + cba$. Equation (7.48) corresponds to the case $A = \mathrm{Mat}_m$. Also, if A has an involution, then the vector space of skew-symmetric elements is a Lie–Jordan subalgebra in A.

Open problem 7.5.7. *Construct examples of algebraic structures that satisfy the identities of* (7.50), (7.51) *and are different from the Lie–Jordan algebras.*

7.6 Rational integrable systems

7.6.1 Inverse element as a solution of a system of differential equations

The function $y = \dfrac{1}{x}$ can be defined as a homogeneous solution of the differential equation

$$y' + y^2 = 0.$$

It turns out that the inverse element in triple Jordan systems can be defined in a similar way.

Let $\mathbf{e}_1, \ldots, \mathbf{e}_N$ be a basis of a triple system $\{X, Y, Z\}$. We intend to define an element

$$\phi(U) = \sum_{k=1}^{N} \phi^k(u^1, \ldots, u^N)\, \mathbf{e}_k, \tag{7.52}$$

inverse to $U = \sum_{i=1}^{N} u^i\, \mathbf{e}_i$ as a solution of a suitable system of partial differential equations.

Proposition 7.6.1. *The overdetermined system of differential equations*

$$\frac{\partial \phi}{\partial u^k} = -\{\phi,\, \mathbf{e}_k,\, \phi\}, \tag{7.53}$$

where $k = 1, \ldots, N$, is compatible iff $\{X, Y, Z\}$ is a triple Jordan system.

Definition 7.6.1. For any triple Jordan system, a nonzero homogeneous[5] solution (7.52) of system (7.53) is called the *inverse element* to the element U.

---

[5]This means that the Euler equation (7.54) is fulfilled.

Remark 7.6.2. *From the Euler equation*

$$\sum u^i \frac{\partial \phi}{\partial u^i} = -\phi \tag{7.54}$$

it follows that a homogeneous solution of system (7.53) *satisfies the identity*

$$\phi(U) = \{\phi(U),\, U,\, \phi(U)\}. \tag{7.55}$$

Example 7.6.3. For the matrix triple Jordan system (7.6) the homogeneous solution $\phi(\mathbf{U})$ of the system (7.53) is exactly the matrix inverse element $\mathbf{U}^{-1}$.

Example 7.6.4. For the triple Jordan system (7.7) we have

$$\phi(\mathbf{u}) = \frac{\mathbf{u}}{|\mathbf{u}|^2}.$$

The following algebraic definition of the element U^{-1} is well known in the theory of triple Jordan systems. Define the linear operator P_X by the formula $P_X(Y) = \{X, Y, X\}$. If the operator P_U is nondegenerate, then, by definition, $U^{-1} = P_U^{-1}(U)$.

Proposition 7.6.2. *If the operator P_U is nondegenerate, then*

$$\phi(U) = P_U^{-1}(U) \tag{7.56}$$

is the unique nonzero homogeneous solution of the system (7.53).

Example 7.6.5. For the triple system (7.8), the operator $P_{\mathbf{u}}$ is degenerate for any $\mathbf{u}$ and formula (7.56) is not applicable. The general solution of the system (7.53) is given by

$$\phi(\mathbf{u}) = \frac{\mathbf{c}}{2\langle \mathbf{c},\, \mathbf{u} \rangle},$$

where $\mathbf{c}$ is an arbitrary constant vector. This is a special case of the formula

$$\phi(U) = \frac{1}{2} C \, (U^T C)^{-1},$$

where C is a constant matrix of dimension $n \times m$, which gives the solution of system (7.53) in the case of triple system (7.9).

Remark 7.6.6. *Example 7.6.5 shows that system* (7.53) *can have many homogeneous solutions. In this case, we call the inverse element any of them.*

Open problem 7.6.7. *Find the element $\phi(U)$ for the triple Jordan system of skew-symmetric $m \times m$ matrices for odd m.*

7.6.2   Several classes of integrable rational Jordan models

The results of this section were published in the paper [212].

In all "rational" integrable models described below, as the element $\phi(U)$ one can take *any* (not necessarily homogeneous) solution of system (7.53).

We introduce the following notation: for any triple Jordan system $\{X, Y, Z\}$ we set

$$\alpha_U(X, Y) \overset{def}{=} \{X, \phi(U), Y\} \tag{7.57}$$

and

$$\sigma_U(X, Y, Z) \overset{def}{=} \{X, \{\phi(U), Y, \phi(U)\}, Z\}.$$

Remark 7.6.8. *According to Propositions 7.1.1 and 7.1.3, for any fixed U the multiplication $\alpha_U(X, Y)$ defines a Jordan algebra, while $\sigma_U(X, Y, Z)$ defines a triple Jordan system. The coefficients $u^1, \ldots, u^N$ of the element U can be regarded as the deformation parameters of these Jordan structures.*

Class 1

For any triple Jordan system, consider a hyperbolic equation

$$U_{xy} = \alpha_U(U_x, U_y). \tag{7.58}$$

In the matrix case, (7.58) coincides with the equation of principle chiral field (6.93). For this reason, we call equation (7.58) *Jordan chiral model.*

It is easy to verify that (7.58) has the following zero-curvature representation (see (1.15))

$$\Psi_x = \frac{2}{(1 - \lambda)} L_{U_x} \Psi, \qquad \Psi_y = \frac{2}{(1 + \lambda)} L_{U_y} \Psi.$$

By L_X we denote the multiplication operator in the algebra with the product $X \circ Y = \alpha_U(X, Y)$.

Remark 7.6.9. *This formula gives us the zero-curvature representation for the matrix σ-model (6.93):*

$$\Psi_x = \frac{1}{(1 - \lambda)} M \Psi, \qquad \Psi_y = \frac{1}{(1 + \lambda)} N \Psi,$$

where Ψ is a matrix and

$$M\Psi \overset{def}{=} -\mathbf{U}_x \mathbf{U}^{-1}\Psi - \Psi \mathbf{U}^{-1}\mathbf{U}_x, \qquad N\Psi \overset{def}{=} -\mathbf{U}_y \mathbf{U}^{-1}\Psi - \Psi \mathbf{U}^{-1}\mathbf{U}_y.$$

This representation is different from the well-known one.

Open problem 7.6.10. *Describe higher symmetries of the Jordan chiral models* (7.58).

Class 2

The following evolution equation

$$U_t = U_{xxx} - 3\,\alpha_U(U_x, U_{xx}) + \frac{3}{2}\,\sigma_U(U_x, U_x, U_x) \tag{7.59}$$

possesses an infinite hierarchy of higher symmetries for any triple Jordan system. The matrix and the two vector equations corresponding to the triple systems (7.6), (7.7) and (7.8) have the form

$$\mathbf{U}_t = \mathbf{U}_{xxx} - \frac{3}{2}\,\mathbf{U}_x\mathbf{U}^{-1}\mathbf{U}_{xx} - \frac{3}{2}\,\mathbf{U}_{xx}\mathbf{U}^{-1}\mathbf{U}_x + \frac{3}{2}\,\mathbf{U}_x\mathbf{U}^{-1}\mathbf{U}_x\mathbf{U}^{-1}\mathbf{U}_x, \tag{7.60}$$

$$\mathbf{u}_t = \mathbf{u}_{xxx} - 3\frac{\langle \mathbf{u}, \mathbf{u}_x \rangle}{|\mathbf{u}|^2}\mathbf{u}_{xx} - 3\frac{\langle \mathbf{u}, \mathbf{u}_{xx} \rangle}{|\mathbf{u}|^2}\mathbf{u}_x - \frac{3}{2}\frac{|\mathbf{u}_x|^2}{|\mathbf{u}|^2}\mathbf{u}_x + 6\frac{\langle \mathbf{u}, \mathbf{u}_x \rangle^2}{|\mathbf{u}|^4}\mathbf{u}_x +$$
$$3\frac{\langle \mathbf{u}_x, \mathbf{u}_{xx} \rangle}{|\mathbf{u}|^2}\mathbf{u} - 3\frac{\langle \mathbf{u}, \mathbf{u}_x \rangle|\mathbf{u}_x|^2}{|\mathbf{u}|^4}\mathbf{u},$$

and

$$\mathbf{u}_t = \mathbf{u}_{xxx} - \frac{3}{2}\frac{\langle \mathbf{c}, \mathbf{u}_x \rangle}{\langle \mathbf{c}, \mathbf{u} \rangle}\mathbf{u}_{xx} - \frac{3}{2}\frac{\langle \mathbf{c}, \mathbf{u}_{xx} \rangle}{\langle \mathbf{c}, \mathbf{u} \rangle}\mathbf{u}_x + \frac{3}{2}\frac{\langle \mathbf{c}, \mathbf{u}_x \rangle^2}{\langle \mathbf{c}, \mathbf{u} \rangle^2}\mathbf{u}_x.$$

Open problem 7.6.11. *Find Lax representations, recursion and Hamiltonian operators for equations* (7.59).

Class 3

The equation

$$V_t = V_{xxx} - \frac{3}{2}\,\alpha_{V_x}(V_{xx}, V_{xx}),$$

generalizing Schwartz–KdV equation (1.71), has an infinite hierarchy of higher symmetries for any triple Jordan system. This equation is related by the potentiation $U = V_x$ with equation (7.59) from Class 2. The matrix equation is given by formula (6.91). The two vector Schwartz–KdV equations have the form

$$\mathbf{u}_t = \mathbf{u}_{xxx} - 3\frac{\langle \mathbf{u}_x, \mathbf{u}_{xx} \rangle}{|\mathbf{u}_x|^2}\mathbf{u}_{xx} + \frac{3}{2}\frac{|\mathbf{u}_{xx}|^2}{\|\mathbf{u}_x\|^2}\mathbf{u}_x,$$

and

$$\mathbf{u}_t = \mathbf{u}_{xxx} - \frac{3}{2}\frac{\langle \mathbf{c}, \mathbf{u}_{xx} \rangle}{\langle \mathbf{c}, \mathbf{u}_x \rangle}\mathbf{u}_{xx}.$$

Class 4

The scalar origin of this class is the Heisenberg model

$$u_t = u_{xx} - \frac{2}{u+v}u_x^2, \qquad v_t = -v_{xx} + \frac{2}{u+v}v_x^2.$$

Its Jordan generalization

$$U_t = U_{xx} - 2\alpha_{U+V}(U_x, U_x), \qquad V_t = -V_{xx} + 2\alpha_{U+V}(V_x, V_x) \qquad (7.61)$$

has the third order symmetry

$$\begin{cases} U_t = U_{xxx} - 6\,\alpha_{U+V}(U_x, U_{xx}) + 6\,\sigma_{U+V}(U_x, U_x, U_x), \\ V_t = V_{xxx} - 6\,\alpha_{U+V}(V_x, V_{xx}) + 6\,\sigma_{U+V}(V_x, V_x, V_x). \end{cases}$$

Remark 7.6.12. *The reduction $V = U$ reduces this symmetry to an equation from Class 2.*

We present here one of the two vector systems:

$$\begin{cases} \mathbf{u}_t = \mathbf{u}_{xx} - 4\dfrac{\langle \mathbf{u}_x, \mathbf{u} + \mathbf{v}\rangle}{|\mathbf{u} + \mathbf{v}|^2}\mathbf{u}_x + 2\dfrac{|\mathbf{u}_x|^2}{|\mathbf{u} + \mathbf{v}|^2}(\mathbf{u} + \mathbf{v}) \\ \mathbf{v}_t = -\mathbf{v}_{xx} + 4\dfrac{\langle \mathbf{v}_x, \mathbf{u} + \mathbf{v}\rangle}{|\mathbf{u} + \mathbf{v}|^2}\mathbf{v}_x - 2\dfrac{|\mathbf{v}_x|^2}{|\mathbf{u} + \mathbf{v}|^2}(\mathbf{u} + \mathbf{v}). \end{cases}$$

Open problem 7.6.13. *Find a Lax representation, recursion and Hamiltonian operators for systems* (7.61).

Remark 7.6.14. *Possibly, there exist Lax representations for equations from Classes 2–4 related to the superstructural Lie algebra* [213] *of the corresponding triple Jordan systems.*

7.7 Deformations of nonassociative algebras and integrable systems of geometric type

The structural constants of Jordan multiplication (7.57) depend on the parameters $u^1, \ldots, u^N$ and, thus, we are dealing with N-parametric deformations of N-dimensional Jordan algebras.

In this section, we consider more general N-parametric deformations of nonassociative algebras which can be related to integrable systems.

7.7.1   Geometric description of deformations

Let E be an Euclidean connection on a N-dimensional manifold $\mathcal{M}$, and $\mathbf{u} = (u^l, \cdots, u^N)$ be local coordinates on $\mathcal{M}$. Denote by $E^i_{jk}(\mathbf{u})$ the components of the connection E.

Consider the connection Γ with components $\Gamma^i_{jk}(\mathbf{u}) = E^i_{jk}(\mathbf{u}) + C^i_{jk}(\mathbf{u})$, where $C^i_{jk}(\mathbf{u})$ are the components of some tensor field C on $\mathcal{M}$.

Definition 7.7.1. [77] The connection Γ is called *covariantly constant deformation of the Euclidean connection* E if the deformation tensor C is covariantly constant with respect to the connection Γ.

From the standard recalculation formulas for the torsion and curvature tensors under a connection deformation (see, for example, [214]), the torsion and curvature of the connection Γ can be expressed in terms of the tensor C as follows:

$$T^i_{jk} = C^i_{jk} - C^i_{kj}, \tag{7.62}$$

$$R^i_{mjk} = \sum_r C^i_{rm} C^r_{jk} - C^i_{rm} C^r_{kj} + C^i_{kr} C^r_{jm} - C^i_{jr} C^r_{km}. \tag{7.63}$$

Remark 7.7.2. *Since the tensor C is covariantly constant, Γ is a connection with covariantly constant torsion and curvature.*

Rewriting the fact that C is covariantly constant in terms of the Euclidean connection E, we obtain

$$\nabla_m\left(C^i_{jk}\right) = \sum_r C^i_{rk} C^r_{mj} + C^i_{jr} C^r_{mk} - C^i_{mr} C^r_{jk}, \tag{7.64}$$

where ∇_m denotes the covariant u^m-derivative with respect to E.

The relations (7.64) form an overdetermined system of first order partial differential equations with respect to unknown functions $C^i_{jk}(u^1, \ldots, u^N)$. In the coordinates, where all components of E are equal to zero, system (7.64) takes the form

$$\frac{\partial C^i_{jk}}{\partial u^m} = \sum_r C^i_{rk} C^r_{mj} + C^i_{jr} C^r_{mk} - C^i_{mr} C^r_{jk}. \tag{7.65}$$

7.7.2   Algebraic description of deformations

Let $\mathbf{V}$ be a vector space over $\mathbb{C}$ with a basis $\mathbf{e}_l, \ldots, \mathbf{e}_N$. The tensor $C(\mathbf{u})$ defines the N-parameter family of multiplications

$$\mathbf{e}_j \circ \mathbf{e}_k = \sum_i C^i_{jk}(\mathbf{u})\, \mathbf{e}_i \tag{7.66}$$

on $\mathbf{V}$.

In terms of product (7.66), the deformation equation (7.65) takes the form

$$\partial_X (Y \circ Z) = (X \circ Y) \circ Z + Y \circ (X \circ Z) - X \circ (Y \circ Z).$$

Here, for any vector field $X = \sum x^i \mathbf{e}_i$

$$\partial_X \overset{def}{=} \sum x^i \frac{\partial}{\partial u^i}.$$

Note that (7.62) and (7.63) can be written in a compact form:

$$T(X, Y) = X \circ Y - Y \circ X,$$

and

$$R(X, Y, Z) = [Y, Z, X], \qquad X, Y, Z \in \mathbf{V},$$

where the operation $[\cdot, \cdot, \cdot]$ is defined by formulas (1.5) and (1.6), and the notation like $T(X, Y)$ is used for the value of the tensor T on vectors X and Y.

Theorem 7.7.3. *The system (7.65) is compatible iff for any* $\mathbf{u} = (u^1, \ldots, u^N)$ *the product (7.66) satisfies identity (2.88).*[6]

Theorem 7.7.4. [77]   **(a)** *The class of algebras with the identity (2.88) contains:*

(1) *Associative algebras;*

(2) *Left-symmetric algebras;*

(3) *Lie algebras;*

(4) *Jordan algebras;*

(5) *LT-algebras*[7] *(see [215]).*

(b) *If the initial data* $C^i_{jk}(0)$ *for system (7.65) defines an algebra of one of Classes 1–5, then the algebra with multiplication (7.66) belongs to the same class for sufficiently small* $\mathbf{u}$. *In other words, these classes of algebras are invariant with respect to deformation (7.65).*

Open problem 7.7.5. *Suppose that the functions* $C^i_{jk}(\mathbf{u})$ *satisfy (7.65). Prove that for small* $\mathbf{u}$ *the algebra with the structural constants* $C^i_{jk}(\mathbf{u})$ *is isomorphic to the algebra with the structural constants* $C^i_{jk}(0)$.

Proposition 7.7.1. *Let* $\{X, Y, Z\}$ *be a triple Jordan system, and* $\phi(\mathbf{u})$ *be any solution of system (7.53). Then, the structural constants of the product*

$$X \circ Y = \{X, \phi, Y\}$$

satisfy system (7.65).

---

[6]That is, this product defines a SS-algebra (see Definition 2.4.3).

[7]Any commutative algebra with the identity (2.88) is an LT-algebra.

7.7.3   Evolution equations of geometric type

Consider evolution systems of the form

$$u_t^i = u_{xxx}^i + A_{jk}^i(\mathbf{u})\, u_x^j\, u_{xx}^k + B_{jks}^i(\mathbf{u})\, u_x^j\, u_x^k u_x^s, \qquad i = 1, \ldots, N. \qquad (7.67)$$

Henceforth, the summation is meant over repeated indices. For such systems, the coefficients are functions of the variables $u^1, \ldots, u^N$.

It is useful to rewrite formula (7.67) as

$$u_t^i = u_{xxx}^i + 3\,\alpha_{jk}^i\, u_x^j u_{xx}^k + \left(\frac{\partial \alpha_{km}^i}{\partial u^j} + 2\alpha_{jr}^i \alpha_{km}^r - \alpha_{rj}^i \alpha_{km}^r + \beta_{jkm}^i\right) u_x^j u_x^k u_x^m. \qquad (7.68)$$

The class of systems (7.68) is invariant with respect to arbitrary point transformations $\mathbf{u} \mapsto \mathbf{\Phi}(\mathbf{u})$, where $\mathbf{u} = (u^1, \ldots, u^N)$. It is easy to verify that under such changes of coordinates the functions α_{jk}^i and β_{jkm}^i are transformed as the components of an affine connection Γ and of a tensor β, respectively.

Example 7.7.6. In the case $N = 1$, equation (7.68) has the form

$$u_t = u_{xxx} + 3\,\alpha(u)\, u_x u_{xx} + \Big(\alpha'(u) + \alpha(u)^2 + \beta(u)\Big)\, u_x^3.$$

Using the symmetry approach (see Section 4.2.5), one can show that this equation possesses higher symmetries iff $\beta' = 2\alpha\beta$. By a suitable point transformation of the form $u \to \Phi(u)$ the function α can be reduced to zero (for $N = 1$ any affine connection is flat). Then the function β becomes a constant and the equation is reduced to $u_t = u_{xxx} + \text{const}\, u_x^3$ (see (4.52) and Example 5.3.17).

Without loss of generality, we assume that the tensor β is totally symmetric:

$$\beta(X, Y, Z) = \beta(Y, X, Z) = \beta(X, Z, Y)$$

for any vectors X, Y, Z.

Question 7.7.7. *For which affine connections Γ and tensors β is the system (7.68) integrable?*

Let R and T be the curvature and torsion tensors of the connection Γ. In order to formulate some necessary integrability conditions, we introduce the following tensor:

$$\sigma(X, Y, Z) \overset{def}{=} \beta(X, Y, Z) - \frac{1}{3}\delta(X, Y, Z) + \frac{1}{3}\delta(Z, X, Y),$$

where

$$\delta(X, Y, Z) \overset{def}{=} T(X, T(Y, Z)) + R(X, Y, Z) - \nabla_X(T(Y, Z)).$$

From the condition $\delta(X, Y, Z) = -\delta(X, Z, Y)$ it follows that

$$\sigma(X, Y, Z) = \sigma(Z, Y, X).$$

Theorem 7.7.8.[8] *If the system (7.68) has a symmetry of the form*

$$\mathbf{u}_\tau = \mathbf{u}_n + \mathbf{G}(\mathbf{u}, \mathbf{u}_x, \cdots, \mathbf{u}_{n-1}), \qquad n \geq 5,$$

then the following conditions are fulfilled:

$$\nabla_X(R(Y, Z, V)) = R(Y, X, T(Z, V)), \tag{7.69}$$

$$\nabla_X(\delta(Y, Z, V)) = 0, \tag{7.70}$$

$$\nabla_X(\sigma(Y, Z, V)) = 0, \tag{7.71}$$

$$\delta(X, Y, Z) = \sigma(X, Z, Y) - \sigma(X, Y, Z). \tag{7.72}$$

Open problem 7.7.9. *Find necessary algebraic integrability conditions for the tensors T, R and σ (cf. formula (7.77) of Theorem 7.7.10).*

Integrable equations in the case $T = 0$

Theorem 7.7.10. [216, 217] *If a system (7.68) has a higher symmetry of order ≥ 5 and $T = 0$, then the conditions*

$$\nabla_X(R(Y, Z, V)) = 0, \tag{7.73}$$

$$\nabla_X(\sigma(Y, Z, V)) = 0, \tag{7.74}$$

$$R(X, Y, Z) = \sigma(X, Z, Y) - \sigma(X, Y, Z), \tag{7.75}$$

$$\sigma(X, Y, Z) = \sigma(Z, Y, X), \tag{7.76}$$

and

$$\sigma(X, \sigma(Y, Z, V), W) - \sigma(W, V, \sigma(X, Y, Z)) + \\ \sigma(Z, Y, \sigma(X, V, W)) - \sigma(X, V, \sigma(Z, Y, W)) = 0 \tag{7.77}$$

hold.

Remark 7.7.11. *The identities (7.76) and (7.77) mean that the functions $\sigma^i_{jkm}(\mathbf{u})$ are structural constants of a triple Jordan system for any fixed $\mathbf{u}$.*

---

[8]This statement was proved by S. Svinolupov and the author, and was announced in the survey [216] dedicated to the memory of Sergei Svinolupov. Unfortunately, the survey was written in a hurry and contains both misprints and significant inaccuracies.

Remark 7.7.12. *According to formula (7.73), in the case $T = 0$, the affine connected space is symmetric. The identity (7.71) shows that we are dealing with a covariantly constant deformation of the triple Jordan system, which is related to the curvature tensor by formula (7.75).*

Conjecture 7.7.13. *If a system (7.68) with $T = 0$ satisfies the conditions (7.73)–(7.77), then it has an infinite set of higher symmetries and nondegenerate conservation laws.*

Conjecture 7.7.14. *For any triple Jordan system with structural constants s^i_{jkm} there exists a unique (up to point transformations) affine connection with $T = 0$ and a tensor $\sigma(\mathbf{u})$ satisfying the conditions (7.73)–(7.77) and such that*

$$\sigma^i_{jkm}(0) = s^i_{jkm}.$$

Open problem 7.7.15. *For systems (7.68) with $T = 0$, satisfying the conditions (7.73)–(7.77), find an universal formula for a recursion operator.*

Examples of equations with $T = 0$

Equations (7.59) have the form (7.68) with $T = 0$. One can verify that the conditions (7.73)–(7.77) are fulfilled for them. Moreover, any equation

$$U_t = U_{xxx} - 3\, A_U(U_x, U_{xx}) + \frac{3}{2}\, B_U(U_x, U_x, U_x), \tag{7.78}$$

where the product $X \circ Y = -A_U(X, Y)$ is defined by the deformation system (7.65) for any Jordan algebra (see part **b** of Theorem 7.7.4), and B_U is the corresponding triple system (7.10), satisfies the integrability conditions (7.73)–(7.77).

The following two vector equations

$$\mathbf{u}_t = \mathbf{u}_{xxx} + \frac{3}{2}\left(P(\mathbf{u}, \mathbf{u}_x)(\mathbf{c} - |\mathbf{c}|^2\mathbf{u})\right)_x + 3\frac{\lambda - 1}{\lambda + 1}|\mathbf{c}|^2 P(\mathbf{u}, \mathbf{u}_x)\mathbf{u}_x, \tag{7.79}$$

where $\lambda = 1$ or $\lambda = 0$, $\mathbf{u}(x, t)$ is a N-dimensional vector, $\mathbf{c}$ is a fixed constant vector and

$$P(\mathbf{u}, \mathbf{u}_x) = \left|\, \mathbf{u}_x + \frac{\langle \mathbf{c}, \mathbf{u}_x\rangle \mathbf{u}}{1 - \langle \mathbf{c}, \mathbf{u}\rangle}\,\right|^2,$$

satisfy the integrability conditions (7.73)–(7.77) but do not belong to the class of equations (7.78) described above.

Several integrable vector examples with $T = 0$ are contained in Section 8.2.4.

Open problem 7.7.16. *Clarify a geometric nature of the deformations $\sigma(\mathbf{u})$ for equations (7.79).*

Remarks on the case $T \neq 0$

For all known examples of integrable equations with $T \neq 0$ the condition $R = 0$ is satisfied.

Example 7.7.17. It is easy to verify that for the integrable matrix equation [194]

$$\mathbf{U}_t = \mathbf{U}_{xxx} - 3\,\mathbf{U}_x\mathbf{U}^{-1}\mathbf{U}_{xx} \tag{7.80}$$

$T \neq 0$ and the curvature tensor equals zero.

This equation, as well as equation (7.60), belongs to the class of matrix equations of the form

$$\mathbf{U}_t = \mathbf{U}_{xxx} + 3\,c_1\,\mathbf{U}_x\mathbf{U}^{-1}\mathbf{U}_{xx} + 3\,c_2\,\mathbf{U}_{xx}\mathbf{U}^{-1}\mathbf{U}_x + c_3\,\mathbf{U}_x\mathbf{U}^{-1}\mathbf{U}_x\mathbf{U}^{-1}\mathbf{U}_x.$$

Open problem 7.7.18. *Find all sets of constants c_1, c_2, c_3 for which this equation is integrable.*

Section 8.2.4 contains several more examples [218] of integrable systems (7.68) with $T \neq 0$, $R = 0$.

Conjecture 7.7.19. *For the affine connected space corresponding to any integrable system (7.68), either the torsion tensor or the curvature tensor is equal to zero.*

It follows from (7.70) that the affine connected spaces with $R = 0$ that correspond to integrable systems (7.68) are generated by some Bol loops [75, Theorem 5.5], and that the binary and ternary operations

$$X \circ Y = T(X,Y), \qquad (X,Y,Z) = \nabla_Z\big(T(X,Y)\big) - T\big(Z, T(X,Y)\big)$$

satisfy the identities of left Sabinin algebra [75]:

$$X \circ X = 0, \qquad (Y,X,X) = 0, \qquad (X,Y,Z) + (Y,Z,X) + (Z,X,Y) = 0,$$

$$(X,Y,(Z,U,V)) = (Z,U,(X,Y,V)) + ((X,Y,Z),U,V) + (Z,(X,Y,U),V),$$

$$(X,Y,Z) \circ U - (X,Y,U) \circ Z + (Z,U,X \circ Y) -$$

$$(X,Y,Z \circ U) + (X \circ Y) \circ (Z \circ U) = 0.$$

In addition, the ternary operation is connected with the tensor σ by $(X,Y,Z) = \sigma(X,Y,Z) - \sigma(X,Z,Y)$. For matrix equation (7.80) these two operations and the triple system σ are generated by the associative matrix multiplication:

$$X \circ Y = XY - YX, \qquad (X,Y,Z) = XYZ - XZY + ZYX - YZX,$$

$$\sigma(X,Y,Z) = XYZ + ZYX.$$

In this example the tensor σ defines a triple Jordan system. It is still unknown whether this is always the case for integrable systems with a nonzero torsion.

Chapter 8

Integrable vector evolution equations

In Chapter 7, we have observed that many polynomial integrable systems associated with nonassociative algebras have a vector of arbitrary dimension N as an unknown variable. The first attempt to classify such integrable systems was made in [219].

8.1 Examples of integrable polynomial vector systems

In this section, we present examples of integrable vector λ-homogeneous (see Theorem 1.4.9) polynomial systems. The vector equations (7.34), (7.35), (7.38), and (7.39) provide examples of such integrable systems.

In what follows, we present several integrable vector systems taken from the paper [219]. By analogy with the scalar case, the homogeneity parameter λ was assumed there to be equal to 2, 1 or $\frac{1}{2}$.

Example 8.1.1. A vector generalization of the Ibragimov–Shabat equation (1.39) has the form

$$\mathbf{u}_t = \mathbf{u}_{xxx} + 3\langle \mathbf{u}, \mathbf{u}\rangle\, \mathbf{u}_{xx} + 6\langle \mathbf{u}, \mathbf{u}_x\rangle\, \mathbf{u}_x + 3\langle \mathbf{u}, \mathbf{u}\rangle^2\, \mathbf{u}_x + 3\langle \mathbf{u}_x, \mathbf{u}_x\rangle\, \mathbf{u}.$$

Example 8.1.2. There are two of the following systems of the derivative NLS type for vectors $\mathbf{u}$ and $\mathbf{v}$ (cf. the corresponding scalar equations from Appendix 9.4):

$$\begin{cases} \mathbf{u}_t = \mathbf{u}_{xx} + 2\alpha\langle \mathbf{u}, \mathbf{v}\rangle\, \mathbf{u}_x + 2\alpha\langle \mathbf{u}, \mathbf{v}_x\rangle\, \mathbf{u} - \alpha\beta\langle \mathbf{u}, \mathbf{v}\rangle^2\, \mathbf{u}, \\ \mathbf{v}_t = -\mathbf{v}_{xx} + 2\beta\langle \mathbf{u}, \mathbf{v}\rangle\, \mathbf{v}_x + 2\beta\langle \mathbf{v}, \mathbf{u}_x\rangle\, \mathbf{v} + \alpha\beta\langle \mathbf{u}, \mathbf{v}\rangle^2\, \mathbf{v} \end{cases}$$

and [220]

$$\begin{cases} \mathbf{u}_t = \mathbf{u}_{xx} + 2\alpha\langle \mathbf{u}, \mathbf{v}\rangle\mathbf{u} + 2\beta\langle \mathbf{u}, \mathbf{v}_x\rangle\mathbf{u} + \beta(\alpha - 2\beta)\langle \mathbf{u}, \mathbf{v}\rangle^2\mathbf{u}, \\ \mathbf{v}_t = -\mathbf{v}_{xx} + 2\alpha\langle \mathbf{u}, \mathbf{v}\rangle\mathbf{v}_x + 2\beta\langle \mathbf{v}, \mathbf{u}_x\rangle\mathbf{v} - \beta(\alpha - 2\beta)\langle \mathbf{u}, \mathbf{v}\rangle^2\mathbf{v}. \end{cases}$$

Here, α and β are arbitrary constants, one of which can be normalized by a scaling.

In the examples above, the exponent of homogeneity λ equals $\dfrac{1}{2}$.

Example 8.1.3. The following four integrable systems with respect to a vector $\mathbf{u}$ and a scalar function u were found in [219]:

$$\begin{cases} u_t = u_{xxx} + uu_x - \langle \mathbf{u}, \mathbf{u}_x\rangle, \\ \mathbf{u}_t = \mathbf{u}_{xxx} + u\mathbf{u}_x + u_x\mathbf{u}, \end{cases} \tag{8.1}$$

$$\begin{cases} u_t = u_{xxx} + 3uu_x + 3\langle \mathbf{u}, \mathbf{u}_x\rangle, \\ \mathbf{u}_t = u\,\mathbf{u}_x + u_x\,\mathbf{u}, \end{cases} \tag{8.2}$$

$$\begin{cases} u_t = \langle \mathbf{u}, \mathbf{u}_x\rangle, \\ \mathbf{u}_t = \mathbf{u}_{xxx} + 2u\,\mathbf{u}_x + u_x\,\mathbf{u}, \end{cases} \tag{8.3}$$

$$\begin{cases} u_t = u_{xxx} + uu_x + \langle \mathbf{u}, \mathbf{u}_x\rangle, \\ \mathbf{u}_t = -2\,\mathbf{u}_{xxx} - u\,\mathbf{u}_x. \end{cases} \tag{8.4}$$

The system (8.1) is exactly (7.18), where $\mathbf{c} = (1, 0, \ldots, 0)^T$. A vector generalization of the Ito equation (8.2) is contained in the paper [221]. The systems (8.3) and (8.4) are generalizations of the corresponding scalar systems from [37]. For all these systems, $\lambda = 2$.

Open problem 8.1.4. *In the framework of the general scheme from* [222], *find Lax representations for systems* (8.1)–(8.4).

Remark 8.1.5. *The fifth order symmetry*

$$\begin{cases} u_\tau = u_{xxxxx} + 10uu_{xxx} + 25u_xu_{xx} + 20u^2u_x - \\ \qquad 10\langle \mathbf{u}, \mathbf{u}_{xxx}\rangle - 15\langle \mathbf{u}_x, \mathbf{u}_{xx}\rangle - 10u_x\langle \mathbf{u}, \mathbf{u}\rangle - 20u\langle \mathbf{u}, \mathbf{u}_x\rangle, \\ \mathbf{u}_\tau = -9\,\mathbf{u}_{xxxxx} - 30u\,\mathbf{u}_{xxx} - 45u_x\,\mathbf{u}_{xx} - (35u_{xx} + 20u^2 + 5\langle \mathbf{u}, \mathbf{u}\rangle)\,\mathbf{u}_x \\ \qquad -(10u_{xxx} + 20uu_x + 5\langle \mathbf{u}, \mathbf{u}_x\rangle)\,\mathbf{u} \end{cases} \tag{8.5}$$

of system (8.3) is a vector extension of equation (1.43). Indeed, if the vector part is missing (i.e. $\mathbf{u} = 0$), then system (8.3) becomes trivial, and (8.5) turns into the Kaup–Kupershmidt equation (1.43).

An extensive list of integrable systems with $\lambda = 1$ for one scalar and one vector variable is contained in [223].

8.2 Symmetry approach to classification of integrable vector equations

It turns out that in the vector case the symmetry approach can be developed [224] in the same generality as for scalar evolution equations (see Section 4.2). In particular, we will no longer assume that the right-hand side of the equation is polynomial.

Example 8.2.1. The Harry–Dym vector equation

$$\mathbf{u}_t = \langle \mathbf{u}, \mathbf{u} \rangle^{3/2} \, \mathbf{u}_{xxx}, \tag{8.6}$$

where $\mathbf{u}$ is a N-component vector, has infinitely many symmetries and conservation laws for any N.

Equations (7.34), (7.35) and (8.6) belong to the class of vector equations of the form

$$\mathbf{u}_t = f_n \, \mathbf{u}_n + f_{n-1} \, \mathbf{u}_{n-1} + \cdots + f_1 \, \mathbf{u}_1 + f_0 \, \mathbf{u}, \qquad \mathbf{u}_i = \frac{\partial^i \mathbf{u}}{\partial x^i}. \tag{8.7}$$

Here, the coefficients f_i depend on scalar products between the vectors $\mathbf{u}, \mathbf{u}_x, \ldots,$ $\mathbf{u}_{n-1}$. The assumption that the vector $\mathbf{u}$ is finite-dimensional is not essential for us. For example, the vector space can consist of functions of variables t, x and y, and the scalar product can have the form

$$\langle U, V \rangle = \int_{-\infty}^{\infty} U(x, t, y) \, V(x, t, y) \, dy.$$

In this case, we arrive at a special class of $(2 + 1)$-dimensional nonlocal evolution equations.

A key point is that we are considering the scalar products

$$u_{[i,j]} = \langle \mathbf{u}_i, \mathbf{u}_j \rangle, \qquad i, j = 0, 1, \ldots \qquad j \geq i \tag{8.8}$$

as independent variables.[1] Denote by $\mathcal{F}$ the differential field of functions depending on a finite number of variables (8.8).

---

[1]This means that equations under study are integrable for an arbitrary dimension N of the vector $\mathbf{u}$. For a fixed N there are algebraic relations between the variables (8.8).

Clearly, in the case of an N-dimensional Euclidean space, equations (8.7) are isotropic, that is, are invariant with respect to the group $SO(N)$. For this reason we will call equations of the form (8.7) with coefficients from the field $\mathcal{F}$ *isotropic*.

Examples of integrable nonisotropic vector equations

Nonisotropic vector equations, as a rule, contain a constant vector or an operator, which alters under orthogonal transformations.

Example 8.2.2. The vector KdV equation (7.18) (see [195] and Section 7.2) and equations (7.79) contain an arbitrary constant vector $\mathbf{c}$.

Example 8.2.3. Equation (2.44) contains an arbitrary operator R such that $R^T = R$.

In the case of Example 8.2.2, we have to take for $\mathcal{F}$ the field of functions of independent variables

$$\langle \mathbf{u}_i, \mathbf{u}_j \rangle, \quad 0 \leq i \leq j, \qquad \langle \mathbf{c}, \mathbf{c} \rangle, \qquad \langle \mathbf{u}_i, \mathbf{c} \rangle. \tag{8.9}$$

Open problem 8.2.4. *Find all integrable cases for several simplest classes of vector equations with an arbitrary vector* $\mathbf{c}$.

For equations of the type (2.44), we consider another scalar product $(X, Y) = \langle X, R(Y) \rangle$ and take for $\mathcal{F}$ the field of function of independent variables

$$u_{[i,j]} = \langle \mathbf{u}_i, \mathbf{u}_j \rangle, \qquad v_{[i,j]} = (\mathbf{u}_i, \mathbf{u}_j), \qquad i, j = 0, 1, \ldots \qquad j \geq i. \tag{8.10}$$

Remark 8.2.5. *It is clear that linear transformations of the pair of scalar products of the form*

$$\bar{u}_{[i,j]} = c_1 u_{[i,j]} + c_2 v_{[i,j]}, \qquad \bar{v}_{[i,j]} = c_3 u_{[i,j]} + c_4 v_{[i,j]}, \qquad c_1 c_4 \neq c_2 c_3$$

preserve the class of all integrable equations of this type.

The symmetry approach, which we develop in the next section for the isotropic equations (8.7), can be almost literally generalized to the case of anisotropic equations, whose coefficients depend on variables (8.10).

8.2.1 Canonical densities

Consider equations (8.7) with coefficients from $\mathcal{F}$, where $\mathcal{F}$ is a field of functions depending on the variables (8.8).

Theorem 8.2.6. [224] *If equation* (8.7) *has an infinite series of vector symmetries of the form*

$$\mathbf{u}_\tau = g_m \, \mathbf{u}_m + g_{m-1} \, \mathbf{u}_{m-1} + \cdots + g_1 \, \mathbf{u}_1 + g_0 \, \mathbf{u}, \qquad g_i \in \mathcal{F}, \qquad (8.11)$$

then

(i) *The equation possesses a formal Lax pair* $L_t = [A, \, L]$, *where*

$$L = a_1 \, D + a_0 + a_{-1} \, D^{-1} + \cdots, \qquad A = \sum_0^n f_i \, D^i. \qquad (8.12)$$

Here, $a_i \in \mathcal{F}$, *and* f_i *are the coefficients of equation* (8.7);

(ii) *The following functions*

$$\rho_{-1} = \frac{1}{a_1}, \qquad \rho_0 = \frac{a_0}{a_1}, \qquad \rho_i = \operatorname{res} L^i, \qquad i \in \mathbb{N} \qquad (8.13)$$

are densities of local (*see Definition 2.0.4*) *conservation laws for equation* (8.7).

The conservation laws with densities (8.13) are called *canonical*.

Remark 8.2.7. *Unlike* (4.21), *the differential operator* A *in* (8.12) *is not the Fréchet derivative of the right-hand side of the equation. For this reason, we avoid the names "formal symmetry" or "formal recursion operator" for the series* L.

Proof. Let us outline a proof of Theorem (8.2.6). (i) We rewrite equation (8.7) and its symmetry (8.11) as

$$\mathbf{u}_t = A(\mathbf{u}), \qquad \mathbf{u}_\tau = B(\mathbf{u}), \qquad \text{where} \quad B = \sum_0^m g_i \, D^i. \qquad (8.14)$$

The compatibility of equations (8.14) leads to the operator identity

$$B_t - [A, \, B] = A_\tau.$$

In the case of large m, we can "ignore" the right-hand side, since it has a small order comparing to the other terms. That is, the operator $\bar{L} = B$ approximately satisfies the equation $\bar{L}_t = [A, \bar{L}]$. Then $L_m = B^{1/m}$ is also an approximate solution. The procedure of gluing together the approximate first order solutions L_i corresponding to different symmetries into one formal Lax operator L is similar to the scalar case [3].

The statement (ii) follows from Adler's Theorem 2.1.5. $\square$

Theorem 8.2.8. [224] *If equation* (8.7) *has an infinite series of nontrivial local conservation laws, then*

(i) *There exist a formal Lax operator L and a series S of the form*

$$S = s_1 D + s_0 + s_{-1} D^{-1} + s_{-2} D^{-2} + \cdots,$$

such that

$$S_t + A^+ S + S A = 0, \qquad S^+ = -S, \qquad L^+ = -S^{-1} L S.$$

Here the superscript $^+$ means the formal conjugation (see Section 2.1.1);

(ii) *If the conditions of Item* (i) *are fulfilled, the densities* (8.13) *with $i = 2k$ are trivial, that is, they have the form $\rho_{2k} = D(\sigma_k)$ for some functions $\sigma_k \in \mathcal{F}$.*

It is not difficult to prove this theorem by following the arguments from [3, 224].

8.2.2 Euler operator and Fréchet derivative

Defining the canonical densities in the previous section, we have considered differential operators and pseudo-differential series with scalar coefficients from $\mathcal{F}$. However, similarly to the nonabelian case (see Section 6.3.2), this is not enough if we are going to define the Fréchet derivative and construct Hamiltonian and recursion operators. To do that we have to change the algebra of coefficients.

Let us denote by $R_{i,j}$ the $\mathcal{F}$-linear operator acting on vectors by the rule

$$R_{i,j}(\mathbf{v}) = \langle \mathbf{u_j}, \mathbf{v} \rangle \, \mathbf{u_i}.$$

It easy to see that

$$R_{i,j} R_{p,q} = \langle \mathbf{u}_j, \mathbf{u}_p \rangle R_{i,q}, \qquad R_{i,j}^T = R_{j,i}, \qquad \mathrm{tr}\,(R_{i,j}) = \langle \mathbf{u}_i, \mathbf{u}_j \rangle,$$

$$D \circ R_{i,j} = R_{i,j} D + R_{i+1,j} + R_{i,j+1}.$$

Denote by $\mathcal{X}$ the associative algebra over $\mathcal{F}$ generated by the operators $R_{i,j}$ and the identity operator.

The Fréchet derivative of an element from $\mathcal{F}$ is a differential operator with coefficients from $\mathcal{X}$ defined in a usual way (see Section 6.3.2). For example, the Fréchet derivative of the right-hand side F of equation (8.7) is equal to

$$F_* = \sum_k f_k D^k + \sum_{i,j,k} \frac{\partial f_k}{\partial u_{[i,j]}} \left(R_{k,i} D^j + R_{k,j} D^i \right), \qquad (8.15)$$

where $i, j, k = 0, \ldots, n$. We call differential operators with coefficients from $\mathcal{X}$ *local*.

The Euler operator (or the variational derivative) is given by the formula

$$\frac{\delta}{\delta \mathbf{u}} = \sum_{i \leq j} (-D)^i \circ \mathbf{u}_j \left(\frac{\partial}{\partial u_{[i,j]}} \right) + (-D)^j \circ \mathbf{u}_i \left(\frac{\partial}{\partial u_{[i,j]}} \right).$$

Apparently, most of the Hamiltonian structures for vector integrable equations are nonlocal. For example, the Hamiltonian operator $\mathcal{H}$ and the symplectic operator $\mathcal{T}$ for the vector mKdV equation

$$\mathbf{u}_t = \mathbf{u}_{xxx} + \langle \mathbf{u}, \mathbf{u} \rangle \, \mathbf{u}_x$$

have the form [225]

$$\mathcal{H}(\mathbf{w}) = D(\mathbf{w}) + \langle \mathbf{u}, D^{-1} \circ \mathbf{u} \rangle \, \mathbf{w} - \langle \mathbf{u}, D^{-1} \circ \mathbf{w} \rangle \mathbf{u},$$

$$\mathcal{T}(\mathbf{w}) = D(\mathbf{w}) + \mathbf{u} \, D^{-1} \circ \langle \mathbf{u}, \mathbf{w} \rangle.$$

It is easy to verify that the coefficients of these operators, written in the form of pseudo-differential series, belong to $\mathcal{X}$.

Remark 8.2.9. *One could define a formal symmetry as a pseudo-differential series with coefficients from $\mathcal{X}$, satisfying the relation (4.21) and develop a symmetry approach based on the existence of such a formal symmetry. However, the complexity of the Fréchet derivative (8.15), comparing with the operator A of the form (8.12), makes the computations very laborious. In addition, Theorem 8.2.6 is also valid for various types of nonisotropic equations (see Examples 8.2.2 and 8.2.3), while the definition of the algebra $\mathcal{X}$ essentially depends on the choice of $\mathcal{F}$.*

8.2.3   Vector isotropic equations of KdV type

Consider vector equations of the form

$$\mathbf{u}_t = \mathbf{u}_{xxx} + f_2 \mathbf{u}_{xx} + f_1 \mathbf{u}_x + f_0 \mathbf{u}, \tag{8.16}$$

where the coefficients f_i are scalar functions of six independent variables

$$\langle \mathbf{u}, \mathbf{u} \rangle, \quad \langle \mathbf{u}, \mathbf{u}_x \rangle, \quad \langle \mathbf{u}_x, \mathbf{u}_x \rangle, \quad \langle \mathbf{u}, \mathbf{u}_{xx} \rangle, \quad \langle \mathbf{u}_x, \mathbf{u}_{xx} \rangle, \quad \langle \mathbf{u}_{xx}, \mathbf{u}_{xx} \rangle. \tag{8.17}$$

Theorem 8.2.10. [224] *For equations of the form (8.16), the canonical densities are defined by the following recurrent formula:*

$$\rho_{n+2} = \frac{1}{3}\left[\sigma_n - f_0\,\delta_{n,0} - 2\,f_2\,\rho_{n+1} - f_2\,D\rho_n - f_1\,\rho_n\right]$$

$$-\frac{1}{3}\left[f_2\sum_{s=0}^{n}\rho_s\,\rho_{ns} + \sum_{0\le s+k\le n}\rho_s\,\rho_k\,\rho_{ns-k} + 3\sum_{s=0}^{n+1}\rho_s\,\rho_{ns+1}\right]$$

$$-D\left[\rho_{n+1} + \frac{1}{2}\sum_{s=0}^{n}\rho_s\,\rho_{ns} + \frac{1}{3}D\rho_n\right],\qquad n\ge 0,\tag{8.18}$$

where ρ_0 and ρ_1 are defined by formulas (cf. (4.41) and (4.42))

$$\rho_0 = -\frac{1}{3}\,f_2,$$

$$\rho_1 = \frac{1}{9}\,f_2^2 - \frac{1}{3}\,f_1 + \frac{1}{3}\,D(f_2).$$

Remark 8.2.11. *The same formula for canonical densities is also true in the nonisotropic case. A similar recurrence for equations (8.7) of the most general form in the case $n = 3$ is given in Section 8.2.8.*

Open problem 8.2.12. *Perform a complete classification of integrable equations (8.16) in the isotropic and/or nonisotropic cases.*

One of the main obstacles in solving this problem is the existence of *triangular equations*. Some equations (8.16) allow a partial separation of variables in the spherical coordinates

$$\boldsymbol{u} = R\boldsymbol{v},\qquad |\boldsymbol{v}| = 1,\qquad \text{where}\quad R = |\boldsymbol{u}|.$$

Let

$$v_{[i,j]} = \langle \partial_x^i \boldsymbol{v},\, \partial_x^j \boldsymbol{v}\rangle,\qquad i \le j.$$

Since $v_{[0,0]} = 1$, we have $D(v_{[0,0]}) = 2v_{[0,1]} = 0$. In addition, $Dv_{[0,1]} = v_{[0,2]} + v_{[1,1]} = 0$, i.e. $v_{[0,2]} = -v_{[1,1]}$ and so on. It is clear that all variables $v_{[0,k]}$ can be expressed through $v_{[i,k]}$, $1 \le i \le k < \infty$.

If in the spherical coordinates equation (8.16) has the form

$$\begin{cases} \boldsymbol{v}_t = \boldsymbol{v}_{xxx} + g_2\boldsymbol{v}_{xx} + g_1\boldsymbol{v}_x + g_0\boldsymbol{v}, \\[2mm] R_t = R_{xxx} + S(v_{[1,1]}, v_{[1,2]}, v_{[2,2]}, R, R_x, R_{xx}), \end{cases}$$

where the coefficients g_i depend only on $v_{[1,1]}, v_{[1,2]}, v_{[2,2]}$, we call equation (8.16) *triangular.*

If the equation for $\boldsymbol{v}$ is an integrable equation on the unit sphere (see Section 8.2.6), then often all necessary integrability conditions for equation (8.16) are fulfilled, and all canonical densities, after the transition to spherical coordinates, depend only on the variables $v_{[i,j]}$. At the same time, in the right-hand side of equation (8.16), there remain arbitrary functions originating from the function S.

Example 8.2.13. Consider the equation

$$\boldsymbol{u}_t = \boldsymbol{u}_{xxx} - 3\frac{u_{[0,1]}}{u_{[0,0]}}\boldsymbol{u}_{xx} + \left(6\frac{u_{[0,1]}^2}{u_{[0,0]}^2} - 3\frac{u_{[0,2]}}{u_{[0,0]}}\right)\boldsymbol{u}_x +$$

$$\left(a_1 u_{[1,2]} + a_2 u_{[0,1]} u_{[0,2]} + a_3 u_{[0,1]} u_{[1,1]} + a_4 u_{[0,1]}^3\right)\boldsymbol{u}. \tag{8.19}$$

For any functions $a_i(u_{[0,0]})$ all integrability conditions from Theorem 8.2.6 for this equation are fulfilled. In the spherical coordinates, the equation takes the form

$$\boldsymbol{v}_t = \boldsymbol{v}_{xxx} + 3v_{[1,1]}\boldsymbol{v}_x + 3v_{[1,2]}\boldsymbol{v},$$

$$R_t = R_{xxx} + \frac{R_x R_{xx}}{R}(R^4 b_2 + R^2 b_1 - 6) - R_x v_{[1,1]}(R^4 b_2 - R^4 b_3 - R^2 b_1 - 3) +$$

$$\frac{R_x^3}{R^2}(R^6 b_4 + R^4 b_3 + 6) + R v_{[1,2]}(R^2 b_1 - 3),$$

where $b_i(R) = a_i(u_{[0,0]}) \equiv a_i(R^2)$. The equation for the vector $\boldsymbol{v}$ is integrable on the sphere $|\boldsymbol{v}| = 1$ (see Theorem 8.2.24).

The classification results for some particular subclasses of equation (8.16) were obtained in [226, 227, 218]. Here we formulate some of them.

Equations invariant with respect to shift of vector

Consider equations of the form

$$\mathbf{u}_t = \mathbf{u}_{xxx} + f_2 \mathbf{u}_{xx} + f_1 \mathbf{u}_x, \tag{8.20}$$

where f_i depends only on $\langle \mathbf{u}_x, \mathbf{u}_x \rangle$, $\langle \mathbf{u}_x, \mathbf{u}_{xx} \rangle$, $\langle \mathbf{u}_{xx}, \mathbf{u}_{xx} \rangle$. It is clear that such equations are invariant with respect to the shift $\mathbf{u} \mapsto \mathbf{u} + \mathbf{b}$, where $\mathbf{b}$ is a constant vector.

Theorem 8.2.14. [226] *Any equation* (8.20) *with an infinite series of higher symmetries or local conservation laws belongs to the following list:*

$$\mathbf{u}_t = \mathbf{u}_{xxx} + \frac{3}{2}\left(\frac{a^2\, u_{[1,2]}^2}{1 + a\,u_{[1,1]}} - a\,u_{[2,2]}\right)\mathbf{u}_x,$$

$$\mathbf{u}_t = \mathbf{u}_{xxx} - 3\,\frac{u_{[1,2]}}{u_{[1,1]}}\,\mathbf{u}_{xx} + \frac{3}{2}\,\frac{u_{[2,2]}}{u_{[1,1]}}\,\mathbf{u}_x,$$

$$\mathbf{u}_t = \mathbf{u}_{xxx} - 3\,\frac{u_{[1,2]}}{u_{[1,1]}}\,\mathbf{u}_{xx} + \frac{3}{2}\left(\frac{u_{[2,2]}}{u_{[1,1]}} + \frac{u_{[1,2]}^2}{u_{[1,1]}^2(1 + a\,u_{[1,1]})}\right)\mathbf{u}_x,$$

$$\mathbf{u}_t = \mathbf{u}_{xxx} - \frac{3}{2}\,(p+1)\,\frac{u_{[1,2]}}{p\,u_{[1,1]}}\,\mathbf{u}_{xx} + \frac{3}{2}\,(p+1)\left(\frac{u_{[2,2]}}{u_{[1,1]}} - \frac{a\,u_{[1,2]}^2}{p^2\,u_{[1,1]}}\right)\mathbf{u}_x.$$

Here a is an arbitrary constant and $p = \sqrt{1 + a\,u_{[1,1]}}$. Note that if $a = 0$, then the last equation of the list reduces to

$$\mathbf{u}_t = \mathbf{u}_{xxx} - 3\,\frac{u_{[1,2]}}{u_{[1,1]}}\,\mathbf{u}_{xx} + 3\,\frac{u_{[2,2]}}{u_{[1,1]}}\,\mathbf{u}_x.$$

Remark 8.2.15. *It was verified that each of these equations has a fifth order symmetry. To prove integrability, for all equations auto-Bäcklund transformations with an arbitrary parameter (see Section 8.2.5) were found* [226].

8.2.4 Vector equations of geometric type

Consider equations (8.16) whose component form belongs to the class of systems (7.67). It is easy to see that such equations have the following structure:

$$\begin{aligned}
\mathbf{u}_t = \mathbf{u}_{xxx} + a_1 u_{[0,1]}\,\mathbf{u}_{xx} + (a_2 u_{[0,2]} + a_3 u_{[1,1]} + a_4 u_{[0,1]}^2)\,\mathbf{u}_x \\
+ (a_5 u_{[1,2]} + a_6 u_{[0,2]} u_{[0,1]} + a_7 u_{[1,1]} u_{[0,1]} + a_8 u_{[0,1]}^3)\,\mathbf{u},
\end{aligned} \tag{8.21}$$

where the coefficients a_i are functions of one variable: $a_i = a_i(u_{[0,0]})$. The components of the corresponding affine connection (see Section 7.7.3) and its torsion and curvature tensors[2] are given by the formulas

$$A(X,Y) = \frac{1}{3}\Big(a_1\langle \mathbf{u}, X\rangle\,Y + a_2\langle \mathbf{u}, Y\rangle\,X + \big(a_5\langle X, Y\rangle + a_6\langle \mathbf{u}, X\rangle\,\langle \mathbf{u}, Y\rangle\big)\mathbf{u}\Big),$$

$$T(X,Y) = \frac{1}{3}(a_1 - a_2)\Big(\langle \mathbf{u}, X\rangle\,Y - \langle \mathbf{u}, Y\rangle\,X\Big)$$

---

[2]The curvature tensor is defined as

$$R^i_{jkl} = \frac{\partial}{\partial u_k}A^i_{jl} - \frac{\partial}{\partial u_l}A^i_{jk} + A^i_{sk}A^s_{jl} - A^i_{sl}A^s_{jk}.$$

and

$$R(X,Y,Z) = \frac{1}{9}\Big(q\,\langle \boldsymbol{u}, X\rangle\langle \boldsymbol{u}, Z\rangle + p\,\langle X, Z\rangle\Big)Y - \frac{1}{9}\Big(q\,\langle \boldsymbol{u}, X\rangle\langle \boldsymbol{u}, Y\rangle + p\,\langle X, Y\rangle\Big)Z + \frac{r}{9}\Big(\langle \boldsymbol{u}, Y\rangle\langle X, Z\rangle - \langle \boldsymbol{u}, Z\rangle\langle X, Y\rangle\Big)\boldsymbol{u},$$

where

$$p = a_2 a_5 \boldsymbol{u}^2 - 3a_2 + 3a_5, \qquad q = a_2 a_6 \boldsymbol{u}^2 + a_2^2 + 3a_6 - 6a_2', \qquad r = a_5 a_6 \boldsymbol{u}^2 + a_5^2 - 3a_6 + 6a_5'.$$

The class of equations (8.21) is invariant with respect to point transformations of the form

$$\boldsymbol{u} = f(v_{[0,0]})\,\boldsymbol{v}. \tag{8.22}$$

Under a transformation (8.22), the coefficient a_1 changes according to the rule

$$\tilde{a}_1(v_{[0,0]}) = 2\,f^{-1}f'\big(v_{[0,0]}\,a_1 f^2 + 3\big) + a_1\,f^2, \quad \text{where} \quad a_1(u_{[0,0]}) = a_1\big(v_{[0,0]}f^2\big).$$

It is easy to see that in the case $a_1 = -3\,u_{[0,0]}^{-1}$ we obtain $\tilde{a}_1 = -3\,v_{[0,0]}^{-1}$. For any other function a_1 we can find a function f such that $\tilde{a}_1 = 0$.

Thus, we arrive at two nonequivalent with respect to transformation (8.22) classes of equations (8.21):

$$1. \quad a_1 = 0 \qquad \text{and} \qquad 2. \quad a_1 = -\frac{3}{u_{[0,0]}}.$$

Theorem 8.2.16. [218] *Any nontriangular (see Section 8.2.3) equation* (8.21) *with $a_1 = 0$ that has higher symmetries reduces to one of the equations of the following list 1 by a scaling $\boldsymbol{u} \mapsto \text{const }\boldsymbol{u}$.*

List 1.

$$\boldsymbol{u}_t = \boldsymbol{u}_{xxx} + 3\lambda\,\boldsymbol{u}_x \left(\frac{u_{[0,1]}^2}{1 + u_{[0,0]}} - u_{[1,1]} \right) + 3\boldsymbol{u}\,F, \quad \text{where } \lambda = 1 \text{ or } \lambda = \frac{1}{2}, \tag{8.23}$$

$$\boldsymbol{u}_t = \boldsymbol{u}_{xxx} - 3\,\boldsymbol{u}_x \left(\frac{u_{[0,2]}}{1 + u_{[0,0]}} - \frac{u_{[0,0]}u_{[1,1]}}{1 + u_{[0,0]}} + \frac{u_{[0,0]}u_{[0,1]}^2}{(1 + u_{[0,0]})^2} \right) + 3\,\boldsymbol{u}F, \tag{8.24}$$

$$\boldsymbol{u}_t = \boldsymbol{u}_{xxx} - 3\,\boldsymbol{u}_x \left(\frac{u_{[0,2]}}{1 + u_{[0,0]}} + \frac{(1 - u_{[0,0]})u_{[1,1]}}{2\,(1 + u_{[0,0]})} - \frac{(2 - u_{[0,0]})u_{[0,1]}^2}{2\,(1 + u_{[0,0]})^2} \right) + 3\,\boldsymbol{u}F, \tag{8.25}$$

where

$$F = u_{[0,1]}\frac{u_{[0,2]} + u_{[1,1]}}{1 + u_{[0,0]}} - u_{[1,2]} - \frac{u_{[0,1]}^3}{(1 + u_{[0,0]})^2}.$$

Theorem 8.2.17. *Any nontriangular equation of the form* (8.21) *with* $a_1 = -\dfrac{3}{u_{[0,0]}}$, *that has higher symmetries reduces to one of the equations of the list* 2 *by a point transformation of the form* (8.22).

List 2.

$$\boldsymbol{u}_t = \boldsymbol{u}_{xxx} - 3\boldsymbol{u}_{xx}\frac{u_{[0,1]}}{u_{[0,0]}} - 3\boldsymbol{u}_x\left(\frac{u_{[1,1]}}{u_{[0,0]}} - \frac{u_{[0,1]}^2}{u_{[0,0]}^2}\right), \tag{8.26}$$

$$\boldsymbol{u}_t = \boldsymbol{u}_{xxx} - 3\boldsymbol{u}_{xx}\frac{u_{[0,1]}}{u_{[0,0]}} - \frac{3}{2}\boldsymbol{u}_x\left(\frac{u_{[1,1]}}{u_{[0,0]}} - 2\frac{u_{[0,1]}^2}{u_{[0,0]}^2}\right), \tag{8.27}$$

$$\boldsymbol{u}_t = \boldsymbol{u}_{xxx} - 3\boldsymbol{u}_{xx}\frac{u_{[0,1]}}{u_{[0,0]}} - \frac{3}{2}\boldsymbol{u}_x\left(2\frac{u_{[0,2]}}{u_{[0,0]}} + \frac{u_{[1,1]}}{u_{[0,0]}}\right)$$
$$+3\boldsymbol{u}\left(\frac{u_{[1,2]}}{u_{[0,0]}} - \frac{u_{[0,1]}u_{[1,1]}}{u_{[0,0]}^2} + \frac{4}{3}\frac{u_{[0,1]}^3}{u_{[0,0]}^3}\right). \tag{8.28}$$

Remark 8.2.18. *Using the above formulas for T and R, it is easy to verify that for equations* (8.23) *and* (8.28)[3] *the torsion T is equal to zero while for equations* (8.24)–(8.27) *we have $T \neq 0$, $R = 0$.*

To prove the integrability of the equations from Theorems 8.2.16 and 8.2.17, in the paper [218] auto-Bäcklund transformations with an arbitrary parameter were found.

8.2.5   Vector auto-Bäcklund transformations

A first order vector auto-Bäcklund transformation is defined by the formula

$$\mathbf{u}_x = h\,\mathbf{v}_x + f\,\mathbf{u} + g\,\mathbf{v},$$

where $\mathbf{u}$ and $\mathbf{v}$ are solutions of the same vector equation. The coefficients f, g and h are (scalar) functions of variables

$$u_{[0,0]} = \langle \mathbf{u}, \mathbf{u}\rangle, \qquad v_{[i,j]} \overset{def}{=} \langle \mathbf{v}_i, \mathbf{v}_j\rangle, \qquad w_i \overset{def}{=} \langle \mathbf{u}, \mathbf{v}_i\rangle, \qquad 0 \leq i \leq j \leq 1.$$

Example 8.2.19. An auto-Bäcklund transformation for the vector Schwartz–KdV equation

---

[3]Equation (8.28) is point equivalent to the isotropic vector equation from Class 2 (see Section 7.6.2).

$$\mathbf{u}_t = \mathbf{u}_{xxx} - 3\,\frac{u_{[1,2]}}{u_{[1,1]}}\,\mathbf{u}_{xx} + \frac{3}{2}\,\frac{u_{[2,2]}}{u_{[1,1]}}\,\mathbf{u}_x$$

has the form

$$\mathbf{u}_x = \frac{2\,\mu}{\mathbf{v}_x^2}\,\langle \mathbf{u} - \mathbf{v},\,\mathbf{v}_\mathbf{x}\rangle\,(\mathbf{u} - \mathbf{v}) - \frac{\mu}{\mathbf{v}_\mathbf{x}^2}\,|\mathbf{u} - \mathbf{v}|^2\,\mathbf{v}_\mathbf{x},$$

where μ is an arbitrary parameter.

The superposition formula [137]

$$\mathbf{z} = \mathbf{u} + (\mu - \nu)\,\frac{\nu\,(\mathbf{u} - \mathbf{v}')^2\,(\mathbf{u} - \mathbf{v}) - \mu\,(\mathbf{u} - \mathbf{v})^2\,(\mathbf{u} - \mathbf{v}')}{\big(\mu\,(\mathbf{u} - \mathbf{v}) - \nu\,(\mathbf{u} - \mathbf{v}')\big)^2},$$

corresponding to this auto-Bäcklund transformation, connects four different solutions

$$
\begin{array}{ccc}
\mathbf{v}' & \xrightarrow{\ \mu\ } & \mathbf{z} \\
{\scriptstyle \nu}\big\uparrow & & \big\uparrow{\scriptstyle \nu} \\
\mathbf{u} & \xrightarrow[\ \mu\ ]{} & \mathbf{v}
\end{array}
$$

of the vector Schwartz–KdV equation.

Remark 8.2.20. *For an integrable equation, the auto-Bäcklund transformation is an integrable vector semi-discrete equation with the continuous variable x and the discrete shift* $\mathbf{u} \mapsto \mathbf{v}$. *The superposition formula is a fully discrete integrable model.*

Remark 8.2.21. *In the presence of the arbitrary parameter, the multiple applying of an auto-Bäcklund transformation allows one to construct exact solutions of the solitonic type with an arbitrary number of arbitrary constants.*

Remark 8.2.22. *The existence of a vector auto-Bäcklund transformation with the arbitrary essential parameter is the most easily verifiable evidence of the integrability of a vector equation.*

Open problem 8.2.23. *Classify integrable vector semi-discrete and discrete equations of the simplest form in the spirit of the papers* [228] *and* [137].

8.2.6 Integrable equations on the sphere

Isotropic equations

Consider equations (8.16) on the sphere $\langle \mathbf{u},\,\mathbf{u}\rangle = 1$. In this case, the coefficients are functions of only three independent variables

$$u_{[1,1]} = \langle \mathbf{u}_x,\,\mathbf{u}_x\rangle, \qquad u_{[1,2]} = \langle \mathbf{u}_x,\,\mathbf{u}_{xx}\rangle, \qquad u_{[2,2]} = \langle \mathbf{u}_{xx},\,\mathbf{u}_{xx}\rangle \qquad (8.29)$$

instead of six variables (8.17). Indeed, differentiating the constraint $\langle \mathbf{u}, \mathbf{u} \rangle = 1$, we can express the remaining scalar products in terms of (8.29). Moreover, the relation $\langle \mathbf{u}, \mathbf{u}_t \rangle = 0$ gives

$$f_0 = f_2\, u_{[1,1]} + 3\, u_{[1,2]}$$

and therefore every equation (8.16) on a sphere has the form

$$\mathbf{u}_t = \mathbf{u}_3 + f_2\, \mathbf{u}_2 + f_1\, \mathbf{u}_1 + (f_2\, u_{[1,1]} + 3\, u_{[1,2]})\, \mathbf{u}, \qquad |\mathbf{u}| = 1. \qquad (8.30)$$

Theorem 8.2.24. [224] *Any equation (8.30) with an infinite sequence of higher symmetries or local conservation laws is contained in the following list:*

$$\mathbf{u}_t = \mathbf{u}_{xxx} - 3\,\frac{u_{[1,2]}}{u_{[1,1]}}\,\mathbf{u}_{xx} + \frac{3}{2}\left(\frac{u_{[2,2]}}{u_{[1,1]}} + \frac{u_{[1,2]}^2}{u_{[1,1]}^2\,(1 + a\, u_{[1,1]})}\right)\mathbf{u}_x,$$

$$\mathbf{u}_t = \mathbf{u}_{xxx} + \frac{3}{2}\left(\frac{a^2\, u_{[1,2]}^2}{1 + a\, u_{[1,1]}} - a\,(u_{[2,2]} - u_{[1,1]}^2) + u_{[1,1]}\right)\mathbf{u}_x + 3\, u_{[1,2]}\,\mathbf{u},$$

$$\mathbf{u}_t = \mathbf{u}_{xxx} - 3\,\frac{u_{[1,2]}}{u_{[1,1]}}\,\mathbf{u}_{xx} + \frac{3}{2}\,\frac{u_{[2,2]}}{u_{[1,1]}}\,\mathbf{u}_x,$$

$$\mathbf{u}_u = \mathbf{u}_{xxx} - 3\,\frac{u_{[1,2]}}{u_{[1,1]}}\,\mathbf{u}_{xx} + 3\,\frac{u_{[2,2]}}{u_{[1,1]}}\,\mathbf{u}_x,$$

$$\mathbf{u}_t = \mathbf{u}_{xxx} - 3\,\frac{(q + 1)\, u_{[1,2]}}{2\, q\, u_{[1,1]}}\,\mathbf{u}_{xx} + 3\,\frac{(q - 1)\, u_{[1,2]}}{2\, q}\,\mathbf{u}$$

$$+ \frac{3}{2}\left(\frac{(q + 1)\, u_{[2,2]}}{u_{[1,1]}} - \frac{(q + 1)\, a\, u_{[1,2]}}{2}\, q^2 u_{[1,1]} + u_{[1,1]}\,(1 - q)\right)\mathbf{u}_x,$$

$$\mathbf{u}_t = \mathbf{u}_3 + 3\, u_{[1,1]}\mathbf{u}_1 + 3\, u_{[1,2]}\,\mathbf{u},$$

$$\mathbf{u}_t = \mathbf{u}_3 + \frac{3}{2}\, u_{[1,1]}\mathbf{u}_1 + 3\, u_{[1,2]}\mathbf{u},$$

where a and c are arbitrary constants, and $q = \sqrt{1 + a\, u_{[1,1]}}$.

8.2.7 Equations with two scalar products on the sphere

Consider equations (8.16), whose coefficients are functions of the variables (8.10), where $i \le j \le 2$. All such integrable equations on the sphere $|\mathbf{u}| = 1$ were found in [224, 229].

We present here only equations with rational coefficients from the classification lists of these papers:

$$\mathbf{u}_t = \mathbf{u}_{xxx} + \left(\frac{3}{2}\, u_{[1,1]} + v_{[0,0]}\right)\mathbf{u}_x + 3\, u_{[1,2]}\,\mathbf{u}_0,$$

$$\mathbf{u}_t = \mathbf{u}_{xxx} - 3\frac{u_{[1,2]}}{u_{[1,1]}}\,\mathbf{u}_{xx} + \frac{3}{2}\left(\frac{u_{[2,2]}}{u_{[1,1]}} + \frac{u_{[1,2]}^2}{u_{[1,1]}^2} + \frac{v_{[1,1]}}{u_{[1,1]}}\right)\mathbf{u}_x,$$

$$\mathbf{u}_t = \mathbf{u}_{xxx} - 3\frac{u_{[1,2]}}{u_{[1,1]}}\,\mathbf{u}_{xx} + \frac{3}{2}\left(\frac{u_{[2,2]}}{u_{[1,1]}} + \frac{u_{[1,2]}^2}{u_{[1,1]}^2} - \frac{(v_{[0,1]} + u_{[1,2]})^2}{q\,u_{[1,1]}} + \frac{v_{[1,1]}}{u_{[1,1]}}\right)\mathbf{u}_x,$$

where $q = u_{[1,1]} + v_{[0,0]} + a$, and a is an arbitrary constant,

$$\mathbf{u}_t = \mathbf{u}_{xxx} - 3\frac{v_{[0,1]}}{v_{[0,0]}}\,\mathbf{u}_{xx} - 3\left(\frac{v_{[0,2]}}{v_{[0,0]}} - 2\frac{v_{[0,1]}^2}{v_{[0,0]}^2}\right)\mathbf{u}_x + 3\left(u_{[1,2]} - \frac{v_{[0,1]}}{v_{[0,0]}}u_{[1,1]}\right)\mathbf{u}$$

$$\mathbf{u}_t = \mathbf{u}_{xxx} - 3\frac{v_{[0,1]}}{v_{[0,0]}}\,\mathbf{u}_{xx} - 3\left(\frac{2v_{[0,2]} + v_{[1,1]} + a}{2v_{[0,0]}} - \frac{5}{2}\frac{v_{[0,1]}^2}{v_{[0,0]}^2}\right)\mathbf{u}_x + 3\left(u_{[1,2]} - \frac{v_{[0,1]}}{v_{[0,0]}}u_{[1,1]}\right)\mathbf{u},$$

$$\mathbf{u}_t = \mathbf{u}_{xxx} - 3\frac{v_{[0,1]}}{v_{[0,0]}}\left(\mathbf{u}_{xx} + u_{[1,1]}\mathbf{u}\right) + 3u_{[1,2]}\,\mathbf{u} + \frac{3}{2}\Bigg(-\frac{u_{[2,2]}}{v_{[0,0]}} + \frac{(u_{[1,2]} + v_{[0,1]})^2}{v_{[0,0]}(v_{[0,0]} + u_{[1,1]})} +$$

$$+\frac{(v_{[0,0]} + u_{[1,1]})^2}{v_{[0,0]}} + \frac{v_{[0,1]}^2 - v_{[0,0]}\,v_{[1,1]}}{v_{[0,0]}^2}\Bigg)\mathbf{u}_x.$$

The first of these equations is exactly (2.44). To justify the integrability of all these equations, auto-Bäcklund transformations with an arbitrary parameter were found for them.

Hyperbolic vector equations on the sphere with integrable third order symmetries

In the paper [230], hyperbolic vector equations of the form

$$\mathbf{u}_{xy} = h_0\mathbf{u} + h_1\mathbf{u}_x + h_2\mathbf{u}_y, \qquad |\mathbf{u}| = 1$$

were considered. Here h_i are scalar functions depending on two scalar products $(\cdot,\cdot)$ and $\langle\cdot,\cdot\rangle$ between the vectors $\mathbf{u}, \mathbf{u}_x$ and $\mathbf{u}_y$. All such equations with third order integrable vector x and y-symmetries (cf. Section 4.1.1) were found.

Example 8.2.25. The hyperbolic equation on the sphere

$$\mathbf{u_{xy}} = \frac{\mathbf{u_x}}{\langle\mathbf{u},\mathbf{u}\rangle}\left(\langle\mathbf{u},\mathbf{u_y}\rangle + \sqrt{1 + \langle\mathbf{u},\mathbf{u}\rangle(\mathbf{u_x},\mathbf{u_x})^{-2}}\,\phi\right) - (\mathbf{u_x},\mathbf{u_y})\mathbf{u},$$

where $\quad \phi = \sqrt{\langle\mathbf{u},\mathbf{u_y}\rangle^2 + \langle\mathbf{u},\mathbf{u}\rangle(1 - \langle\mathbf{u_y},\mathbf{u_y}\rangle)},$

has integrable vector x and y-symmetries of third order. In the case $N = 2$ this equation, after the stereographic projection, is equivalent to the equation (see [110, Section 1])

$$u_{xy} = \operatorname{sn}(u)\sqrt{u_x^2 + 1}\sqrt{u_y^2 + 1}.$$

8.2.8   Anisotropic equations with constant vector

Consider the vector KdV equation (7.18) (see Section 7.2). This equation has both higher symmetries and local conservation laws. The simplest conserved densities are

$$\rho_1 = \langle \mathbf{c}, \mathbf{u} \rangle, \qquad \rho_2 = \mathbf{c}^2 \mathbf{u}^2 - 2\langle \mathbf{c}, \mathbf{u} \rangle^2,$$

and

$$\rho_3 = 6\langle \mathbf{c}, \mathbf{u}_x \rangle^2 - 3\mathbf{c}^2 \mathbf{u}_x^2 + 3\mathbf{c}^2 \mathbf{u}^2 \langle \mathbf{c}, \mathbf{u} \rangle - 4\langle \mathbf{c}, \mathbf{u} \rangle^3.$$

Equation (7.18) belongs to the class of nonisotropic vector equations of the form

$$\mathbf{u}_t = f_3 \, \mathbf{u}_{xxx} + f_2 \, \mathbf{u}_{xx} + f_1 \, \mathbf{u}_x + f_0 \, \mathbf{u} + h \, \mathbf{c}, \tag{8.31}$$

where the coefficients f_i and h depend on the variables

$$\langle \mathbf{u}_i, \mathbf{u}_j \rangle, \qquad \langle \mathbf{u}_i, \mathbf{c} \rangle, \qquad 0 \le i \le j. \tag{8.32}$$

Without loss of generality, we can assume that $\langle \mathbf{c}, \mathbf{c} \rangle = 1$.

Equations (7.79) give us examples of integrable equations of the form (8.31) with $f_3 = 1$ and $h = 0$.

Theorem 8.2.26. *If an equation*

$$\mathbf{u}_t = f_n \, \mathbf{u}_n + f_{n-1} \, \mathbf{u}_{n-1} + \cdots + f_1 \, \mathbf{u}_1 + f_0 \, \mathbf{u} + h \, \mathbf{c} = A(\mathbf{u}) + h \, \mathbf{c}, \tag{8.33}$$

where the coefficients depend on the scalar products (8.32), has an infinite set of symmetries, then it possesses a formal Lax pair $L_t = [A, L]$. Here, L is a pseudo-differential series of the form

$$L = a_1 \, D + a_0 + a_{-1} \, D^{-1} + \cdots$$

with scalar coefficients depending on variables (8.32) and the differential operator A is defined by (8.33).

The recurrent formula for densities of the canonical conservation laws

$$D_t \, \rho_i = D \, \theta_i, \qquad i = -1, 0, 1, 2, \ldots$$

for equation (8.31) has the form

$$\rho_{n+2} = \frac{a}{3}\left(\theta_n - f_0 \, \delta_{n,0} - 2 \, a f_2 \, \rho_{n+1} - f_2 \frac{d}{dx}\rho_n - f_1 \, \rho_n \right) -$$

$$\frac{a}{3} f_2 \sum_{i+j=n} \rho_i \, \rho_j - \frac{1}{3} a^{-2} \sum_{i+j+k=n} \rho_i \, \rho_j \, \rho_k - a^{-1} \sum_{i+j=n+1} \rho_i \, \rho_j -$$

$$a^{-2} D \, a \, \rho_{n+1} - \frac{1}{2} a^{-2} D \sum_{i+j=n} \rho_i \, \rho_j - \frac{1}{3} a^{-2} \, D^2 \, \rho_n, \qquad n \ge 0,$$

where $f_3 \overset{def}{=} a^{-3}$ and

$$\rho_{-1} = a,$$

$$\rho_0 = -\frac{1}{3} a^3 f_2 - D \ln a,$$

$$\rho_1 = \frac{a}{3} \theta_{-1} + a^{-1} \rho_0^2 - \frac{1}{3} a^2 f_1 + a^{-3} (D a)^2 + 2 a^{-2} \rho_0 D a - a^{-1} D \rho_0 - \frac{1}{3} a^{-2} D^2 a.$$

Remark 8.2.27. *This formula is also valid for integrable equations (8.7) with $n = 3$ in the cases, when the coefficients depend on the variables (8.8) or (8.10). If $f_3 = a = 1$, then it coincides with (8.18).*

Despite the presence of the recurrent formula for the canonical densities and some examples (see (7.79) and (7.18)), so far there are no classification results for integrable equations of the form (8.31).

Chapter 9

Appendices

9.1 Appendix 1. Hyperbolic equations with third order integrable symmetries

Consider equations of the form

$$u_{xy} = \Psi(u, u_x, u_y), \tag{9.1}$$

possessing *integrable* (see Appendix 9.3) x and y-symmetries of the form

$$u_t = u_{xxx} + F(u, u_x, u_{xx}), \qquad u_\tau = u_{yyy} + G(u, u_y, u_{yy}). \tag{9.2}$$

Theorem 9.1.1. *Any equation* (9.1) *that has integrable symmetries of the form* (9.2)*, can be reduced by transformations of the form*

$$x \leftrightarrow y, \qquad u \mapsto \phi(u), \qquad x \mapsto \alpha x, \qquad y \mapsto \beta y,$$

$$u \mapsto u + \gamma x + \delta y, \qquad u \mapsto u \exp(\alpha x + \beta y), \qquad u \mapsto u + c\, x\, y$$

to one of the following:

$$u_{xy} = c_1 e^u + c_2 e^{-u}; \qquad u_{xy} = f(u)\sqrt{u_x^2 + 1}, \quad \text{where} \quad f'' = cf;$$

$$u_{xy} = \sqrt{u_x}\sqrt{u_y^2 + 1}; \qquad u_{xy} = \sqrt{\wp(u) - \mu}\sqrt{u_x^2 + 1}\sqrt{u_y^2 + c};$$

$$u_{xy} = 2\, uu_x; \qquad u_{xy} = 2\, u_x\sqrt{u_y}; \qquad u_{xy} = u_x\sqrt{u_y^2 + 1};$$

$$u_{xy} = \sqrt{u_x u_y}; \qquad u_{xy} = \frac{u_x(u_y + c)}{u}, \qquad c \neq 0;$$

$$u_{xy} = \left(c_1 e^u + c_2 e^{-u}\right) u_x;$$

$$u_{xy} = \frac{u_y \eta}{\sinh(u)} \left(\eta e^u - 1\right); \qquad u_{xy} = \frac{2 u_y \eta}{\sinh(u)} \left(\eta \cosh(u) - 1\right);$$

$$u_{xy} = \frac{2\eta\xi}{\sinh(u)} \left((\eta\xi + 1)\cosh(u) - \xi - \eta\right); \qquad u_{xy} = \frac{u_y}{u}\eta(\eta - 1) + c\,u\,\eta(\eta + 1);$$

$$u_{xy} = \frac{2u_y}{u}\eta(\eta - 1); \qquad u_{xy} = \frac{2\eta\xi}{u}(\eta - 1)(\xi - 1); \qquad u_{xy} = \frac{u_x u_y}{u} - 2u^2 u_y;$$

$$u_{xy} = \frac{u_x(u_y + c)}{u} - u u_y; \qquad u_{xy} = \sqrt{u_y} + c\,u_y; \qquad u_{xy} = c\,u.$$

Here $(\wp')^2 = 4\wp^3 - g_2\wp - g_3$, μ *is any root of the equation* $4\mu^3 - g_2\mu - g_3 = 0$, c, c_1, c_2, g_2, g_3 *are arbitrary constants, and* $\xi = \sqrt{u_y + 1}$, $\eta = \sqrt{u_x + 1}$.

9.2   Appendix 2. Scalar hyperbolic equations of Liouville type

Here are collected [110, 166] *nonlinear* equations (5.3) of Liouville type (see Definition 5.3.1), known to the author. The list is presented up to transformations of the form

$$x \mapsto \zeta(x), \qquad y \mapsto \xi(y), \qquad u \mapsto \theta(x, y, u)$$

and the involution $x \leftrightarrow y$.

Class 1. Equations from this class have the form

$$u_{xy} = -\frac{W_y}{W_{u_x}}.$$

Here, the function $W(x, y, u_x)$ satisfies the equation

$$u_x = q_0(x, y) + \sum_{i=1}^{n} \alpha_i(y)\, q_i(x, W).$$

where α_i and q_i are arbitrary functions.   □

Class 2. (see Section 5.4.1) Equations of the form

$$u_{xy} = -\frac{P_u}{P_{u_x}}\, u_y,$$

where

$$u_x = \alpha(x, P)\, u^2 + \beta(x, P)\, u + \gamma(x, P),$$

and α, β, γ are arbitrary functions.

Remark 9.2.1. *The well-known equation*

$$u_{xy} = e^u \, u_y$$

belongs to this class.

Class 3. Equations of the form

$$u_{xy} = A_n(x, y) \sqrt{u_x \, u_y}, \qquad n \geq 1,$$

where $A_n(x, y)$ is any solution of the equation $h_n = 0$. The expressions h_i, $i > 1$ are defined by the recurrence

$$h_{k+1} = 2h_k - h_{k-1} - \frac{\partial^2}{\partial x \partial y} \ln h_k, \qquad k = 1, \ldots,$$

where

$$h_0 = \frac{1}{4} A_n^2, \qquad h_1 = \frac{1}{4} A_n^2 - \frac{\partial^2}{\partial x \partial y} \ln A_n.$$

In particular, Class 3 contains equations with

$$A_n = \frac{2n\lambda}{\lambda(x + y) - xy}$$

and with their degeneration

$$A_n = \frac{2n}{(x + y)}.$$

This equation has x and y-integrals of order n. $\quad\square$

In addition to the three series given above, Liouville type equations of the form

$$u_{xy} = A(x, y, u) \, B(u_x) \bar{B}(u_y) \tag{9.3}$$

are known. Besides equations (5.2), (5.14), (5.15), these are the equations

$$u_{xy} = \left(\frac{1}{u - x} + \frac{1}{u - y} \right) u_x \, u_y;$$

$$u_{xy} = f(u) \, b(u_x), \qquad \text{where} \qquad f f'' - f'^2 = 0, \qquad bb' + u_1 = 0;$$

$$u_{xy} = f(u) \, b(u_x) \, \bar{b}(u_y), \qquad \text{where} \qquad (\ln f)'' - f^2 = 0, \qquad bb' + u_1 = 0, \qquad \bar{b}\bar{b}' + \bar{u}_1 = 0;$$

$$u_{xy} = \frac{1}{u} b(u_x) \, \bar{b}(u_y), \qquad \text{where} \qquad bb' + cb + u_1 = 0, \qquad \bar{b}\bar{b}' + c\bar{b} + \bar{u}_1 = 0;$$

$$u_{xy} = \frac{1}{(x+y)\,b(u_x)\,\bar{b}(u_y)}, \quad \text{where} \quad b' = b^3 + b^2, \qquad \bar{b}' = \bar{b}^3 + \bar{b}^2;$$

$$u_{xy} = \frac{1}{u}\,b(u_x)\,\bar{b}(u_y), \quad \text{where} \quad bb' + b - 2u_1 = 0, \qquad \bar{b}\bar{b}' + \bar{b} - 2\bar{u}_1 = 0.$$

Equations of more complex structure than (9.3) can be found in [166, 167]. For example, two equations found by Lainé have the form

$$u_{xy} = \left(\frac{1}{u-x} + \frac{1}{u-y}\right)u_x u_y + \frac{u_y}{u-x}\sqrt{u_x}$$

and

$$u_{xy} = 2\left[(u+Y)^2 + u_y + (u+Y)\sqrt{(u+Y)^2 + u_y}\right] \times \left[\frac{\sqrt{u_x} + u_x}{u-x} - \frac{u_x}{\sqrt{(u+Y)^2 + u_y}}\right],$$

where $Y = Y(y)$ is an arbitrary function. One more equation (5.38) was discovered by A. Zhiber and the author.

9.3   Appendix 3. Integrable evolution equations

Admissible point transformations

Let us describe point transformations that are usually used in the classification of integrable equations (4.10).

(1) Point transformations of the form[1]

$$\tilde{u} = \phi(u);$$

(2) The scaling

$$\tilde{x} = a\,x, \qquad \tilde{t} = a^n\,t, \qquad a \in \mathbb{C};$$

(3) The Galilean transformation

$$\tilde{x} = x + c\,t, \qquad c \in \mathbb{C};$$

(4) If the function F does not depend on u, the shift

$$\tilde{u} = u + c_1\,x + c_2\,t, \qquad c_i \in \mathbb{C}$$

is allowed;

---

[1]If the transformation rule for one of the variables t, x or u is not specified, it is assumed that the variable does not change.

(5) if $F(\lambda u, \lambda u_1, \dots, \lambda u_{n-1}) = \lambda F(u, u_1, \dots, u_{n-1})$, then the transformation

$$\tilde{u} = u \exp(a t + b x), \qquad a, b \in \mathbb{C}$$

is admissible.

The equations related by such transformations are called *equivalent*. It is important to note that the classification results are purely algebraic. We are not interested in the properties of real solutions of the equations under consideration, and therefore the functions and constants in the transformations are supposed to be complex. For example, the equations $u_t = u_3 - u_1^3$ and $u_t = u_3 + u_1^3$ are regarded as equivalent.

9.3.1 Third order equations

Theorem 9.3.1. [118, 3, 14] *Any nonlinear equation*

$$u_t = u_3 + F(u, u_1, u_2)$$

that has infinitely many higher symmetries

$$u_{\tau_i} = G_i(u, \dots, u_{n_i}), \qquad i = 1, 2, \dots, \qquad n_{i+1} > n_i,$$

is equivalent to one of the following:

$$u_t = u_{xxx} + u u_x;$$

$$u_t = u_{xxx} + u^2 u_x;$$

$$u_t = u_{xxx} + u_x^2;$$

$$u_t = u_{xxx} - \frac{1}{2} u_x^3 + (c_1 e^{2u} + c_2 e^{-2u}) u_x;$$

$$u_t = u_{xxx} - \frac{3 u_x u_{xx}^2}{2(u_x^2 + 1)} + c_1 (u_x^2 + 1)^{3/2} + c_2 u_x^3;$$

$$u_t = u_{xxx} - \frac{3 u_{xx}^2}{2 u_x} + \frac{1}{u_x} - \frac{3}{2} \wp(u) u_x^3; \tag{9.4}$$

$$u_t = u_{xxx} - \frac{3 u_x u_{xx}^2}{2(u_x^2 + 1)} - \frac{3}{2} \wp(u) u_x (u_x^2 + 1); \tag{9.5}$$

$$u_t = u_{xxx} - \frac{3 u_{xx}^2}{2 u_x};$$

$$u_t = u_{xxx} - \frac{3 u_{xx}^2}{4 u_x} + c_1 u_x^{3/2} + c_2 u_x^2;$$

$$u_t = u_{xxx} - \frac{3}{4}\frac{u_{xx}^2}{u_x+1} + 3\,u_{xx}u^{-1}(\sqrt{u_x+1}-u_x-1)$$
$$- 6\,u^{-2}u_x(u_x+1)^{3/2} + 3\,u^{-2}u_x\,(u_x+1)(u_x+2);$$

$$u_t = u_{xxx} - \frac{3}{4}\frac{u_{xx}^2}{u_x+1} - 3\,\frac{u_{xx}\,(u_x+1)\cosh u}{\sinh u} + 3\,\frac{u_{xx}\sqrt{u_x+1}}{\sinh u}$$
$$- 6\,\frac{u_x(u_x+1)^{3/2}\cosh u}{\sinh^2 u} + 3\,\frac{u_x\,(u_x+1)(u_x+2)}{\sinh^2 u} + u_x^2(u_x+3);$$

$$u_t = u_{xxx} + 3\,u^2u_{xx} + 9\,uu_x^2 + 3\,u^4u_x;$$

$$u_t = u_{xxx} + 3\,uu_{xx} + 3\,u_x^2 + 3\,u^2u_x. \tag{9.6}$$

Here $\wp(u)$ is any solution of the equation $(\wp')^2 = 4\wp^3 - g_2\wp - g_3$, $g_i \in \mathbb{C}$, and c_1, c_2 are arbitrary constants.

Remark 9.3.2. *Sometimes, instead of equations (9.4) and (9.5), the equations (4.7) and (4.6), which are equivalent to them, are considered. It is easy to check that if $Q' \neq 0$, then doing in equations (4.7) and (4.6) the transformation $u = f(v)$, where $(f')^2 = Q(f)$, we obtain for v equations (9.4) and (9.5), respectively.*

9.3.2 Fifth order equations

Theorem 9.3.3. [14] *Suppose that a nonlinear equation*

$$u_t = u_5 + F(u, u_x, u_2, u_3, u_4)$$

satisfies the following two conditions:

(1) *It possesses infinite set of higher symmetries*

$$u_{\tau_i} = G_i(u, \ldots, u_{n_i}), \qquad i = 1, 2, \ldots, \qquad n_{i+1} > n_i;$$

(2) *It does not have higher symmetries of order less than five.*

Then the equation is equivalent to one from the following list:

$$u_t = u_5 + 5uu_3 + 5u_1u_2 + 5u^2u_1;$$

$$u_t = u_5 + 5uu_3 + \frac{25}{2}u_1u_2 + 5u^2u_1;$$

$$u_t = u_5 + 5u_1u_3 + \frac{5}{3}u_1^3;$$

$$u_t = u_5 + 5u_1u_3 + \frac{15}{4}u_2^2 + \frac{5}{3}u_1^3;$$

$$u_t = u_5 + 5(u_1 - u^2)u_3 + 5u_2^2 - 20uu_1u_2 - 5u_1^3 + 5u^4u_1;$$

$$u_t = u_5 + 5(u_2 - u_1^2)u_3 - 5u_1u_2^2 + u_1^5;$$

$$u_t = u_5 + 5(u_2 - u_1^2 + \lambda_1 e^{2u} - \lambda_2^2 e^{-4u})\,u_3 - 5u_1u_2^2 + 15(\lambda_1 e^{2u} + 4\lambda_2^2 e^{-4u})\,u_1u_2$$
$$+ u_1^5 - 90\lambda_2^2 e^{-4u}\,u_1^3 + 5(\lambda_1 e^{2u} - \lambda_2^2 e^{-4u})^2\,u_1;$$

$$u_t = u_5 + 5(u_2 - u_1^2 - \lambda_1^2 e^{2u} + \lambda_2 e^{-u})\,u_3 - 5u_1u_2^2 - 15\lambda_1^2 e^{2u}\,u_1u_2$$
$$+ u_1^5 + 5(\lambda_1^2 e^{2u} - \lambda_2 e^{-u})^2\,u_1, \qquad \lambda_2 \neq 0;$$

$$u_t = u_5 - 5\frac{u_2u_4}{u_1} + 5\frac{u_2^2u_3}{u_1^2} + 5\left(\frac{\mu_1}{u_1} + \mu_2u_1^2\right)u_3 - 5\left(\frac{\mu_1}{u_1^2} + \mu_2u_1\right)u_2^2$$
$$- 5\frac{\mu_1^2}{u_1} + 5\mu_1\mu_2u_1^2 + \mu_2^2u_1^5;$$

$$u_t = u_5 - 5\frac{u_2u_4}{u_1} - \frac{15}{4}\frac{u_3^2}{u_1} + \frac{65}{4}\frac{u_2^2u_3}{u_1^2} + 5\left(\frac{\mu_1}{u_1} + \mu_2u_1^2\right)u_3 - \frac{135}{16}\frac{u_2^4}{u_1^3}$$
$$- 5\left(\frac{7\mu_1}{4u_1^2} - \frac{\mu_2u_1}{2}\right)u_2^2 - 5\frac{\mu_1^2}{u_1} + 5\mu_1\mu_2u_1^2 + \mu_2^2u_1^5;$$

$$u_t = u_5 - \frac{5}{2}\frac{u_2u_4}{u_1} - \frac{5}{4}\frac{u_3^2}{u_1} + 5\frac{u_2^2u_3}{u_1^2} + \frac{5}{2}\frac{u_2u_3}{\sqrt{u_1}} - 5(u_1 - 2\mu u_1^{1/2} + \mu^2)u_3 - \frac{35}{16}\frac{u_2^4}{u_1^3}$$
$$- \frac{5}{3}\frac{u_2^3}{u_1^{3/2}} + 5\left(\frac{3\mu^2}{4u_1} - \frac{\mu}{\sqrt{u_1}} + \frac{1}{4}\right)u_2^2 + \frac{5}{3}u_1^3 - 8\mu u_1^{5/2} + 15\mu^2u_1^2 - \frac{40}{3}\mu^3u_1^{3/2};$$

$$u_t = u_5 + \frac{5}{2}\frac{f - u_1}{f^2}u_2u_4 + \frac{5}{4}\frac{2f - u_1}{f^2}u_3^2 + 5\mu(u_1 + f)^2u_3 + \frac{5}{4}\frac{4u_1^2 - 8u_1f + f^2}{f^4}u_2^2u_3 +$$
$$\frac{5}{16}\frac{2 - 9u_1^3 + 18u_1^2f}{f^6}u_2^4 + \frac{5\mu}{4}\frac{(4f - 3u_1)(u_1 + f)^2}{f^2}u_2^2 + \mu^2(u_1 + f)^2\left(2f(u_1 + f)^2 - 1\right);$$

$$u_t = u_5 + \frac{5}{2}\frac{f - u_1}{f^2}u_2u_4 + \frac{5}{4}\frac{2f - u_1}{f^2}u_3^2 - 5\omega(f^2 + u_1^2)u_3 + \frac{5}{4}\frac{4u_1^2 - 8u_1f + f^2}{f^4}u_2^2u_3 +$$
$$\frac{5}{16}\frac{2 - 9u_1^3 + 18u_1^2f}{f^6}u_2^4 + \frac{5}{4}\omega\frac{5u_1^3 - 2u_1^2f - 11u_1f^2 - 2}{f^2}u_2^2 - \frac{5}{2}\omega'(u_1^2 - 2u_1f + 5f^2)u_1u_2$$
$$+ 5\omega^2u_1f^2(3u_1 + f)(f - u_1);$$

$$u_t = u_5 + \frac{5}{2}\frac{f - u_1}{f^2}u_2u_4 + \frac{5}{4}\frac{2f - u_1}{f^2}u_3^2 + \frac{5}{4}\frac{4u_1^2 - 8u_1f + f^2}{f^4}u_2^2u_3$$
$$+ \frac{5}{16}\frac{2 - 9u_1^3 + 18u_1^2f}{f^6}u_2^4 + 5\omega\frac{2u_1^3 + u_1^2f - 2u_1f^2 + 1}{f^2}u_2^2$$
$$- 10\omega u_3(3u_1f + 2u_1^2 + 2f^2) - 10\omega'(2f^2 + u_1f + u_1^2)u_1u_2$$
$$+ 20\omega^2u_1(u_1^3 - 1)(u_1 + 2f);$$

$$u_t = u_5 + \frac{5}{2}\frac{f - u_1}{f^2}u_2 u_4 + \frac{5}{4}\frac{2f - u_1}{f^2}u_3^2 - 5c\frac{f^2 + u_1^2}{\omega^2}u_3$$

$$+ \frac{5}{4}\frac{4u_1^2 - 8u_1 f + f^2}{f^4}u_2^2 u_3 + \frac{5}{16}\frac{2 - 9u_1^3 + 18u_1^2 f}{f^6}u_2^4$$

$$- 10\omega\left(3u_1 f + 2u_1^2 + 2f^2\right)u_3 - \frac{5}{4}c\frac{11u_1 f^2 + 2u_1^2 f + 2 - 5u_1^3}{\omega^2 f^2}u_2^2$$

$$+ 5\omega\frac{2u_1^3 + u_1^2 f - 2u_1 f^2 + 1}{f^2}u_2^2 + 5c\omega'\frac{u_1^2 + 5f^2 - 2u_1 f}{\omega^3}u_1 u_2$$

$$- 10\omega'\left(2f^2 + u_1 f + u_1^2\right)u_1 u_2 + 20\omega^2 u_1(u_1^3 - 1)(u_1 + 2f)$$

$$+ 40\frac{cu_1 f^3(2u_1 + f)}{\omega} + 5\frac{c^2 u_1 f^2(3u_1 + f)(f - u_1)}{\omega^4}, \qquad c \neq 0.$$

Here λ_1, λ_2, μ, μ_1, μ_2, *and* c *are arbitrary constants, the function* $f(u_1)$ *is defined by the algebraic equation*

$$(f + u_1)^2(2f - u_1) + 1 = 0,$$

and $\omega(u)$ *is nonconstant solution of the differential equation*

$$\omega'^2 = 4\omega^3 + c.$$

Differential substitutions connecting the integrable equations of the third or fifth order, can be found in [14].

9.4 Appendix 4. Quasilinear systems of two equations of second order

Theorem 9.4.1. [4, 132] *Any nonlinear nontriangular quasilinear system of the form* (4.16), *having a symmetry* (4.15), *up to scalings of* t, x, u, v, *shifts of* u *and* v *and the involution*

$$u \leftrightarrow v, \qquad t \leftrightarrow -t$$

belongs to the following list:

$$\begin{cases} u_t = u_{xx} + (u + v)u_x + uv_x, \\ v_t = -v_{xx} + (u + v)v_x + vu_x; \end{cases}$$

$$\begin{cases} u_t = u_{xx} - 2(u + v)u_x - 2uv_x + 2u^2 v + 2uv^2 + \alpha u + \beta v + \gamma, \\ v_t = -v_{xx} + 2(u + v)v_x + 2vu_x - 2u^2 v - 2uv^2 - \alpha u - \beta v - \gamma; \end{cases}$$

$$\begin{cases} u_t = u_{xx} + vu_x + uv_x, \\ v_t = -v_{xx} + vv_x + u_x; \end{cases}$$

$$\begin{cases} u_t = u_{xx} + 2vu_x + 2uv_x + 2uv^2 + u^2 + \alpha u + \beta v + \gamma, \\ v_t = -v_{xx} - 2vv_x - u_x; \end{cases}$$

$$\begin{cases} u_t = u_{xx} + \alpha v_x + (u+v)^2 + \beta(u+v) + \gamma, \\ v_t = -v_{xx} + \alpha u_x - (u+v)^2 - \beta(u+v) - \gamma; \end{cases}$$

$$\begin{cases} u_t = u_{xx} + (u+v)u_x + 4\alpha v_x + \alpha(u+v)^2 + \beta(u+v) + \gamma, \\ v_t = -v_{xx} + (u+v)v_x + 4\alpha u_x - \alpha(u+v)^2 - \beta(u+v) - \gamma; \end{cases}$$

$$\begin{cases} u_t = u_{xx} + 2\alpha u^2 v_x + 2\beta uvu_x + \alpha(\beta - 2\alpha)u^3 v^2 + \gamma u^2 v + \delta u, \\ v_t = -v_{xx} + 2\alpha v^2 u_x + 2\beta uvv_x - \alpha(\beta - 2\alpha)u^2 v^3 - \gamma uv^2 - \delta v; \end{cases}$$

$$\begin{cases} u_t = u_{xx} + 2uvu_x + (\alpha + u^2)v_x, \\ v_t = -v_{xx} + 2uvv_x + (\beta + v^2)u_x; \end{cases}$$

$$\begin{cases} u_t = u_{xx} + 2\alpha uvu_x + 2\alpha u^2 v_x - \alpha\beta u^3 v^2 + \gamma u, \\ v_t = -v_{xx} + 2\beta v^2 u_x + 2\beta uvv_x + \alpha\beta u^2 v^3 - \gamma v; \end{cases}$$

$$\begin{cases} u_t = u_{xx} + 2uvu_x + 2(\alpha + u^2)v_x + u^3 v^2 + \beta u^3 + \alpha uv^2 + \gamma u, \\ v_t = -v_{xx} - 2uvv_x - 2(\beta + v^2)u_x - u^2 v^3 - \beta u^2 v - \alpha v^3 - \gamma v; \end{cases}$$

$$\begin{cases} u_t = u_{xx} + 4uvu_x + 4u^2 v_x + 3vv_x + 2u^3 v^2 + uv^3 + \alpha u, \\ v_t = -v_{xx} - 2v^2 u_x - 2uvv_x - 2u^2 v^3 - v^4 - \alpha v; \end{cases}$$

$$\begin{cases} u_t = u_{xx} + 4uu_x + 2vv_x, \\ v_t = -v_{xx} - 2vu_x - 2uv_x - 3u^2 v - v^3 + \alpha v; \end{cases}$$

$$\begin{cases} u_t = u_{xx} + vv_x, \\ v_t = -v_{xx} + u_x; \end{cases}$$

$$\begin{cases} u_t = u_{xx} + 6(u+v)v_x - 6(u+v)^3 - \alpha(u+v)^2 - \beta(u+v) - \gamma, \\ v_t = -v_{xx} + 6(u+v)u_x + 6(u+v)^3 + \alpha(u+v)^2 + \beta(u+v) + \gamma; \end{cases}$$

$$\begin{cases} u_t = u_{xx} + vv_x, \\ v_t = -v_{xx} + uu_x. \end{cases}$$

Here α, β, γ, δ are arbitrary constants. We omit the term $(c\,u_x, c\,v_x)^T$ in the right-hand side of all systems. It is a classical symmetry generated by the invariance of systems with respect to the shift of x.

Bibliography

References for Introduction

[1] Faddeev L. D. and Takhtajan L. A., *The Hamiltonian Methods in the Theory of Solitons*, 2007, Berlin, Springer, 592 pp.

[2] Reyman A. G. and Semenov-Tian-Shansky M. A., *Integrable systems. Group theory approach*, 2003, Izhevsk, ICS Publ., 350 pp.

[3] Sokolov V. V. and Shabat A. B., *Classification of integrable evolution equations*, Soviet Scientific Reviews, Section C, 1984, **4**, 221–280.

[4] Mikhailov A. V., Shabat A. B., and Yamilov R. I., *Symmetry approach to classification of nonlinear equations. Complete lists of integrable systems.* Russian Math. Surveys, 1987, **42**(4), 1–63.

[5] Mikhailov A. V., Shabat A. B., and Sokolov V. V., *Symmetry Approach to Classification of Integrable Equations*, In *What is integrability?* Ed. V.E. Zakharov, Springer Series in Nonlinear Dynamics, Springer-Verlag, 1991, 115–184.

[6] Adler V. E., Shabat A. B., and Yamilov R. I., *Symmetry approach to the integrability problem*, Theoret. and Math. Phys., 2000, **125**(3), 1603–1661.

[7] Adler V. E., Marikhin V. G., and Shabat A. B., *Lagrangian chains and canonical Bäcklund transformations*, Theoret. and Math. Phys., 2001, **129**(2), 1448–1465.

[8] Mikhailov A. V. and Sokolov V. V., *Symmetries of differebshntial equations and the problem of Integrability*, Ed. A.V. Mikhailov, *Lecture Notes in Physics*, Springer, 2009, **767**, 19–88.

[9] Ibragimov N. H., *Transformation Groups Applied to Mathematical Physics Mathematics*, Dordrecht: D. Reidel, 1985, 394 pp.

[10] Fokas A. S., *Symmetries and integrability*, Studies in Applied Mathematics, 1987, 77(3), 253–299.

[11] Krasilchchik I. S., Lychagin V. V., and Vinogradov A. M., *Geometry of jet spaces and nonlinear partial differential equations*, Series (Adv. Stud. Contemp. Math., **1**, 1996, Publication New York, NY : Gordon and Breach, 441 pp.

[12] Olver P. J., *Applications of Lie Groups to Differential Equations*, (2nd edition), *Graduate Texts in Mathematics*, 1993, **107**, Springer–Verlag, New York.

[13] Zhiber A. V. and Shabat A. B., *Klein-Gordon equations with a nontrivial group*, Sov. Phys. Dokl., 1979, **24**(8), 608–609.

[14] Meshkov A. G. and Sokolov V. V., *Integrable evolution equations with constant separant*, Ufa Mathematical Journal, 2012, **4**(3), 104–154, arXiv nlin.SI 1302.6010.

[15] Kaplansky I., *An introduction to differential algebra*, 1957, Paris, Hermann, 63 pp.

[16] Novikov S. P., Manakov S. V., Pitaevskii L. P., and Zakharov V. E., *Theory of Solitons: The Inverse Scattering Method*, 1984, Springer-Verlag, 320 pp.

[17] Ablowitz M. J. and Segur H., *Solitons and the Inverse Scattering Transform*, 1981, SIAM, Philadelphia, 424 pp.

[18] Manakov S. V., *Note on the integration of Euler's equations of the dynamics of an n-dimensional rigid body*, Funct. Anal. Appl., 1976, **10**(4), 328–329.

[19] Golubchik I. Z. and Sokolov V. V., *Factorization of the loop algebra and integrable top-like systems*, Theoret. and Math. Phys., 2004, **141**(1), 1329–1347. arXiv:nlin.SI/0403023 v1.

[20] Lax P., *Integrals of nonlinear equations of evolution and solitary waves*, Comm. Pure Applied Math., 1968, **21**(5), 467–490.

[21] Zakharov V. E. and Shabat A. B., *Exact theory of two-dimensional self-focusing and one-dimensional self-modulation of waves in nonlinear media*, Soviet Physics JETP, 1972, **34**(1), 62–69.

[22] Zakharov V. E. and Shabat A. B., *Integration of nonlinear equations of mathematical physics by the method of inverse scattering. II*, Funct. Anal. Appl., 1979, **13**(3), 166–174.

[23] Borisov A. V. and Mamaev I. S., *Rigid body dynamics*, Izhevsk, Chaotic and Regular Dynamics, 2001, 384 pp.

[24] Audin M., *Spinning Tops: A Course on Integrable Systems*, Cambridge University Press, 1999, 148 pp.

[25] Sklyanin E., *Some algebraic structures connected withthe Yang–Baxter equation*, Funct. Anal. Appl., 1983, **16**(4), 263–270.

[26] Odesskii A. and Feigin B., *Sklyanin's elliptic algebras*, Funct. Anal. Appl., 1989, **23**(3), 207–214.

[27] Odesskii A. V., *Elliptic algebras*, Russian Math. Surveys, 2002, **57**(6), 1127–1162.

[28] De Sole A., Kac V. G., and Valeri D., *Double Poisson vertex algebras and noncommutative Hamiltonian equations*, Adv. Math., 2015, **281**, 1025–1099.

[29] Svinolupov S. I. and Sokolov V. V., *Factorization of evolution equations*, Russian Math. Surveys, 1992, **47**(3), 127 - 162.

[30] Sanders J. A. and Wang J. P., *On the integrability of homogeneous scalar evolution equations*, J. Differential Equations, 1998, **147**, 410–434.

[31] Sawada S. and Kotera T., *A method for findind N-soliton solutions of the KdV and KdV-like equation*, Prog. Theor. Phys., 1974, **51**, 1355–1367.

[32] Kaup D. J., *On the inverse scattering problem for the cubic eigenvalue problem of the class* $\varphi_{xxx} + 6Q\phi_x + 6R\phi = \lambda\phi$, Stud. Appl. Math., 1980, **62**, 189–216.

[33] Bäcklund A. V., *Einiges uber Curven und Flachentransformationen*, Lund Universitets Arsskrift, 1875, **10**, 1–12.

[34] Miura R. M., *Korteweg-de Vries equation and generalization, I. A remarkable explicit nonlinear transformation*, 1968, J. Math. Phys., **9**, 1202–1204.

[35] Drinfeld V. G. and Sokolov V. V., *Equations that are related to the Korteweg-de Vries equation*, Sov. Math. Dokl., 1985, **32**, 361–365.

[36] Arnold V. I., *Geometrical Methods in the Theory of Ordinary Differential Equations*, Springer, NY, 1988, 250 pp.

References for Part 1

[37] Drinfeld V. G. and Sokolov V. V., *Lie algebras and equations of Korteweg de Vries type*, J. Soviet Math., 1985, **30**, 1975–2036.

[38] Krichever I. M. and Phong D. H., *Symplectic forms in the theory of solitons*, 1997, arXiv preprint hep-th/9708170.

[39] Adler M., *On a trace functional for formal pseudo-differential operators and the symplectic structure of the Korteweg de Vries type equations*, Invent. Math., 1979, **50**(3), 219–248,

[40] Gelfand I. and Dikii L., *Asymptotic behavior of the resolvent of Sturm-Liouville equations and the algebra of the Korteweg-de Vries equations*, Russian Math. Surveys, 1975, **30**(5), 77–113.

[41] Sokolov V. V., *Hamiltonian property of the Krichever-Novikov equation*, Sov. Math. Dokl., 1984, **30**, 44–47.

[42] Gurses M., Karasu A., and Sokolov V. V., *On construction of recursion operator from Lax representation*, J. Math. Phys., 1999, **40**(12), 6473–6490.

[43] Ore O., *Theory of noncommutative polynomials*, Ann. Math., 1933, **34**, 480–508.

[44] Gel'fand I. M. and Dikii L. A., *Fractional powers of operator and Hamiltonian systems*, Funct. Anal. Appl., 1976, **10**(4), 259–273.

[45] Sokolov V. V. and Shabat A. B., *$(L-A)$-pairs and Ricatti-type substitutions*, Funct. Anal. Appl., 1980, **14**(2), 148–150.

[46] Fordy A. P. and Gibbons J., *Factorization of operators I : Miura transformations*, J. Math. Phys., 1980, **21**(10), 2508–2510.

[47] Drinfeld V. G. and Sokolov V. V., *New evolution equations possess $(L-A)$-pairs*, Proceedings of S.L. Sobolev seminar, 1981, **2**, 5–9.

[48] Krichever I. M. and Novikov S. P., *Holomorphic bundles over algebraic curves and non-linear equations*, Russian Math. Surveys, 1980, **35**(6), 53–79.

[49] Svinolupov S. I., Sokolov V. V., and Yamilov R. I., *On Bäcklund transformations for integrable evolution equation*, Sov. Math., Dokl., 1983, **28**, 165–168.

[50] Manakov S. V., *On the theory of two-dimensional stationary self-focusing electromagnetic waves*. Sov. Phys. JETP, 1974, **38**(2), 248–253.

[51] De Groot M. F., Hollowood T. J., and Miramontes J. L., *Generalized Drinfeld-Sokolov hierarchies*, Comm. Math. Phys., 1992, **145**, 57–84.

[52] Golubchik I. Z. and Sokolov V. V., *Integrable equations on $\mathbb{Z}$-graded Lie algebras*, Theoret. and Math. Phys., 1997, **112**(3), 1097–1103.

[53] De Sole A, Kac V. G., and D. Valeri D, *Classical affine W-algebras for $\mathfrak{gl}_N$ and associated integrable Hamiltonian hierarchies*, Comm. Math. Phys, 2016, **348**(1), 265–319,   arXiv:1509.06878.

[54] Sklyanin E. K., *On the complete integrability of the Landau—Lifschitz equation*, Preprint No. E-3, Leningrad Branch Math. Inst. (LOMI), 1979, Leningrad, 32 pp. (in Russian).

[55] Zakharov V. E. and Mikhailov A. V., *Method of the inverse scattering problem with spectral parameter on an algebraic curve*, Funct. anal. appl., 1983, **17**(4), 247–251.

[56] Golubchik I. Z. and Sokolov V. V., *Multicomponent generalization of the hierarchy of the Landau—Lifshitz equation*, Theoret. and Math. Phys, 2000, **124**(1), 909–917.

[57] Krichever I. M., *Vector bundles and Lax equations on algebraic curves*, Comm. Math. Phys., 2002, **229**(2), 229–269; http://arxiv.org/abs/hep-th/0108110.

[58] Krichever I. M. and Sheinman O. K., *Lax operator algebras*, Funct. Anal. Appl., 2007, **41**(4), 284–294.

[59] Cherednik I. V., *Functional realizations of basis representations of factoring Lie groups and algebras*, Funct. Anal. Appl., 1985, **19**(3), 193–206.

[60] Skrypnyk T., *Deformations of loop algebras and classical integrable systems: Finite-dimensional Hamiltonian systems*, Reviews in Math. Phys., 2004, **16**(7), 823–849.

[61] Belavin A. A. and Drinfeld V. G., *Solutions of the classical Yang–Baxter equation for simple Lie algebras*, Funct. Anal. Appl., 1982, **16**(3), 159–180.

[62] Sokolov V. V., *On decompositions of the loop algebra over $\mathfrak{so}_3$ into a sum of two subalgebras*, Sov. Math. Dokl., 2004, **70**(1), 568–570.

[63] Ostapenko V., *Endomorphisms of lattices of a Lie algebra over formal power series field*, C.R. Acad. Sci., 1992. Paris, **315**, Serie 1, 669–673.

[64] Efimovskaya O. V. *Factorization of the loop algebra over $\mathfrak{so}(4)$ and integrable nonlinear differential equations*, J. Math. Sci., 2007, **144**(2), 3926–3937.

[65] Sokolov V. V. and Tsyganov A. V., *On Lax pairs for the generalized Kowalewski and Goryachev-Chaplygin tops*, Theoret. and Math. Phys., 2002, **131**(1), 543–549, nlin.SI/0111035.

[66] Sokolov V. V., *A new integrable case for the Kirchhoff equation*, Theoret. and Math. Phys., 2001, **129**(1), 1335–1340.

[67] Golubchik I. Z. and Sokolov V. V., *Generalized Heizenberg equations on Z-graded Lie algebras*, Theoret. and Math. Phys., 1999, **120**(2), 1019–1025.

[68] Mikhailov A. V. and Shabat A. B., *Integrable deformations of the Heisenberg model*, Phys. Lett. A, 1986, **116**(4), 191–194.

[69] Perelomov A. M., *Some remarks on the integrability of the equations of motion of a rigid body in an ideal fluid*, Funct. anal. appl., 1981, **15**(2), 144–146.

[70] Golubchik I. Z. and Sokolov V. V., *Compatible Lie brackets and integrable equations of the principle chiral model type*, Funct. anal. appl., 2002, **36**(3), 172–181.

[71] Efimovskaya O. V. and Sokolov V. V., *Decomposition of the loop algebra over* $\mathfrak{so}(4)$ *and integrable models of the chiral type*, J. Math. Sci., 2006, **136**(6), 4385–4391.

[72] Cherednik I. V., *On integrability of two-dimensional assimetric chiral $O(3)$-field and its quantum analog*, Nuclear Physics, 1981, **33**, 278–282.

[73] Kostant B., *Quantization and representation theory*, in: Lect. Notes, 1979, **34**, 287–316.

[74] Chernyakov Yu. and Sorin A., *Explicit Semi-invariants and Integrals of the Full Symmetric sl_n Toda Lattice*, Lett. in Math. Phys., 2014, **104**, 1045–1052.

[75] Mikheev P. O. and Sabinin L. V., *Smooth quasigroups and geometry*, J. Soviet Math., 1990, **51**(6), 2642–2666.

[76] Vinberg E. B., *Convex homogeneous cones*, Transl. Moscow Math. Soc., 1963, **12**, 340–403.

[77] Sokolov V. V. and Svinolupov S. I., *Deformation of nonassociative algebras and integrable differential equations*, Acta Applicandae Mathematica, 1995, **41**(1-2), 323–339.

[78] Golubchik I. Z., Sokolov V. V., and Svinolupov S. I., *A new class of nonassociative algebras and a generalized factorization method*, Schrödinger Inst. preprint serie, Wien, 1993, Preprint **53**, 19 p.

[79] Golubchik I. Z. and Sokolov V. V., *Generalized operator Yang-Baxter equations, integrable ODEs and nonassociative algebras*, J. Nonlin. Math. Phys., 2000, **7**(2), 1–14.

[80] Semenov-Tian-Shansky M. A., *What is a classical rr-matrix?*, Funct. Anal. Appl., 1983, **17**(4), 259–272.

[81] Newell A., *Solutions in Mathematics and Physics*, SIAM, Philadelphia 1985, 244 pp.

[82] Golubchik I. Z. and Sokolov V. V., *On some generalizations of the factorization method*, Theoret. and Math. Phys., 1997, **110**(3), 267–276.

References for Part 2

[83] Magri F., *A simple model of the integrable Hamiltonian equation*, J. Math. Phys., 1978, **19**, 1156–1162.

[84] Magri F., Casati P., Falqui G., and Pedroni M., *Eight lectures on Integrable Systems, In: Integrability of Nonlinear Systems* (Y. Kosmann-Schwarzbach et al. eds.), Lecture Notes in Physics (2nd edition), 2004, **495**, 209–250.

[85] Gelfand I. M. and Zakharevich I., *Lenard schemes, and the local geometry of bi-Hamiltonian Toda and Lax structures*, Selecta Math. (N.S.), 2000, **6**(2), 131–183.

[86] Mischenko A. S. and Fomenko A. T., *Euler equations on finite-dimensional Lie groups*, Russian Acad. Sci. Izv. Math., 1978, **12**(2), 371–389.

[87] Khesin B., Levin A., and Olshanetsky M., *Bi-Hamiltonian structures and quadratic algebras in hydrodynamics and on noncommutative torus*, Comm. Math. Phys., 2004, **250**, 581–612.

[88] Odesskii A. V. and Sokolov V. V., *Compatible Lie brackets related to elliptic curve*, J. Math. Phys., 2006, **47**(1), 013506, 1–14.

[89] Feigin B. L. and Odesskii A. V., *Functional realization of some elliptic Hamiltonian structures and bozonization of the corresponding quantum algebras*, Integrable Structures of Exactly Solvable Two-Dimensional Models of Quantum Field Theory, (S. Pakuliak et al., eds.), NATO Sci. Ser. II Math. Phys. Chem. Klumer, Dordrecht, 2001, **35**, 109–122.

[90] Sokolov V. V., *Elliptic Calogero–Moser Hamiltonians and compatible Poisson brackets*, in *Recent developments in Integrable Systems and related topics of Mathematical Physics*, V. M. Buchstaber et al. (eds.), PROMS, Springer, 2018, **273**, 38–46   nlin. arXiv: 1608.08511.

[91] Odesskii A. V. and Sokolov V. V., *Pairs of compatible associative algebras, classical Yang–Baxter equation and quiver representations*, Comm. in Math. Phys., 2008, **278**(1), 83–99.

[92] Odesskii A. V. and Sokolov V. V., *Integrable matrix equations related to pairs of compatible associative algebras*, J. Phys. A: Math. Theor., 2006, **39**, 12447–12456.

[93] Odesskii A. V. and Sokolov V. V., *Algebraic structures connected with pairs of compatible associative algebras, International Mathematics Research Notices*, 2006, ID 43734, 1–35.

[94] Golubchik I. Z. and Sokolov V. V., *Factorization of the loop algebras and compatible Lie brackets, J. Nonlin. Math. Phys.*, 2005, **12**(1), 343–350.

[95] Golubchik I. Z. and Sokolov V. V., *Compatible Lie brackets and Yang–Baxter equation, Theoret. and Math. Phys.*, 2006, **146**(2), 159–169.

[96] Sklyanin E. K. and Takebe T., *Separation of variables in the elliptic Gaudin model, Comm. Math. Phys.*, 1999, **204**, 17–38.

[97] Aguiar M., *On the associative analog of Lie bialgebras, J. Algebra*, 2001, **244**(2), 492–532.

[98] Vinberg E. B., *Discrete linear groups generated by reflections, Russian Acad. Sci. Izv. Math.*, 1971, **5**(5), 1083–1119.

[99] Veselov A. P., *On Darboux-Treibich-Verdier potentials, Lett. in Math. Phys.*, 2011, **96**(1-3), 209–216.

[100] Olshanetsky M. A. and Perelomov A. M., *Quantum integrable systems related to Lie algebras, Phys. Repts.*, 1983, **94**, 313–393.

[101] Matushko M. G. and Sokolov V. V., *Polynomial forms for quantum elliptic Calogero–Moser Hamiltonians, Theoret. and Math. Phys.*, 2017, **191**(1), 480–490.

[102] Sokolov V. V. and Turbiner A. V., *Quasi-exact-solvability of the A_2 Elliptic model: algebraic form, $\mathfrak{sl}(3)$ hidden algebra, polynomial eigenfunctions, J. Phys. A: Math. Gen.*, 2015, **48**(15), 155201.   nlin. arXiv:1409.7439.

[103] Rühl W. and Turbiner A. V., *Exact solvability of the Calogero and Sutherland models, Mod. Phys. Lett.*, 1995, **A10**, 2213–2222.   hep-th/9506105.

[104] Sokolov V. V., *Algebraic quantum Hamiltonians on the plane, Theoret. and Math. Phys.*, 2015, **184**(1), 940–952.   nlin. arXiv:1503.05185.

[105] Takemura K., *Quasi-exact solvability of Inozemtsev models, J. Phys. A: Math. Gen.*, 2002, **35**, 8867–8881.

[106] Dubrovin B. A., *Geometry of 2D topological field theories, In Integrable Systems and Quantum Groups*, Lecture Notes in Math., 1996, **1620**, 120–348.

[107] Vinberg E. B., *On certain commutative subalgebras of a universal enveloping algebra, Russian Acad. Sci. Izv. Math.*, 1991, **36**(1), 1–22.

References for Part 3

[108] Zhiber A. V., *Quasilinear hyperbolic equations with an infinite-dimensional symmetry algebra, Russian Acad. Sci. Izv. Math.*, 1995, **45**(1), 33–54.

[109] Borisov A. B., and Zykov S. A., *The dressing chain of discrete symmetries and proliferation of nonlinear equations, Theoret. and Math. Phys.*, 1998, **115**(2), 530–541.

[110] Zhiber A. V. and Sokolov V. V., *Exactly integrable hyperbolic equations of Liouville type, Russian Math. Surveys*, 2001, **56**(1), 61–101.

[111] Meshkov A. G. and Sokolov V. V., *Hyperbolic equations with symmetries of third order, Theoret. and Math. Phys.*, 2011, **166**(1), 43–57, arXiv:0912.5092.

[112] Zhiber A. V. and Shabat A. B., *Systems of equations $u_x = p(u,v)$, $v_y = q(u,v)$ that possess symmetries, Sov. Math. Dokl.*, 1984, **30**(1), 23–26.

[113] Ibragimov N. Kh. and Shabat A. B., *Infinite Lie-Bäcklund algebras, Funct. anal. appl.*, 1980, **14**(4), 313–315.

[114] Babela M. and Odesskii A., *A family of integrable evolution equations of third order, J. Nonlin. Math. Phys.*, 2017, **24**(1), 73–78.

[115] Svinolupov S. I. and Sokolov V. V., *Evolution equations with nontrivial conservative laws, Funct. anal. appl.*, 1982, **16**(4), 317–319.

[116] Svinolupov S. I., *Second-order evolution equations with symmetries, Russian Math. Surveys*, 1985, **40**(5), 241–242.

[117] Sokolov V. V. and Svinolupov S. I., *Weak nonlocalities in evolution equations, Math. Notes*, 1990, **48**(6), 1234–1239.

[118] Svinolupov S. I. and Sokolov V. V., *On conservations laws for the equations possessing nontrivial Lie-Bäcklund algebra, Integrable systems: collection of the papers.* Ed. A.B. Shabat, BB AS USSR, Ufa. 1982, 53–67, (in Russian).

[119] Hernández Heredero R., Sokolov V. V., and Svinolupov S. I., *Toward the classification of third order integrable evolution equations, J. Phys. A.*, 1994. **13**, 4557–4568.

[120] Hernández Heredero R., *Classification of fully nonlinear integrable evolution equations of third order*, J. Nonlin. Math. Phys., 2005, **12**(4), 567–585.

[121] Drinfeld V. G., Svinolupov S. I., and Sokolov V. V., *Classification of fifth order evolution equations possessing infinite series of conservation laws*, Dokl. AN USSR, 1985, **A10**, 7–10, (in Russian).

[122] Fujimoto A. and Watanabe J., *Classification of fifth-order evolution equations with nontrivial symmetries*, Math. Japonicae, 1983, **28**(1), 43–65.

[123] Sokolov V. V. *On the symmetries of evolution equations*, Russian Math. Surveys, 1988, **43**(5), 165–204.

[124] Olver P. and Wang J. P., *Classification of integrable one-component systems on associative algebras*, Proc. London Math. Soc., 2000, **81**(3), 566–586.

[125] Chen H. H., Lee Y. C., and Liu C. S., *Integrability of nonlinear Hamiltonian systems by inverse scattering method*, Phys. Scr., 1979, **20**(3-4), 490–492.

[126] Fokas A. S., *A symmetry approach to exactly solvable evolution equations*, J. Math. Phys., 1980, **21**(6), 1318–1325.

[127] Kaptsov O. V., *Classification of evolution equations by conservation laws*, Funct. Anal. Appl., 1982, **16**(1), 59–61.

[128] Abellanas L. and Galindo A. J., *Evolution equations with high order conservation laws*, J. Math. Phys., 1983, **24**(3), 504–509.

[129] Mikhailov A. V. and Shabat A. B., *Integrability conditions for systems of two equations $u_t = A(u)u_{xx} + B(u, u_x)$. I*, Theoret. and Math. Phys., 1985, **62**(2), 107–122 .

[130] Mikhailov A. V. and Shabat A. B., *Integrability conditions for systems of two equations $u_t = A(u)u_{xx} + B(u, u_x)$. II*, Theoret. and Math. Phys., 1986, **66**(1), 31–44.

[131] Mikhailov A. V., Shabat A. B., and Yamilov R. I., *Extension of the module of invertible transformations. Classification of integrable systems*, Comm. Math. Phys., 1988, **115**(1), 1–19.

[132] Sokolov V. V. and Wolf T., *A symmetry test for quasilinear coupled systems*, Inverse Problems, 1999, **15**, L5–L11.

[133] Borovik A. E., Popkov V. Yu, and Robuk V. N., *Generation of nonlinear structures in exactly solvable dissipative systems*, Sov. Math. Dokl., 1989, **305**, 841–843.

[134] Gel'fand I. M., Manin Yu. I., and Shubin M. A., *Poisson brackets and kernel of variational derivative in formal variational calculus*, Funct. Anal. Appl., 1976, **10**(4), 30–34.

[135] Dorfman I. Ya., *Dirac Structures and Integrability of Nonlinear Evolution Equations*, John Wiley&Sons, Chichester, 1993, 176 pp.

[136] Meshkov A. G., *Necessary conditions of the integrability*, Inverse Problems, 1994, **10**, 635–653.

[137] Adler A., Bobenko A., and Suris Yu., *Classification of integrable equations on quad-graphs. The consistency approach*, Comm. Math. Phys., 2003, **233**(3), 513–543.

[138] Ibragimov N. Kh. and Shabat A. B., *Evolutionary equations with nontrivial Lie–Bäcklund group*, Funct. anal. appl., 1980, **14**(1), 19–28.

[139] Calogero F. and Degasperis A., *Reduction technique for matrix nonlinear evolutions equations solvable by the spectral transform*, Preprint 151, Istituto di Fisica G. Marconi Univ. di Roma, 1979, 1–37.

[140] Cartan E., *Les systemes differentiels exterieurs et leurs applications geometriques*, Hermann, Paris, 1945, 180 pp.

[141] Muckminov F. Kh. and Sokolov V. V., *Integrable evolution equations with constrains*, Mathematics of the USSR-Sbornik, 1988, **61**(2), 389–410.

[142] Carpentier S., *Compatible Hamiltonian operators for the Krichever-Novikov equation*, Comptes Rendus Mathematique, 2017, **355**(7), 744–747.

[143] Fuchssteiner B., *Application of Hereditary Symmetries to Nonlinear Evolution Equations*, Nonlinear Anal. Theory Meth. Appl., 1979, **3**, 849–862.

[144] Krasil'shchik I. S. and Kersten P. H., *Symmetries and Recursion Operators for Classical and Supersymmetric Differential Equations*, Kluwer academic publishers, London, 2000, 384 pp.

[145] Demskoi D. K. and Sokolov V. V., *On recursion operators for elliptic models*, Nonlinearity, 2008, **21**, 1253–1264,   arxiv nlin. SI/0607071.

[146] Mokhov O. I. and Ferapontov E. V., *Nonlocal Hamiltonian operators of hydrodynamic type associated with constant curvature metrics*, Russian Math. Surveys, 1990, **45**(3), 218–219.

[147] Maltsev A. Ya. and Novikov S. P., *On the local systems Hamiltonian in the weakly nonlocal Poisson brackets*, Phys. D, 2001, **156**(1-2), 53–80.

[148] Hernández Heredero R., Shabat A. B., and Sokolov V. V., *A new class of linearizable equations*, J. Phys. A: Math. Theor., 2003, **36**(47), L605–L614.

[149] Hernández Heredero R., and Sokolov V. V., *The symmetry approach to integrability: recent advances*, to be published in *Nonlinear Systems and Their Remarkable Mathematical Structures*, N. Euler and M.C. Nucci (editors), CRC Press, Boca Raton, 2019, **2**, 119–157,   nlin. arXiv 1904.01953,

[150] Mikhailov A. V., Novikov V. S. and Wang J. P., *On classification of integrable nonevolutionary equations*, *Studies in Applied Mathematics*, 2007, **118**(4), 419–457.

[151] Novikov V. S. and Wang J. P., *Symmetry structure of integrable nonevolutionary equations*, *Studies in Applied Mathematics*, 2007, **119**(4), 393–428.

[152] Martínez Alonso L. and Shabat A. B., *Towards a theory of differential constraints of a hydrodynamic hierarchy*, J. Nonlin. Math. Phys., 2003, **10**, 229–242.

[153] Caparrós Quintero C. and Hernández Heredero R., *Formal recursion operators of integrable nonevolutionary equations and Lagrangian Systems*, J. Phys. A: Math. Theor., 2018, **51**, 385201.

[154] Hirota R. and Satsuma J., *Nonlinear evolution equations generated from the Bäklund transformation for the Boussinesq equation*, Progr. Theor. Phys., 1977, **57**, 797–807.

[155] Leznov A. N. and Savel'ev M. V., *Group-theoretical methods for integration of nonlinear dynamical systems*, 2012, Birkhauser Boston, 292 pp.

[156] Darboux G., *Sur les équations aux dérivées partielles du second ordre*, Ann. Ecole Normale Sup., 1980, **VII**, 163–173.

[157] Goursat E., *Leçons sur l'intégration des équations aux dérivées partielles du second ordre à deux variables indépendantes*, Tome I, Tome II, Hermann, Paris, 1896, 1898.

[158] Gau E., *Sur l'intégration des équations aux dérivées partielles du second ordre par la métode de M. Darboux*, Journ. de Math., 1911, **6**(7), 123–240.

[159] Vessiot E., *Sur les équations aux dérivées partielles du second ordre, $F(x, y, p, q, r, s, t) = 0$, intégrables par la methode de Darboux*, J. Math. Pure Appl., 1939, **18**, 1–61.

[160] Gosse R., *Des certaines équations aux dérivées partielles du second ordre intégrables par la métode de Darboux*, Journ. de Math., 1925, **9**(4), 381–399.

[161] Sokolov V. V. and Zhiber A. V., *On the Darboux integrable hyperbolic equation*, Phys. Lett. A, 1885, **208**, 303–308.

[162] Anderson I. M. and Kamran N., *The variational bicomplex for hyperbolic second-order scalar partial differential equations in the plane*, Duke Math. J., 1997, **87**(2), 265–319.

[163] Sokolov V. V. and Startsev S. Ya., *Symmetries of nonlinear hyperbolic systems of the Toda chain type*, Theoret. and Math. Phys., 2008, **155**(2), 802–811.

[164] Kapsov O. V., *On Goursat classification problem*, Programming, 2012, **2**, 68–71, (in Russian).

[165] Startsev S. Ya., *Differential substitutions of the Miura transformation type*, Theoret. and Math. Phys., 1998, **116**(3), 1001–1010.

[166] Lainé M. E., *Sur l'application de la methode de Darboux aux equations* $s = f(x, y, z, p, q)$, Comptes Rendus, 1926, **182**, 1126–1127.

[167] Zhiber A. V. and Sokolov V. V., *New example of a nonlinear hyperbolic equation, possessing integrals*, Theoret. and Math. Phys., 1999, **120**(1), 834–839.

[168] Sanders J. A. and Wang J. P, *Family of operators and their Lie algebras*, J. Lie Theory, 2002, **12**(2), 503–514.

[169] Startsev S. Ya., *Pre-Hamiltonian operators related to hyperbolic equations of Liouville type*, 2020,   nlin. arXiv 2002.10442

[170] Kiselev A. V. and van de Leur J. W., *Symmetry algebras of Lagrangian Liouville-type systems*, Theoret. and Math. Phys., 2010, **162**(2), 149–162.

[171] Startsev S. Ya., *Laplace invariants for integrable hyperbolic equations and differential substitutions of Miura type*, PhD Thesis, Ufa Mathematical Institute, 1997 (in Russian),   arXiv:1710.11068 [nlin.SI]

[172] Zhiber A. V. and Startsev S. Ya., *Integrals, solutions, and existence problems for Laplace transformations of linear hyperbolic systems*, Math. Notes, 2003, **74**(6), 803–811.

[173] Guryeva A. M. and Zhiber A. V., *Laplace Invariants of Two-Dimensional Open Toda Lattices*, Theoret. and Math. Phys., 2004, **138**(3), 338–355.

[174] Guryeva A. M. *Laplace method of cascade integration and nonlinear systems of hyperbolic equations*, PhD Thesis, Ufa Mathematical Institute, 2005, (in Russian).

[175] Mikhailov A. V. and Sokolov V. V., *Integrable ODEs on Associative Algebras*, *Comm. in Math. Phys.*, 2000, **211**(1), 231–251.

[176] Calogero F., *An integrable many-body problem*, *J. Math. Phys.*, 2011, **52**(10), 102702.

[177] Odesskii A. V., Roubtsov V. N., and Sokolov V. V. *Bi-Hamiltonian ODEs with matrix variables*, *Theoret. and Math. Phys.* 2012, **171**(1), 442–447, arXiv nlin.SI 1105.1740.

[178] Wolf T., *The program CRACK for solving PDEs in general relativity*, *Relativity and Scientific Computing*, Ed. F. W. Hehl, R. A. Puntigam and H. Ruder, Berlin: Springer, 1996, 241–257.

[179] Sokolov V. V. and Wolf T., *On nonabelization of integrable polynomial ODEs*, *Lett. in Math. Phys.*, 2020, **110**(3), 533–553,   arXiv nlin. 1809.03030, 1807.05583.

[180] Wolf T. and Efimovskaya O., *On integrability of the Kontsevich nonabelian ODE system*, *Lett. in Math. Phys.*, 2012, **100**(2), 161–170.

[181] Rota G.-C., *Baxter operators, an introduction*, in *"Gian-Carlo Rota on combinatorics"*, *Contemp. Mathematicians*, 1995, Birkhauser Boston, Boston, MA, 504–512.

[182] Schedler T., *Trigonometric solutions of the associative Yang–Baxter equation*, *Math. Res. Lett.*, 2003, **10**(2-3), 301–321.

[183] Elashvili A. G., *Frobenius Lie algebras*, *Funct. Anal. Appl.*, 1982, **16**(4), 326–328.

[184] Zobnin A. I., *Anti-Frobenius algebras and associative Yang–Baxter equation*, *Matem. Mod.*, 2014, **26**(11), 51–56.

[185] Bielawski R., *Quivers and Poisson structures*, *Manuscripta Mathematica*, 2013, **141**(1-2), 29–49.

[186] Sokolov V. V., *Classification of constant solutions for associative Yang–Baxter equation on Mat_3*, *Theoret. and Math. Phys.*, 2013, **176**(3), 1156–1162.   nlin. arXiv:1212.6421.

[187] Farkas D. and Letzter G., *Ring theory from symplectic geometry*, *J. Pure Appl. Algebra*, 1998, **125**(1-3), 155–190.

[188] Crawley-Boevey W., *Poisson structures on moduli spaces of representations*, *J. Algebra*, 2011, **325**, 205–215.

[189] Odesskii A. V. and Sokolov V. V., *Nonabelian elliptic Poisson structures on projective spaces*, math.QA. arXiv: 1911.03320

[190] Van den Bergh M., *Double Poisson algebras, Trans. Amer. Math. Soc.*, 2008, **360**(11), 555–603.

[191] Odesskii A. V., Roubtsov V. N., and Sokolov V. V. *Double Poisson brackets on free associative algebras, Contemporary Mathematics*, 2013, **592**, 225–241, arXiv nlin. 1208.2935.

[192] Olver P. J. and Sokolov V. V., *Integrable Evolution Equations on Associative Algebras, Comm. Math. Phys.*, 1998, **193**, 245–268.

[193] Marchenko V. A., *Nonlinear equations and operator algebras*, Kiev, Naukova dumka, 1986, 152 pp.

[194] Kupershmidt B. A., *KP or mKP*, Providence, R.I. : AMS, 2000, 600 pp.

[195] Svinolupov S. I. and Sokolov V. V., *Vector–matrix generalizations of classical integrable equations, Theoret. and Math. Phys.*, 1994, **100**(2), 959–962.

[196] Olver P. J. and Sokolov V. V., *Non-abelian integrable systems of the derivative nonlinear Schrodinger type, Inverse Problems.*, 1998, **6**, L5–L8.

[197] Svinolupov S. I., *On the analogues of the Burgers equation, Phys. Lett. A*, 1989, **135**(1), 32–36.

[198] Svinolupov S. I., *Jordan algebras and generalized Korteweg-de Vries equations, Theoret. and Math. Phys.*, 1991, **87**(3), 611–620.

[199] Svinolupov S. I., *Generalized Schrödinger equations and Jordan pairs, Comm. Math. Phys.*, 1992, **143**(1), 559–575.

[200] Fordy A. P. and Kulish P., *Nonlinear Schrödinger equations and simple Lie algebras, Comm. Math. Phys.*, 1983, **89**, 427–443.

[201] Fordy A. P., *Derivative nonlinear Schrödinger equations and Hermitian symmetric spaces, J. Phys. A.: Math. Gen.*, 1984, **17**, 1235–1245.

[202] Athorne C. and Fordy A. P., *Generalized KdV and mKdV equations associated with symmetric spaces, J. Phys. A: Math. Gen.*, 1987, **20**, 1377–1386.

[203] Jacobson N., *Structure and representations of Jordan algebras*, Providence, 1968, 461 pp.

[204] Zelmanov E., *Primary Jordan Triple Systems. III.*, *Sib. Math. J.*, 1985, **26**(1), 55–64.

[205] Kazarian M. E. and Lando S. K., *An algebro-geometric proof of Witten's conjecture*, J. Amer. Math. Soc., 2007, **20**(4), 1079–1089.

[206] Svinolupov S. I., *Jordan algebras and integrable systems*, Funct. Anal. Appl., 1993, **27**(4), 257–265.

[207] Shestakov I. P. and Sokolov V. V., *Multi-component generalizations of mKdV equation and non-associative algebraic structures*, Journal of Algebra and Applications, 2020,   nlin. arXiv 1905.01016 [math-ph].

[208] Sanders J. A. and Wang J. P., *Integrable systems in n-dimensional Riemannian Geometry*, Moscow Mathematical Journal, 2003, **3**(4), 1369–1393.

[209] Kulish P. P. and Sklyanin E. K., *O(N)-invariant nonlinear Schodinger equation. - A new completely integrable system*, Phys. Lett. A, 1981, **84**(7), 349–352.

[210] Loos O., *Assoziative Tripelsysteme*, Manuscr. Math., 1972, **7**, 103–112, (in German).

[211] Grishkov A. N. and Shestakov I. P., *Speciality of Lie-Jordan Algebras*, J. Algebra, 2001, **237**, 621–636.

[212] Svinolupov S. I. and Sokolov V. V., *Deformations of triple Jordan systems and integrable equations*, Theoret. and Math. Phys., 1997, **108**(3), 1160–1163.

[213] Meyberg K., *Lectures on algebras and triple systems*, Lecture Notes, University of Virginia, Charlottesville, 1972, 266 pp.

[214] Helgason S., *Differential Geometry, Lie Groups, and Symmetric Spaces*, Academic Press, New York, 1978, 634 pp.

[215] Zhevlakov K. A., Slin'ko A. M., Shestakov I. E, and Shirshov A.I., *Rings that are nearly associative*, Academic Press, New York, 1982, 385 pp.

[216] Habibullin I. V., Sokolov V. V., and Yamilov R. I., *Multi-component integrable systems and nonassociative structures*, in *Nonlinear Physics: theory and experiment*, Ed. E. Alfinito, M. Boiti, L. Martina, F. Pempinelli, World Scientific Publisher. Singapore, 1996, 139–168.

[217] Sokolov V. V., *Integrable evolution systems of geometric type*, Theoret. and Math. Phys., 2020, **202**(3), 428–436.

[218] Meshkov A. G. and Sokolov V. V., *Classification of integrable vector equations of geometric type*, J. Geom. and Phys., 2020, **149**,   nlin. arXiv:1904.09351v2 [math-ph]

[219] Sokolov V. V. and Wolf T., *Classification of integrable polynomial vector evolution equations*, J. Phys. A., 2001, **34**, 11139–11148.

[220] Tsuchida T. and Wadati M., *New integrable systems of derivative nonlinear Schrödinger equations with multiple components*, Phys. Lett. A, 1999, **257**, 53–64.

[221] Kupershmidt B. A., *A coupled Korteweg-de Vries equation with dispersion*, J. Phys. A, 1985, **18**, L571–L573.

[222] De Sole A., Kac V. G., and Valeri D., *Classical affine W-algebras and the associated integrable Hamiltonian hierarchies for classical Lie algebras*, Comm. Math. Phys., 2018, **360**, 851–918, arXiv:1705.10103.

[223] Tsuchida T. and Wolf T., *Classification of polynomial integrable systems of mixed scalar and vector evolution equations: I*, J. Phys. A.: Math. Gen., 2005, **38**(35), 7691–7733.

[224] Meshkov A. G. and Sokolov V. V. *Integrable evolution equations on the N-dimensional sphere*, Comm. Math. Phys., 2002, **232**(1), 1–18.

[225] Sanders J. A. and Wang J. P., *Integrable systems in n-dimensional Riemannian Geometry*, Moscow Mathematical Journal, 2003, **3**(4), 1369–1393.

[226] Meshkov A. G. and Sokolov V. V., *Classification of integrable divergent N-component evolution systems*, Theoret. and Math. Phys., 2004, **139**(2), 609–622.

[227] Balakhnev M. Ju. and Meshkov A. G., *On a classification of integrable vectorial evolutionary equations*, J. Nonlin. Math. Phys., 2008, **15**(2), 212–226.

[228] Yamilov R., *Symmetries as integrability criteria for differential difference equations*, J. Phys. A: Math. Theor., 2006, **39**(45), 541–623.

[229] Balakhnev M. Ju. and Meshkov A. G., *Integrable anisotropic evolution equations on a sphere*, SIGMA, 2005, **1** (paper 027), 11 pp.

[230] Meshkov A. G. and Sokolov V. V., *Vector hyperbolic equations on the sphere possessing integrable third-order symmetries*, Lett. in Math. Phys., 2014, **104**(3), 341–360.

Index